Critical Mass

Formerly on the staff of *Nature*, Philip Ball is now a full-time writer. His books include *Critical Mass*, winner of the Aventis Prize for Science Books, *H2O* and *Bright Earth*. He lives in London.

'[Ball] is as comfortable referring to Tolstoy or Vonnegut's fiction as to Newtonian mechanics, Keynesian economics or a chanting football crowd. His scope is immense – and valuable'
Daily Telegraph

There's something for everyone – the causes of crime, how cities develop, cold war brinkmanship, how panics spread through a crowd, the shape of the internet, and that old favorite, the 'six degrees to Kevin Bacon' networking game. A rewarding comucopia of topics . . . elegantly written throughout.' *Fortean Times*

Ball delves far beyond today's headlines . . . Substantial, impeccably researched and . . . persuasive. For anyone who would like to learn about the intellectual ferment at the surprising junction of physics and social science, *Critical Mass* is the place to start.'
Nature

'Lucid, accessible and engaging . . . Ball makes a persuasive, comprehensive case and it's a welcome antidote to popular individualistic thought.' *Glasgow Herald*

'Exquisitely produced and painstakingly researched . . . Ball writes patiently and eloquently . . . Exciting . . . A rousing call-to-arms, and an elegant answer to the shallow tradition of British empiricism.' *Independent*

Also by Philip Ball

The Ingredients: A Guided Tour of the Elements

Bright Earth: The Invention of Colour

Stories of the Invisible: A Guided Tour of Molecules

H_2O: A Biography of Water

The Self-made Tapestry: Pattern Formation in Nature

Made to Measure: New Materials for the 21st Century

Designing the Molecular World: Chemistry at the Frontier

Critical Mass

HOW ONE THING LEADS TO ANOTHER

*Being an enquiry into the interplay of chance
and necessity in the way that human culture,
customs, institutions, cooperation and
conflict arise*

PHILIP BALL

arrow books

Published in the United Kingdom in 2005 by Arrow Books

14 16 18 20 19 17 15 13

First published in the United Kingdom in 2004 by William Heinemann

Arrow Books
The Random House Group Limited
20 Vauxhall Bridge Road, London SW1V 2SA

www.randomhouse.co.uk

Addresses for companies within The Random House Group Limited
can be found at:
www.randomhouse.co.uk/offices.htm

The Random House Group Limited Reg. No. 954009

A CIP catalogue record for this book
is available from the British Library

ISBN 9780099457862

The Random House Group Limited makes every effort to ensure
that the papers used in its books are made from trees that
have been legally sourced from well-managed and credibly
certified forests. Our paper procurement policy can be found at:
www.randomhouse.co.uk/paper.htm

Typeset by SX Composing DTP, Rayleigh, Essex
Printed and bound in Great Britain by
Cox & Wyman Ltd, Reading, Berkshire

Contents

Acknowledgements   vii

Picture credits   ix

Introduction   1
Political Arithmetick

1   Raising Leviathan   7
The brutish world of Thomas Hobbes

2   Lesser Forces   38
The mechanical philosophy of matter

3   The Law of Large Numbers   58
Regularities from randomness

4   The Grand Ah-Whoom   99
Why some things happen all at once

5   On Growth and Form   121
The emergence of shape and organization

6   The March of Reason   145
Chance and necessity in collective motion

7   On the Road   193
The inexorable dynamics of traffic

8   Rhythms of the Marketplace   220
The shaky hidden hand of economics

9   Agents of Fortune   253
Why interaction matters to the economy

10   Uncommon Proportions   281
Critical states and the power of the straight line

11   The Work of Many Hands   311
The growth of firms

12   Join the Club   337
Alliances in business and politics

13   Multitudes in the Valley of Decision   369
Collective influence and social change

14   The Colonization of Culture   423
Globalization, diversity and synthetic societies

15   Small Worlds   443
Networks that bring us together

16   Weaving the Web   467
The shape of cyberspace

17   Order in Eden   505
Learning to cooperate

18   Pavlov's Victory   538
Is reciprocity good for us?

19   Towards Utopia?   564
Heaven, hell and social planning

Epilogue   586
Curtain Call

Notes   589
Bibliography   615
Index   635

handwritten notes: for notebook
chronology
bibliographical
index: names all for
want for follow-up
honest introspection

from Ben: respect for the spines of books

Acknowledgements

Writing this book has often felt like a collaborative exercise, for which I am eager to spread some credit (while of course remaining accountable for all blame). For wise, perceptive and tolerant suggestions and advice, and for the provision of materials and references, I am deeply indebted to many real experts in this arena, in particular Robert Axelrod, Robert Axtell, Albert-Laszló Barabási, Eshel Ben-Jacob, Rama Cont, Dirk Helbing, Steve Keen, Thomas Lux, Mitsugu Matsushita, Joe McCauley, Mark Newman, Paul Ormerod, Craig Reynolds, Sorin Solomon, Gene Stanley, Alessandro Vespignani and Tamás Vicsek. The support of my editors, Ravi Mirchandani, Caroline Knight and John Glusman, has been vital, and many rough edges were knocked off the text by the careful attention of the copyeditor, John Woodruff. I have been encouraged as ever by the good judgement of my agent Peter Robinson and the support of my wife Julia.

Philip Ball
London, October 2003

Picture credits

1.1   Thomas Hobbes by Isaac Fuller (1606–72), Burghley House Collection, Lincolnshire, UK/Bridgeman Art Library.

1.2   Title page from Hobbes' *Leviathan*, 1651, London; © Bettman/CORBIS.

2.1   Engraving of James Clerk Maxwell after a photograph by Fergus; © CORBIS.

3.1   Alphonse Quetelet; courtesy AKG Images.

5.1   (a) Manuel Velarde, Universidad Complutense, Madrid. (b) From M. C. Cross and P. Hohenberg (1993). *Reviews of Modern Physics* **65**, 851–1112. (c) David Cannell, University of California at Santa Barbara.

5.3   Mitsugu Matsushita, Chuo University.

5.4   Photographs: Mitsugu Matsushita.

5.5   From W. A. Bentley and W. J. Humphreys (1931). *Snow Crystals*. Reprinted by Dover, New York, 1962.

5.6, 6.1   Eshel Ben-Jacob, Tel-Aviv University.

6.2   (b) Norbert Wu.

6.3   Tamás Vicsek, Eötvös-Loránd University.

6.8, 6.9   Dirk Helbing, Technical University of Dresden.

6.11, 6.12   Michael Batty, University College London.

6.13   Hernán Makse, City University of New York.

6.14   Rui Carvalho and Alan Penn, University College London.

7.3, 7.4   From D. E. Wolf (1999). *Physica A* **263**, 438–51.

10.1   Alastair Bruce, University of Edinburgh.

13.2, 13.3   From J. M. Epstein (2001). *Computational Economics* **18**(1), 9–24.

13.4   Paul Ormerod, Volterra Consulting.

16.3   Hawoong Jeong, Korea Advanced Institute of Science and Technology.

16.4   Kindly provided by Hawoong Jeong.

16.6   Hawoong Jeong.

18.5   (a, b) Martin Nowak, Oxford University. (c) From M. A. Nowak and R. M. May (1993). *International Journal of Bifurcation and Chaos* **3**(1), 35–78.

Note on trademarks

All trademarks referred to in this book are the properties of their owners. Although trademarks, and their owners, are not systematically listed or marked in the text, all trademarks are acknowledged, and there is no intention of infringing upon any trademark.

Introduction

Political Arithmetick

On the 7th of November 1690 a manuscript was delivered to England's new king, William III. William, the Prince of Orange, had only the previous year deposed the unpopular, Catholic James II in a bloodless coup; and in that time of turmoil the book's message might have provided some solace to the monarch, for it set out to show that England was a solid and secure force in the world.

The book's author was Sir William Petty, once a professor of anatomy at Oxford University and physician general to the English army in Ireland. He had died in 1687, but his work was delivered to the royal court by his son, the Earl of Shelburne. Petty claimed to prove:

That a small Country, and few People, may by their Situation, Trade, and Policy, be equivalent in Wealth and Strength, to afar greater People, and Territory . . .

That France cannot by reason of Natural and Perpetual Impediments, be more powerful at Sea, than the English, or Hollanders.

That the People, and Territories of the King of England, are Naturally near as considerable, for Wealth, and Strength, as those of France.

That the Impediments of England's Greatness, are but contingent and removeable.

That the Power and Wealth of England, hath increased above this forty years.

That one tenth part, of the whole Expence, of the King of England's Subjects; is sufficient to maintain one hundred thousand Foot, thirty thousand Horse, and forty thousand Men at Sea, and to defray all other Charges, of the Government: both Ordinary and Extraordinary, if the same were regularly Taxed and Raised.

That there are spare Hands enough among the King of England's Subjects, to earn two Millions per annum, more than they now do, and there are Employments, ready, proper, and sufficient, for that purpose.

That there is Mony sufficient to drive the Trade of the Nation.

That the King of England's Subjects, have Stock, competent, and convenient to drive the Trade of the whole Commercial World.[1]

England, in other words, was styled for greatness. On what grounds did Petty make these bold assertions? The book was called *Political Arithmetick*, and it claimed to make a science of politics. Just as Isaac Newton's law of gravity ultimately rested on the quantitative measurements and deductions of the astronomers Tycho Brahe and Johannes Kepler, so Petty used numbers to derive proofs of the healthy state of English society.

'The Method I take to do this', he explained,

is not yet very usual; for instead of using only comparative and superlative Words, and intellectual Arguments, I have taken the course (as a Specimen of the Political Arithmetick I have long aimed at) to express my self in Terms of Number, Weight, or Measure; to use only Arguments of Sense, and to consider only such Causes, as have visible Foundations in Nature; leaving those that depend upon the mutable Minds, Opinions, Appetites, and Passions of particular Men, to the Consideration of others: Really professing my self as unable to speak satisfactorily upon those Grounds (if they may be call'd Grounds), as to foretel the cast of a Dye; to play well at Tennis, Billiards, or Bowles, (without long practice,) by virtue of the most elaborate Conceptions that ever have been written *De Projectilibus & Missilibus*, or of the Angles of Incidence and Reflection.[2]

In other words, while Petty professed to know little about mutable human nature, he believed that society could be understood to the extent that it could be measured and quantified. The science of political arithmetic, he argued, could free a nation's leaders from man's irrationality, and be used to fashion sound and verifiable principles of governance.

How dismayed Petty would have been to find that, three hundred years later, political scientists were still lamenting the fact that human affairs are dominated by whim and prejudice rather than being led by reason and logic. In *Man, the State, and War* (1954), Kenneth Waltz voices the hope that dealings between nations might one day be conducted by the use of rational theory rather than by dogma and polemic. 'In the absence of an elaborated theory of international politics,' he says, 'the causes one finds and the remedies one proposes are often more closely related to temper and training than to the objects and events of the world about us.'[3]

Waltz certainly does not envisage anything as simple as what Petty had in mind – a kind of Newtonian physics of society. But Petty's efforts, which now look woefully naive, nevertheless find an echo in contemporary physics. Over the past two decades, something extra-ordinary has been happening in this field of science. Tools, methods and ideas developed to understand how the blind material fabric of the universe behaves are finding application in arenas for which they were never designed, and for which they might at first glance appear ridiculously inappropriate. Physics is finding its place in a science of society.

This book is all about how that happened, why it is worth taking seriously and where it might lead. It is also about the limits and caveats of a physics of society, whose potential for misapplication is considerable.

We have been here before. In the 1970s, the catastrophe theory of René Thom seemed to promise an understanding of how sudden changes in society might be provoked by small effects. This initiative atrophied rather quickly, since Thom's phenomenological and qualitative theory did not really offer fundamental explanations and mechanisms for the processes it described. Chaos theory, which matured in the 1980s, has so far proved rather more robust, supplying insights into how complicated and ever-shifting ('dynamical') systems rapidly cease to be precisely predictable even if their initial states are known in great detail. Chaos theory has been advocated as a model for market economics, and its notion of stable dynamical states, called attractors, seemed to provide some explanation for why certain modes of social behaviour or organization remain immune to small perturbations. But this theory has not delivered anything remotely resembling a science of society.

The current vogue is for the third of these three C's: complexity. The buzzwords are now 'emergence' and 'self-organization', as complexity theory seeks to understand how order and stability arise from the interactions of many agents according to a few simple rules.

The physics I shall discuss in this book is not unrelated to the idea of complexity – indeed, the two often overlap. But very often what passes today for 'complexity science' is really something much older, dressed up in fashionable apparel. The main themes in complexity theory have been studied by physicists for over a hundred years, and these scientists have evolved a toolkit of concepts and techniques to which complexity studies have added barely a handful of new items. At the root of this sort of physics is a phenomenon which immediately

4

explains why the discipline may have something to say about society: it is a science of *collective behaviour*. At face value it is not obvious how the bulk properties of insensate particles of matter should bear any relation to how humans behave en masse. Yet physicists have discovered that systems whose component parts have a capacity to act collectively often show recurrent features, even though they might seem to have nothing at all in common with one another.

With that in mind, I hope to show that the new physics of society is able to accommodate just those characteristics of humankind that Petty felt it expedient to exclude: the 'mutable Minds, Opinions, Appetites, and Passions of particular Men'. I want to suggest that, even with our woeful ignorance of why humans behave the way they do, it is possible to make some predictions about how they behave collectively. That is to say, we can make predictions about society even in the face of individual free will, and perhaps even illuminate the limits of that free will.

William Petty thought that quantification alone was enough to qualify his Political Arithmetick as a science. But his contemporary Thomas Hobbes had a rather deeper appreciation of what a science of society should be about. It must look beyond mere numbers, Hobbes implied, and grapple with the difficult question of mechanism. We must ask not just *how* things happen in society, but *why*. In the first part of the book we shall see where Hobbes's mechanistic approach and Petty's arithmetic led in attempts to understand society, and how – most curiously – they fed back into physics itself in the nineteenth century. We shall see how physics deals with systems of many components, all interacting with one another at once, and why regular and predictable behaviour emerges in statistical form from such seeming chaos.

Treating people as though they were just so much insensate matter (or rather, *appearing* perhaps to do so) is a contentious business, which is why we shall approach the physics-based modelling of society with cautious steps, showing first why life (I am tempted to say 'mere life') need not in itself present a boundary to the application of statistical physics. First the bacterium; later the world.

Yet you should not expect to find a 'theory of society' expounded in this book. Indeed, the modern trend towards 'unified theories' – grand, over-arching frameworks – in science, while having its uses, is arguably unhealthy. If there is such a thing as a physics of society, it does not come in the form of some universal equation into which we feed numbers and out of which emerges a deterministic description of social behaviour. The case must be constructed by example, and the tools subtly adapted to each specific purpose. The survey presented in this book is by no means exhaustive; but we shall look at what physics has to say about how people move around in open spaces, how they make decisions and cast votes, and form alliances and groups and companies. We shall see physics used to understand some aspects of the behaviour of economic markets and to reveal the hidden structure in networks of social and business contacts. We shall uncover physics of a sort in the politics of conflict and cooperation.

Underlying all of this is a more difficult question: does physics simply help us to explain and understand, or can we use it to anticipate and thereby to avoid problems, to improve our societies, to make a better and safer world? Or is that merely another dream destined for the already overflowing graveyard of utopias past?

1

Raising Leviathan

The brutish world of Thomas Hobbes

A work on politics, on morals, a piece of criticism, even a manual on the art of public speaking would, other things being equal, be all the better for having been written by a geometrician.

Bernard Fontenelle, Secretary of the Académie Française

(late seventeenth century)

'I perceive', says I, 'the world has become so mechanical that I fear we shall quickly become ashamed of it; they will have the world be in large what a watch is in small, which is very regular, and depends only on the just disposing of the several parts of the movement. But pray tell me, madame, had you not formerly a more sublime idea of the universe?'

Bernard Fontenelle (1686)

The most complete exposition of a social myth often comes when the myth itself is waning.

Robert M. MacIver (1947)

It is no longer very useful to ask the question, 'Who governs Britain?' Discuss.

Exercise in Stephen Cotgrove (1967)

7

*

> Brothers will fight
> And kill each other . . .
> Men will know misery . . .
> An axe-age, a sword-age,
> Shields will be cloven,
> A wind-age, a wolf-age,
> Before the world ends.[1]

This is how the Norsemen envisaged Ragnarok, the Twilight of the Gods; but in political exile in France in 1651, Thomas Hobbes must have thought that he had lived through it already. At Naseby and Marston Moor, Newbury and Edgehill, the stout yeomanry of England had hacked one another to bloody ruin. Oliver Cromwell reigned as Lord Protector of a country shocked to find itself a republic, its line of monarchial succession severed by the executioner's axe.

The combatants in the English Civil War, unlike those in France's revolution or America's blood-soaked battle of North against South, had few clear ideological distinctions. The Royalists fought under the king's banner, but the Roundheads also claimed allegiance to 'King and Parliament'. For all his presumptuous arrogance, Charles I had no desire to live outside the constitution and laws of the land. Both sides were Anglican and wary of Papists. There were aristocrats in the Parliamentarian ranks and common folk among the Cavaliers. Many of those who slaughtered one another might have found little to dispute had they wielded words instead of swords.

Such a conflict could be nothing but a prescription for confusion, once the beheading of the king brought it to an end. Embarrassed by the power with which fate had invested him, Cromwell searched in vain for a constitutional solution that would guarantee stability. Such was the might the Ironsides gave him that as Lord Protector he could experiment more freely than any British ruler before or since – though this was a freedom Cromwell would happily have relinquished had he felt able. Time and again he created parliaments on which to shed some burden of authority, only to dissolve them once he found them unworthy of the responsibility.

In the turmoil of those times, none could be certain that friends would not become foes, or old opponents emerge as allies. The Presbyterian Scottish Parliament, whose fierce antagonism to Charles I had precipitated the conflict between Parliament and Crown in the 1630s, was by 1653 fighting against Cromwell with Charles II as its figurehead. Cromwell himself expelled from the House of Commons the Parliament he had fought to instate, and struggled to maintain control of the monstrous army he had created. After Cromwell's death in 1658 this militia restored Parliament and craved an end to the Protectorate. John Lambert led the troops to victory over a Royalist uprising in 1659 even as he plotted to restore Charles II to the throne (and conveniently make him brother-in-law to Lambert's daughter). Yet in the end it was by defeating Lambert that George Monk, a former Royalist, restored a Parliament in 1660 that he knew would crown the exiled king.

What could the common people have possibly craved more than stability? Twenty years of war and changing fortunes had convinced them that only a monarchy would supply it; and Charles II, who had narrowly escaped the tender mercies of the Ironsides just eight years previously, found a loyal army and a joyous population awaiting his return from France.

There is no way to understand the extraordinary quest on which Thomas Hobbes (1588–1679) embarked without acknowledging its historical setting. Centuries of monarchical rule over a hierarchical society had been graphically dismembered with the fall of the axe on

30 January 1649. A system of governance previously upheld by divine and moral imperatives had been revealed as arbitrary and contingent. Almost every political idea that was to follow in later centuries was voiced in seventeenth-century England, and many of them were put into practice. Soldiers and labourers became Levellers and Diggers, advocating socialist principles of equality and an end to individual ownership of land. Cromwell himself seems to have toyed with the notion of a freely elected democratic government, yet he spent much of his Protectorate heading a military dictatorship. Charles I dissolved Parliament and instigated an absolute monarchy in the years before the Civil War.

Which of these systems should a society adopt? The issue was a burning one. Although war between nations was regarded almost as a natural state of affairs, it might hardly pain the common person beyond the imposition of new taxes and levies. But internal strife was agonizing. The Civil War in England, conducted on the whole with restraint toward civilians, was bad enough; but the Thirty Years' War, which ravaged Europe from the early part of the century, killed one in every three of the inhabitants of many German states. For Hobbes and many of his contemporaries, civil peace was worth almost any price.

England's miseries were a symptom of broader changes in the Western world. The feudal system of the Middle Ages was waning before the rise of a prosperous middle class, and from the ranks of this vigorous and ambitious sector came many of the Parliamentarians who no longer felt obliged to submit to every royal whim. The monarchy, with its counsellors and Star Chamber, harked back to the medieval-ism of Elizabethan society, but the spirit of the age cleaved to some-thing more democratic, however limited in scope. The Reformation of Luther and Calvin had split Europe asunder; no longer did a single Church rule all of Christendom. The backlash to the assault on ecclesi-astical tradition – prompted not only by Luther's heresy but by the

Humanism of the Renaissance – gave birth to the Counter-Reformation, the Council of Trent, the Jesuits and the persecutory zeal of the Inquisition. The greater the religious diversity, it seemed, the greater the intolerance.

And emerging from this ferment were ideas about the nature of the world that were ultimately to prove as challenging as any of the proclamations that Luther nailed to the church door in Wittenberg. Copernicus had been fortunate to develop his heliocentric theory – the idea that the Earth revolves around the Sun – in the early sixteenth century, before the Counter-Reformation; his first manuscript, circulated around 1530, even received papal sanction. But by 1543, when the full treatise was published after his death, it was prefaced with a note through which the editor, Andreas Osiander, hoped to evade ecclesiastical condemnation by indicating that the new view of the celestial bodies should be regarded simply as a convenient mathematical fiction. How Galileo fared against papal authority when he placed the same idea on firmer footing is the stuff of legend. The Inquisition condemned him in 1616 and forced him to recant in 1633. But by the middle of the seventeenth century, with René Descartes revitalizing the ancient Greek atomic theory and Isaac Newton soon to be admitted to Trinity College in Cambridge, the banishment of magic and superstition by mechanistic science seems in retrospect inevitable.

Hobbes's masterwork, *Leviathan*, was an attempt to develop a political theory out of this mechanical world-view. He set himself a goal that today sounds absurdly ambitious, although at the dawn of the Enlightenment it must have seemed a natural marriage. Hobbes wanted to deduce, by logic and reason no less rigorous than that used by Galileo to understand the laws of motion, how humankind should govern itself. Starting with what he believed to be irreducible and self-evident axioms, he aimed to develop a science of human interactions, politics and society.

It is hard now to appreciate the magnitude not just of this challenge in itself, but of the shift in outlook that it embodied. There has never been any shortage of views on the best means of governance and social organization. Almost without exception, proposals before Hobbes – and many subsequently – were designed to give the proposers the greatest (perceived) advantage. Emperors, kings and queens sought to justify absolute monarchy by appeal to divine covenant. The Roman Catholic Church was hardly the first theocracy to set itself up as the sole conduit of God's authority. In Plato's Republic, one of the earliest of utopian models, cool and self-confident reasoning argued for a state in which philosophers were accorded the highest status. The rebellious English Parliament of the early 1640s demanded that the King transfer virtually all governing power to them. One could always find an argument to put oneself at the top of the pile.

Hobbes was different. What he aimed to do was to apply the method of the theoretical scientist: to stipulate fundamental first principles and see where they led him. In theory, any conclusion was possible. By analysing human nature and how people interact, he might conceivably have found that the most stable society was one based on what we would now call communism, or democracy, or fascism. In practice, Hobbes's reasoning led him towards the con-clusion that he had probably preferred at the outset – from which we may be sure that his method was not as objective as he would have had the world believe. Nonetheless, its claim to have dispensed with bias and to rely only on indisputable logic is what makes *Leviathan* a landmark in the history of political theory.

But it is something more too. Hobbes's great work is seen today as historically and even philosophically important – but political science has become a very different beast, and no one seriously entertains the notion that Hobbes's arguments remain convincing. Nor should they, in one sense – for as we shall see, his basic postulates are very much a product of their times. Yet *Leviathan* is a direct and in many ways an astonishingly prescient antecedent to a revolutionary development now taking place at the forefront of modern physics. Scientists are beginning to realize that the theoretical framework that underpins contemporary physics can be adapted to describe social structures and behaviour, ranging from how traffic flows to how the economy fluctuates and how businesses are organized.

This framework is not as daunting as it might sound. Contrary to what one might imagine from the popular perception of modern physics, we do not have to delve into the imponderable paradoxes of quantum theory, or the mind-stretching revelations of relativity, or the origins of the universe in the Big Bang in order to understand the basic ideas behind these theories. No, this is an approach rooted in the behaviour of everyday substances and objects: of water, sand, magnets, crystals. But what can such things possibly have to say about the way societies organize themselves? A great deal, as it happens.

Hobbes had no inkling of any of this, but he shared the faith of today's physicists that human behaviour is not after all so complex that it cannot occasionally be understood on the basis of just a few simple postulates, or by the operation of what we might regard as *natural forces*. For Hobbes, contemplating the tumultuous political landscape of his country, the prime force could not be more plain: the lust for power.

THE LEVIATHAN WAKES

Thomas Hobbes (Figure 1.1) had never been able to take anything for granted. His father was a poorly educated and irascible vicar, a drunkard who left his family when Thomas was sixteen and died 'in obscurity'. This put his son to little inconvenience, since from a young age Thomas was supported and encouraged by his wealthy and

Figure 1.1 Thomas Hobbes was the first to seek a physics of society.

altogether more respectable uncle, Francis, a glover and alderman of Malmesbury. Francis watched over the boy's education, helping to nurture a clearly prodigious intellect: by the time the fourteen-year-old Thomas won admittance to Magdalen College at Oxford, he had already translated Euripides' *Medea* from Greek to Latin. He so excelled at the university that, when he graduated, he was recommended to the Earl of Devonshire as a tutor to the earl's son (himself only three years younger than Thomas). From such a position Hobbes was free to continue his studies of the classics. In his early twenties he acted as secretary to Francis Bacon (1561–1626), whose interests ranged from natural science and philosophy to politics and ethics. During this time, until Bacon's death, Hobbes showed no evident inclinations towards science; but Bacon's rational turn of thought left a clear imprint on his thinking.

It was not until 1629 that the forty-year-old Hobbes, a committed classicist, had his eyes opened to the power of scientific and mathematical reasoning. The story goes that Hobbes happened to glance at a book which lay open in a library, and was transfixed. The book was Euclid's *Elements of Geometry*, and Hobbes began to follow one of the Propositions. 'By God, this is impossible!', he exclaimed – but was soon persuaded otherwise. As Hobbes's contemporary, the gossipy biographer John Aubrey, tells it,

So he reads the Demonstration of it, which referred him back to such a Proposition; which proposition he read: that referred him back to another, which he also read, and *sic deinceps* [so on], that at last he was demonstratively convinced of that trueth. This made him in love with Geometry.[2]

Hobbes was deeply impressed by how this kind of deductive reasoning, working forward from elementary propositions, allowed geometers to reach ineluctable conclusions with which all honest and percipient people would be compelled to agree. It was a prescription for certainty.

The axioms of geometry are, by and large, statements that few people would have trouble supposing. They assert such things as 'Two straight lines cannot enclose a space.' We can often convince ourselves of their validity with simple sketches. Other fields of enquiry struggle to muster analogous self-evident starting points. 'I think, therefore I am' may have convinced Descartes that, as an axiom, it is 'so solid and so certain that all the most extravagant suppositions of the sceptics were incapable of upsetting it'; but in fact every word of the sentence is open to debate, and it has none of the compelling visual power of geometry's first principles.

Hobbes was sufficiently enthused to become a would-be geometer himself, but he was never a master of the subject. Through clumsy errors he persuaded himself that he had solved the old geometric conundrum of 'squaring the circle' (a task that is in fact impossible). But that was not his principal concern. In the 1630s the tensions between Crown and Commons led Charles I to dissolve Parliament and embark on an eleven-year period of 'Personal Rule'. In the midst of an unstable society, Hobbes wanted to find a theory of governance with credentials as unimpeachable as those of Euclid's geometry. First, he needed some fundamental hypothesis about human behaviour, which in turn had to be grounded in the deepest soil of science. And there was one man who had dug deeper than any other. In the spring of 1636, Hobbes travelled to Florence to meet Galileo.

The fundamental laws describing how objects move in space are called Newton's laws, since it was Sir Isaac who first formulated them clearly. But the tallest giant from whose shoulders Newton saw afar was Galileo Galilei (1564–1642), who laid the foundations of modern mechanics. Galileo taught the world about falling bodies, which, he said, accelerate at a constant rate as they descend (if one ignores the effects of air resistance). And with his law of inertia, Galileo went beyond the 'common-sense' view of Aristotle (384–322 B C) that objects must be continually pushed if they are not to slow down: on the contrary, said Galileo, in the absence of any force an object will continue to move indefinitely in a straight line at constant velocity.

Aristotle's view is the 'common-sense' one because it is what we experience in everyday life. If you stop pedalling your bicycle, you will eventually come to a standstill. But Galileo realized that this is because frictional forces act in nature to slow us down. If we can eliminate all the forces acting on a body, including gravity and friction, the natural state of the body is motion in an unchanging direction at unchanging speed. This was a truly profound theory, for it saw beyond the practical limitations of Galileo's age to a beautiful and simple truth. (An air pump that could create a good vacuum and thus eliminate air resistance was not invented until 1654.)

Galileo's law of inertia is without doubt one of the deepest laws of nature. On meeting the great man, Hobbes became convinced that this must be the axiom he was seeking. Constant motion was the natural state of all things – including people. All human sensations and emotions, he concluded, were the result of motion. From this basic principle Hobbes would work upwards to a theory of society.

What, precisely, does Hobbes mean by this assumption? It is, to modern eyes, a cold and soulless (not to mention an obscure) description of human nature. He pictured a person as a sophisticated mechanism acted upon by external forces. This machine consists of not only the body with its nerves, muscles and sense organs, but also the mind with its imagination, memory and reason. The mind is purely a kind of calculating machine – a computer, if you will. Such machines were popular in the seventeenth century: the Scottish mathematician John Napier (1550–1617) devised one, as did the French philosopher and mathematician Blaise Pascal (1623–62). They were mechanical devices for adding and subtracting numbers; and this, said Hobbes, is all the mind does too:

When a man *Reasoneth*, hee does nothing else but conceive a summe totall, from *Addition* of parcels; or conceive a Remainder, from *Subtraction* of one summe from another . . . For REASON . . . is nothing but *Reckoning*.[3]

The body, meanwhile, is merely a system of jointed limbs moved by the strings and pulleys of muscles and nerves. Man is an automaton.

Indeed, Hobbes held that the ingenious mechanical automata created by some inventors of the era were truly possessed of a kind of primitive life. To him there was nothing mysterious or upsetting about such an idea. Others were less sanguine: the Spanish Inquisition imprisoned some makers of automata on the grounds that they were dabbling in witchcraft and black magic.

What impelled Hobbes's mechanical people into action was not just external stimuli relayed to the brain by the apparatus of the senses. They were imbued also with an inner compulsion to remain in motion. For what is death but immobility, and which person did not seek to avoid death? 'Every man . . .', said Hobbes, 'shuns . . . death, and this he doth, by a certain impulsion of nature, no less than that whereby a stone moves downward.'[4]

Mankind's volitions, therefore, are divided by Hobbes into 'appetites' and 'aversions': the desire to seek ways of continuing this motion and to avoid things that obstruct them. Some appetites are innate, such as hunger; others are learnt through experience. To decide on a course of action, we weigh up the relevant appetites and aversions and act accordingly.

What Hobbes means by 'motion' is a little vague, for he clearly does not intend to imply that we are forever seeking to run around at full pelt. Motion is rather a kind of liberty – a freedom to move at will. Those things that impede liberty impede motion. Even if a man sits still, the mechanism of his mind may be in furious motion: the freedom to think is an innate desire too.

What room is there in this mechanical description for free will? According to Hobbes, there is none – he was a strict determinist. Humans are puppets whose strings are pulled by the forces at play in the world. Yet Hobbes saw nothing intolerable in this bleak picture. After all, he believed that he had arrived at this basic, indisputable postulate about human nature by *introspection* – by considering his own nature. The first puppet he saw was himself:

whosoever looketh into himself, and considereth what he doth, when he does *think*, *opine*, *reason*, *hope*, *feare*, &c, and upon what grounds; he shall thereby read and know, what are the thoughts, and Passions of all other men, upon the like occasions.[5]

THE MECHANISTIC PHILOSOPHY

If we shudder at this concept of humanity today, it is partly because we regard mechanical, clockwork devices as crude and clumsy. There are now many materialist scientists and philosophers who believe that the brain is a kind of vast and squishy computer whose secrets reside in nothing more than the extreme interconnectedness of its billions of biological switches. As a superior version of our most advanced cultural artefact, this view of the brain is neither unusual nor eccentric.

To the intellectuals of the seventeenth century the same was true of the clock, which as a reliable timekeeper was still a recent innovation. In that age there was nothing inelegant about a mechanical picture of humanity; on the contrary, it showed just how wonderfully wrought people were. As Descartes said:

And as a clock, composed of wheels and counterweights, observes not the laws of nature when it is ill made, and points out the hours incorrectly, than when it satisfies the desire of the maker in every respect; so likewise if the body of man be considered as a kind of machine, so made up and composed of bones, nerves, muscles, veins, blood, and skin, that although there were in it no mind, it would still exhibit the same motions which it at present manifests voluntarily.[6]

As above, so below. If mankind was a clockwork mechanism, so too was the universe. The planets and stars revolved like the gears of a

19

clock contrived by God, the cosmic clockmaker. This set in train the debate about whether or not God's skill had left him any cause to intervene in the world once it was 'wound', which culminated in an intemperate argument between Gottfried Leibniz and Isaac Newton (who seldom argued temperately).

And if the universe was a clockwork mechanism, the way to understand it was to take it apart, piece by piece: to apply the reductionist methodology of science. It was precisely this approach that Hobbes chose to use to analyse the workings of society: he would resolve it into its constituent parts and descry in their motions the simple causative forces. This was his intention in *Leviathan*'s precursor, *De cive* ('On the Citizen'), published in 1642, which contained many of the same ideas:

For everything is best understood by its constitutive causes. For as in a watch, or some such small engine, the matter, figure, and motion of the wheels cannot well be known except it be taken asunder and viewed in parts.[7]

By this time Hobbes had joined other Royalist sympathizers in exile in Paris. He sensed what was in the air in England in 1640, when Charles I had been forced to reconvene Parliament in order to gather taxes to finance the suppression of rebellion in Scotland. So anti-Royalist was the new 'Short' Parliament, which had smouldered in banished discontent for eleven years, that the king rapidly dissolved it again, only to have to resurrect it once more when the Scottish army reached Durham on its march south. From there it was a downhill slide to the outbreak of civil war in 1642. Fearing that his political writings would draw censure (or worse) from the belligerent Short Parliament of 1640, Hobbes left for France.

So Hobbes had thus formulated most of his ideas on 'civil governments and the duties of subjects' before the war began; but its impending prospect lent his efforts some urgency. He had originally intended to write a three-part thesis that began with traditional physics, extended these ideas to the nature of humankind, and only subsequently developed a 'scientific' theory of government. But as he later explained, *De cive* was hastened by circumstances:

my country some years before the civil war did rage, was so boiling hot with questions concerning the rights of dominion and the obedience due from subjects, the true forerunners of an approaching war; and this was the cause which ripened and plucked from me this third part.[8]

In France, Hobbes joined the circle of mechanistic French philosophers whose acquaintance he had made during his earlier European trip in 1634–7. Among them were Marin Mersenne (1588–1648) and Pierre Gassendi (1592–1655), colleagues of Descartes and two of the most enthusiastic supporters of the mechanical world-view. In this sympathetic environment Hobbes refined his theory of human nature and carried it through to deduce the consequences for civic structure. *Leviathan* was published in 1651, and Hobbes presented it to the fugitive Charles II in exile, to whom he had once taught mathematics. There was to be no one, Royalist or Roundhead, who was pleased by what it said.

THE UTOPIANS

Hobbes was not the first to imagine a utopia based on scientific reasoning. The governing philosophers of Plato's Republic live simply and own no private property, but they have absolute power over the lower classes of soldiers and common people, with whom Plato is little concerned. His is a utopia for aristocrats only; the mob might as well be living in a totalitarian, if benevolent, state. But the word 'utopia' comes from the imaginary land devised by the scholar and lawyer Thomas More (1478–1535). In his book *Utopia* (1518), a sailor named Raphael Hythloday describes the eponymous island where he dwelt for five years after sailing there by chance. The meaning of the name is debated; but the common interpretation renders it as either 'good place' or 'no place'.

In More's Utopia, everything is ideal. There is no ownership: everyone lives in an identical house, but houses are exchanged every ten years to dispel any notion that individuals own their homes. All people of the same sex are dressed alike, and their clothing is simple and immune to fashion. Everyone works – enough but not too much – and they are offered non-compulsory educational lectures. All of the many religions are tolerated, and people live moderately and modestly. It is a vision on the one hand refreshingly liberal, equal and just, and on the other terrifyingly bland and spiritless.

When Francis Bacon drew up his own version of the perfect society, he made science its linchpin. *New Atlantis* was a book he never finished; it was published, incomplete, the year after his death. The title harks back to Plato, who mentions the fabled lost civilization several times in his dialogues. But Bacon employs the same conceit as More: European sailors are driven off course in the Pacific Ocean and find themselves at the previously unknown island of Bensalem (a Hebrew word meaning 'son of peace', although the implication is that this is the 'New Jerusalem'). It is a Christian society that dwells on Bensalem, welcoming, kind and compassionate but also fiercely patriarchal and hierarchical. Central to the culture of Bensalem is Salomon's House, an institution devoted to science and the acquisition of new knowledge. The scientists (Fathers) dress and act rather like priests, and have access to vast resources for research. There are laboratories where nature is not only examined but also imitated and manipulated. Artificial environments resembling mines reproduce the conditions in which metals and minerals are formed; new living species are devised and created. 'Neither do we this by chance,' a Father explains, 'but we know beforehand of what matter and commixture, what kind of those creatures will arise.'[9]

Salomon's House resembles in many ways a modern research institution, albeit one unfettered by any constraints on research ethics. Some might see in it the blueprint for biotechnological laboratories in which the stuff of life is cut up, spliced and reconstituted. The Fathers take an oath of secrecy and reveal their inventions only if it suits them. One cannot imagine Bacon having much difficulty with the modern concept of private companies patenting genes.

But Bacon's Bensalem is an essentially arbitrary society: a vision of what its author considered desirable, and one devoted to, rather than derived from, scientific principles. This is why Hobbes's *Leviathan* is original. He does not describe a society ready-made and shaped by his own preferences, but builds it up, with careful logic, from his mechanistic view of how humans behave.

We should take care with what we mean by that. Hobbes was not especially interested in psychology, or in deducing how people will respond to a particular set of circumstances. He was pursuing a moral philosophy – asking whether a course of behaviour is *right*. In this respect, the ground was prepared for him (and characteristically for the times, he does not acknowledge it) by the Dutch philosopher Hugo Grotius (1583–1645), whose *The Laws of War and Peace* (1625) attempted to find the irreducible characteristics of human social existence. Grotius was not looking for scientific or mathematical laws as we would now understand them, but for 'natural laws', which again might be better regarded as natural rights. With ruthless efficiency, Grotius stripped society of its more pleasant features – benevolence, he said, is all very well, but it is not fundamental. There are only two things that people have a natural right to exercise in the company of their fellows: an expectation that they will not be subjected to unwarranted attack, and the freedom to defend themselves if they are. So long as people confine themselves to self-preservation and refrain from injuring others without cause, society is possible. This, said Grotius, is the 'state of nature', the most basic state of social existence. Civilization generally does rather better than that, encouraging courtesies and friendships and learning and the arts and so forth – but these are all optional extras, and society as such can exist without them.

Thus Grotius's 'minimal society' was a grim affair, and his concept of natural rights was not, as we might suppose today, a precondition of liberalism. But it was not at all obvious how even such a brusque, unfriendly society might be maintained. For who was to say when aggression was warranted and when it was not? If food is short, are you justified in killing your neighbour to preserve yourself? Are you justified in doing that pre-emptively, as an insurance policy against possible famine next year? Even if everyone agrees to recognize their fellows' natural rights, social stability doesn't necessarily follow because there is no consensus about how to exercise them.

In the hierarchical societies of medieval Europe this seldom became a problem because people were accustomed to the idea that they should do as they were told by their superiors. They might resent this inequality, but it was rarely questioned. By the Renaissance those certainties had broken down – partly because of changes to the structure of society, partly because of religious unrest and the Reformation, partly because Humanism had exposed people to new ways of thinking and there was more awareness of the sheer diversity of societies past and present. Society suddenly seemed to lack foundational principles or agreed behavioural norms.

Hobbes realized that this relativism of opinions about how to exercise natural rights meant that in the end a 'state of nature' was all about one thing: power.

HOW TO BUILD A COMMONWEALTH

The person without liberty is without power. Even the most humble and self-effacing of us want a little power – to choose when we eat and sleep, where we live and with whom, what we may say or do. Many millions of people in the world lack some or all of these freedoms, but they are among those acknowledged internationally, in the Universal Declaration of Human Rights, as liberties that everyone deserves simply by virtue of being alive.

Hobbes defined power as the ability to secure well-being or personal advantage, 'to obtain some future apparent Good'. People, he said, have some 'Naturall Power' that enables them to do this, stemming from innate qualities such as strength, eloquence and prudence. And they may use these qualities to acquire 'Instrumentall Power', which is merely 'means and Instruments to acquire more': wealth, reputation, influential friends.

So Hobbes's model of society hinged on the assumption that people (if we say 'men' we are not, in this context, being inaccurate) seek to accumulate power, up to a personal level of satiety that varies between individuals. It is a cold-blooded prescription, for sure. The Scottish political scientist Robert MacIver has complained that it neglects all that is good and worthy in man:

Hobbes ignored all the social bonds that spread out from the life of the family, all the traditions and indoctrinations that hold groups of men together, all the customs and innumerable adjustments that reveal the social-izing tendencies of human nature.[10]

Doubtless that is so, and we may want to make the same complaint. The social historian Lewis Mumford condemns this kind of abstraction of society, saying that it reduces the individual to 'an atom of power, ruthlessly seeking whatever power can command'.[11] It has to be admitted that this is precisely what Hobbes intended.

Yet even the nineteenth-century Romantic Ralph Waldo Emerson seems to agree with the Hobbesian interpretation of human nature when he says, 'Life is a search after power.' And in any event, we can agree or disagree with Hobbes's wolfish view of humanity while nevertheless phrasing the valid question: given these postulates, what follows? *If* men behave in this way, what kind of society can arise and be maintained?

Power is relative: the true measure is the amount by which one man's power exceeds that of the others around him. It follows, Hobbes said, that the quest for power is in fact a quest for command over the powers of other men. But how does one command the power of another? In the bourgeois market society that had come to dominate the cultural landscape of the mid-seventeenth century, the answer was simple: he buys it. One man pays another to act on his behalf and to submit to his will.

This does not necessarily mean, as it might sound, simply that a powerful man may hire others to act as bullies, henchmen and mercenaries. Rather, Hobbes had in mind the way a rich merchant employs workers to make and distribute his goods, or a craftsman takes on assistants to execute a contract. Yet his formulation is as icy as his model of man as machine: 'The Value, or WORTH of a man, is as of all other things, his Price; that is to say, so much as would be given for the use of his Power.'[12] It is the ethic of the free market – buy out the competition.

It is not obvious that a society in which appetites for power vary need in itself be an unsettled one, for those with moderate ambitions might be happy enough to work for those with stronger desires. But Hobbes maintained that some men's appetites know no limits. Such power-hungry individuals destabilize a society in which less ambitious men might otherwise labour in harmony. 'I put for a generall inclination of all mankind', he said,

a perpetuall and restlesse desire of Power after power, that ceaseth only in Death. And the cause of this, is not alwayes that a man hopes for a more intensive delight, than he has already attained to; or that he cannot be content with a moderate power: but because he cannot assure the power and means to live well, which he hath present, without the acquisition of more.[13]

And so all are sucked into a perpetual power struggle. Unchecked, this leads to Hobbes's own vision of a State of Nature, in comparison to which Grotius's version – a crabbed, surly society – might sound positively idyllic. It is as bleak and frightening as you can get.

Without any law or law enforcers, every man is open to violent exploitation by others. When everyone seeks to dominate his neighbour without restraint, says Hobbes, there is

no place for Industry; . . . no Culture of the Earth . . . no Knowledge of the face of the Earth . . . no Arts; no Letters; no Society; and which is worst of all, continuall feare, and danger of violent death; And the life of man, solitary, poore, nasty, brutish, and short.[14]

Who would not do all they could to escape such a state? But to proceed logically to a better way, Hobbes found it necessary to introduce two more postulates, which he elevated to the status of Laws of Nature. The first says that a man will not seek actively to harm himself or endanger his life, or to overlook ways of making it safer. Reasonable enough at first glance, this in fact accords us extraordinary percipience in seeing the consequences of the actions we choose, so that we will always make the one most favourable to our self-preservation. But the second law is still more debatable:

That a man be willing, when others are so too, as farre-forth, as for Peace, and defence of himselfe he shall think it necessary, to lay down this right to all things; and be contented with so much liberty against other men, as he would allow other men against himselfe.[15]

In other words, men will, as a corollary of their instinct for self-preservation, be prepared to suppress their exploitative impulses and cooperate with one another. Only thus can peace and stability be brought to the State of Nature.

But cooperation is not enough. For men's unceasing appetites for power will make them liable to defect from this contract the moment they see any advantage in doing so. We shall see later that Hobbes here essentially formulates, three hundred years ahead of time, one of the most influential behavioural dilemmas of the modern era. The solution, he reasons, is for men not simply to give up some of their natural rights to do as they please, but to *transfer* these rights to some authority which is then granted the mandate to impose the contract – by force if necessary.

In whom should this authority reside? Hobbes felt that it did not greatly matter, so long as the authority was there. His fundamental postulates assume a degree of equality among men rarely voiced in seventeenth-century Europe: in the State of Nature, no man's status is greater than another's, although some have the advantage of greater 'Naturall Power'. But then the community *elects* some individual and confers on them absolute power. In effect, they choose a monarch and thenceforth defer to him or her without question.

This resolution is a peculiar mixture, for it amounts to the creation of a despotism by democratic means out of an anarchic state. Hobbes admits that the supreme authority could conceivably be an elected body, not an individual – a Parliament, in effect. But he suspects (and who can dispute it?) that with more than one head of state, internal power conflicts will arise sooner or later.

The powers of Hobbes's elected monarch are absolute, stopping only at the right of individuals always to preserve their own lives. It is up to the sovereign, once elected, to decide how much of each man's power he must enlist to maintain the social contract. Even to a tyranny, says Hobbes, citizens owe an obligation of duty and submission. At the same time this absolutism unites people into a cohesive unit, a Commonwealth: the Leviathan. It was a curious name to give to a supposedly desirable state of society – almost as though Hobbes positively wanted his readers to envisage a dreadful, oppressive regime. Leviathan is a fearsome sea-creature mentioned in the Book of Job:

> If you lay a hand on him,
> You will remember the struggle and never do it again!
> Any hope of subduing him is false;
> The mere sight of him is overpowering . . .
> When he rises up, the mighty are terrified;
> They retreat before his thrashing . . .
> Nothing on earth is his equal –
> A creature without fear.
> He looks down on all that are haughty;
> He is king over all that are proud.[16]

The message is plain – you disobey Leviathan's laws at your peril.

Yet because it has freely elected to be governed this way, the population in some sense shares in the political structure that results.

Figure 1.2 Hobbes's Leviathan is a sovereign who makes society cohere into a Commonwealth – by rule of the sword, if necessary. 'The only way', says Hobbes, 'to erect such a Common Power, as may be able to defend [people] from the invasion of Forraigners, and the injuries of one another . . . is, to conferre all their power and strength upon one Man, . . . that may reduce all their Wills, by plurality of voices, unto one Will . . . This done, the Multitude so united in one Person, is called a COMMON-WEALTH, in latine CIVITAS. This is the Generation of that great LEVIATHAN, or rather (to speak more reverently) of that *Mortall God*, to which wee owe under the *Immortall God*, our peace and defence.'

Leviathan is thus 'one person, of whose acts a great multitude . . . have made themselves every one the author'[17] – an image reinforced by the dramatic frontispiece to the first edition of the book, probably prepared by the artist Wenceslas Hollar (Figure 1.2). In personifying the State in this way, Hobbes was following a long tradition: in the fourteenth century, the Bishop of Rochester Thomas Brinton identified the prince as the head of the 'body politic' and the labourers at the feet. Others took delight in anatomizing every member of society, from priests (chest or ears) to merchants (thighs) to judges (ribs).

The justification for the Leviathan, says Hobbes, is 'the Convenience, or Aptitude to Produce the Peace, and Security of the People'. One can deplore his proposed means of achieving these aims, but the objectives themselves are nevertheless enshrined in all democracies today. In explaining how a mass of selfish individuals can unite to create a sovereign nation, Hobbes gave form to the modern idea of the State. More than this, even: according to historian Frederick Nussbaum, 'Hobbes discovered society.'[18]

And thus Thomas Hobbes believed he had proved monarchy to be the best system of rule, using science and reason alone. He felt that those nations that had enjoyed prolonged civic stability, such as Imperial Rome, had by good luck or judgement hit on the ideal solution that science now revealed with inexorable logic. 'The skill of making, and maintaining Commonwealths,' he said, 'consisteth in certain Rules, as doth Arithmetique and Geometry; not (as Tennis-play) on Practise onely.'[19]

A CALCULUS OF SOCIETY

One might think that Charles II would have been pleased with a treatise claiming to prove scientifically that kings were the best rulers. But he was not at all happy with *Leviathan* – for it proposed that the king comes from the ranks of ordinary men and is instated arbitrarily

31

by election of the masses, like a common parliament! Whereas it was well known that kings ruled by divine decree, deriving their authority not from some social contract but from a heavenly one. To the Royalists, the book was pure treason.

There was no comfort here for supporters of the parliamentary system either. Hobbes's supreme authority, be it an individual or a collective body, subsequently had the right to decide who would succeed them – democracy is exercised once and then relinquished. And to make matters worse, *Leviathan* offended the devout by lambasting those nations who 'acquiesce in the great Mysteries of Christian Religion, which are above Reason'.[20] This was deemed by many to be a declaration of atheism. Hobbes endeared himself to no one.

So it was a dangerous game that Hobbes now played. In the winter of 1651/2, shortly after his book appeared, he retreated from the hostility of the exiled Royalists and returned to Cromwell's England, where the desire for peace and stability under the Protectorate had introduced a degree of tolerance. Hobbes made friends within the new regime, and he fitted in quietly enough until Charles II was restored to the throne in 1660. If there was one thing that the Royalists, new and old, disliked more than Hobbes's political philosophy, it was his views on religion. He had become widely regarded as an atheist, especially by the dominant Anglican Royalists, and he might well have faced imprisonment if the bill to make Christian heresy a criminal offence had been passed by Parliament in 1666. The threat was ever present for the remainder of Hobbes's lifetime; but in spite of this, and decades of ill health notwithstanding, he survived to the truly venerable age of ninety-one.

No nation chose to put the advice in *Leviathan* into practice. Indeed, according to historian Richard Olson, 'because they seemed to inspire both immorality and revolution, Hobbes's theories were generally feared and detested by all respectable persons.'[21] To Scottish philosopher David Hume, 'Hobbes's politics promoted tyranny and his ethics encouraged licentiousness.'[22] But because his ideas were argued with such compelling force and precision, they posed a challenge to all subsequent political philosophers. You could be appalled by Hobbes, but you could not ignore him.

Above all, *Leviathan* established the idea that there was room for reason in politics. Previous utopias were not deductive; their validity was simply asserted. In general they sought either to shore up the status quo or to portray a society conjured into existence from the author's imagination, with no explanation of how things got to be that way. The Leviathan, on the other hand, was at least ostensibly the product of mechanistic science. It was not even necessarily something to be celebrated, but was a necessary evil, the only alternative to grim anarchy.

The social contract proposed by Hobbes might sound like a forerunner of those advocated by John Locke (1623–1704) and Jean-Jacques Rousseau (1712–78), but it is instead the reverse. To Locke and Rousseau, the power conferred upon the head of state comes with an obligation to serve the interests of the populace; for Hobbes, the common people become contracted to serve their ruler. For Hobbes, the principal fear was of anarchy; for Locke it was the abuse of power, which is why he saw the need for safeguards to avoid absolutism.

But although apparently a proponent of autocracy, Hobbes also provides arguments which can be used to support both bourgeois capitalism and liberalism. He expressed an aversion to the way the mercantile society bred men whose 'only glory [is] to grow excessively rich by the wisdom of buying and selling', which they do 'by making poor people sell their labour to them at their own prices.'[23] Yet he saw bourgeois culture as largely inevitable, and sought a system which would accommodate its selfish tendencies without conflict. To this end he left it to the market to assign the value of everything, people included: 'the value of all things contracted for, is measured by the Appetite of the Contractors: and therefore the just value, is that which they be contracted to give.'[24] This free-market philosophy found voice in Adam Smith's *Wealth of Nations* in the following century. Those in Britain and the USA – and elsewhere – who lived through the 1980s will recognize it as an attitude that did not wane with the Age of Enlightenment.

MAN AND MECHANISM

A political scientist taking a chronological approach would track the trajectory of Hobbes's thought via Locke to later thinkers who believed there could be such a thing as a 'calculus of society'. Along this path we would uncover Jeremy Bentham's utilitarianism in the late eighteenth century, an attempt to harmonize the individual's pursuit of personal happiness with the interests of society. Bentham, like Locke, believed that reason alone could show how this might be achieved. His solution was the 'greatest happiness' principle, an optimal state in which the sum total of human happiness was as large as it could possibly be, allowing for the conflicts of interest that inevitably arise when each person seeks their own advantage. Bentham's utopia was quite different from Hobbes's: a democracy with equality for all, including votes for women. Bentham and the Philosophical Radicals, who included John Stuart Mill, paved the way for the socialism of Karl Marx. Marx, of course, was also determined to formulate a 'scientific' political theory – one which in his case was strongly (and misguidedly) influenced by Darwinism.

And so we might go on. But I shall not. These theories indeed seek a foundation in rationality, and we shall revisit them from time to time.

But they are not scientific in the way that the real topic of this book is scientific. There are few political thinkers who have defined a social model with the logical precision of Hobbes, and none who has carried those precepts through to their conclusions in a truly scientific, rather than a suppositional, way. This is not by any means to denigrate such models; rather, it is simply to say that their approach is different. Political theorists tend to concern themselves with what they think *ought* to be; scientists concentrate on the way things *are*. The same is true of the new physics of society: it seeks to find descriptions of observed social phenomena and to understand how they might arise from simple assumptions. Equipped with such models, one can then ask what we would need to do in order to obtain a different result. Such decisions about what is desirable should properly be in the realm of public debate: they are no longer scientific questions. In this sense, the science becomes – as it should be – a servant and guide and not a dictator.

How is it that physics has come to have the confidence, perhaps even the arrogance, to venture into social science? No one in recent decades has set out to construct a physics that would be capable of this. It just so happens that physicists have realized that they have at their disposal tools which can be applied to this new task. These tools were not developed for that purpose; they were first developed to understand atoms.

Carolyn Merchant, in her book *The Death of Nature* (1983), argues that the rise of mechanistic, atomistic philosophy in the seventeenth century sanctioned the manipulations and violations of nature that continue to blight the world today. The utopian society envisaged by Thomas Hobbes, in which people are little more than automata impelled this way and that by mechanical forces, and where scientific reasoning is the arbiter of social justice, sounds like a chilling place to live. It is hard to imagine how any model of society which regards the

35

behaviour of individuals as governed by rigid mathematical rules can offer us a vision of a better way to live, rather than a nightmarish Brave New World.

That, I suspect, is the instinctive objection that many will have to the notion of a 'physics of society'. But I hope to show that the new incursion of physics into the social, political and economic sciences is not like this. It is not an attempt to prescribe systems of control and governance, still less to bolster with scientific reasoning prejudices about how society ought to be run. Neither does it really imagine that people are so many soulless, homogeneous effigies to be shuffled this way and that according to blind mathematics. Instead, what physicists are now trying to do is to gain some understanding of how patterns of behaviour emerge – and patterns undoubtedly do emerge – from the statistical mêlée of many individuals doing their own idiosyncratic thing: helping or swindling one another, cooperating or conflicting, following the crowd or blazing their own trail. By gaining such knowledge we might hope to adapt our social structures to the way things *are* rather than the way some architect or politician or town planner thinks they ought to be. We might identify modes of organization which fit with the way we actually and instinctively act.

These are potential practical benefits of a genuinely inquisitive physics of society; but there emerges from such efforts a broader message too. It is this: collective actions and effects are inevitable. No matter how individualistic we like to think we are, our deeds are often the invisible details of a larger picture. This is not necessarily a description of impotence. Environmentalists and other activists like to entreat us to 'think globally, act locally'. But the physics of society shows that the reverse can take effect too: by concerning ourselves with nothing more than how we interact with our immediate neighbours, by 'thinking locally', we can collectively acquire a coherent, global influence. The consequences of that – good or bad – are worth knowing.

No scientific theory will show us how to build a utopia, but the search for a physics of society will benefit from our acknowledging the lessons of those quixotic attempts, like that of Thomas Hobbes, to do so in the past. These efforts to create a rational utopia show us the dangers of such a rigid programme. Science provides not prescriptions but descriptions. With such understanding, we might hope to make our choices with clearer vision.

2

Lesser Forces

The mechanical philosophy of matter

Nature, it seems, is the popular game
For milliards and milliards and milliards
Of particles playing their infinite game
Of billiards and billiards and billiards.
   Piet Hein (1966)

The Boltzmann is magnificent. I have almost finished it. He is a masterly expounder. I am firmly convinced that the principles of the theory are right, which means that I am convinced that in the case of gases we are really dealing with discrete point masses of definite size, which are moving according to certain conditions . . . This is a step forward in the dynamical explanation of physical phenomena.
   Albert Einstein (1900)

I have endeavoured to show that it is the peculiar function of physical science to lead us to the confines of the incomprehensible, and to bid us behold it and receive it in faith, till such time as the mystery shall open.
   James Clerk Maxwell (1856)

*

Laws make life simpler, and that can be liberating. Immanuel Kant realized this when he said, 'Man is free if he needs to obey no person but solely the laws.'[1]

It is not a trivial matter that science has come to use legal terminology to describe regularities in nature. 'I'm arresting you for breaking the laws of physics', says the policeman to the levitating man in a cartoon. Like many good jokes, this one reveals the snares that language sets. We can break society's laws if we dare, but the laws of physics do not need enforcing, for they are inviolable.

If the Enlightenment enthusiasm for a mechanistic philosophy looks naive to us now, let us not forget what it offered. Such 'natural laws' as Aristotle divined were hardly simplifications; often they were mere tautologies. Objects fell to Earth because they had a downward tendency. The Sun and Moon followed their arcs across the sky because heavenly bodies had a circulating tendency. In contrast, Newton's law of gravity rationalized why cannonballs fall and why the Moon does not. It condensed pages of astronomical data into a concise, simple formula. It helped to fit disparate observations into a single framework. And beyond all this, it suggested that humankind can understand, and not just experience, the hows and whys of existence.

The mechanical laws of Galileo and Newton hold true for planetary orbits and for motes of dust, for a falling apple and a falling star. They are deep and elegant truths, so far as truth can ever be discerned, about the way the universe works. Maybe we can therefore forgive Hobbes and his contemporaries their propensity to use mechanics to explain everything – even the mysteries of the human mind. But in the two centuries that followed the publication of *Leviathan*, delight in mechanics did not diminish. On the contrary, scientists saw ever more reason to believe that they had grasped the central governing principles of all matter, and that explanations for all phenomena simply required the right mechanical description.

It is this account of matter at the fundamental level, hatched in the nineteenth century, that underpins the physics of society. In this chapter we shall see where it came from and what it consists of. It is a theory which invokes many players, and each of them is too small to see. That the whole world can be constructed from a small variety of atoms is an astonishing thing. Understanding what they do when they get together is one of science's greatest triumphs. But no one could have expected this understanding to lead where it has.

PIECES OF EVERYTHING

As the fundamental, irreducible constituent particles of all things, atoms (meaning 'uncuttable') were postulated around 440 BC by the Greek philosopher Leucippus. His pupil Democritus worked out the implications of the hypothesis in great detail. The idea of atoms led to controversy about whether or not there was space (void) between the particles. Anaxagoras (c.500–428 BC) dismissed the notion of void, but the Athenian Epicurus (341–270 BC) questioned how anything could move if all space were packed full of atoms.

Democritus' atomism fell out of favour for two millennia, largely because Aristotle did not like it. Medieval theologians rejected the hypothesis because it could not be made to fit the Christian belief in transubstantiation. Interest was revived in the fifteenth century by the rediscovery of the poem *De rerum natura* ('On the Nature of Things') by the Roman philosopher Lucretius (99–55 BC), a follower of the atomistic doctrine of Epicurus.

Galileo, Francis Bacon, Pierre Gassendi and Isaac Newton believed in atoms, but many other great thinkers did not. While accepting that matter might be made up of small particles, René Descartes saw no reason to assume they could not be divided indefinitely. He asserted that they were borne along like grains and dust in swirling vortices of some all-pervading fluid.

It was generally agreed that the microscopic realm was a world in motion – which implied that mechanics could be used to understand the everyday properties of matter. The idea was first enunciated clearly by Daniel Bernoulli (1700–1782), a mathematician of Flemish descent born in Basle, Switzerland. In 1738 he proposed that gases are composed of tiny particles rushing around and colliding. The pressure exerted by a gas – on the side of an inflated balloon, say – was the result of all the little impacts of the particles hitting the surface.

In 1763 a Serbian Jesuit named Roger Joseph Boscovich (1711–87) identified the ultimate implication of this mechanical atomic theory. A crucial aspect of Isaac Newton's laws of motion is their predictive capability. If we know how an object is moving at any instant – how fast, and in which direction – and if we also know the forces acting on it, we can calculate its future trajectory exactly. This predictability made it possible for astronomers to use Newton's laws of motion and gravity to calculate when, for example, future lunar and solar eclipses would happen.

Boscovich realized that if all the world is just atoms in motion and collision, then an all-seeing mind

could, from a continuous arc described in an interval of time, no matter how small, by all points of matter, derive the law [that is, a universal map] of forces itself . . . Now, if the law of forces were known, & the position, velocity & direction of all the points at any given instant, it would be possible for a mind of this type to foresee all the necessary subsequent motions & states, & to predict all the phenomena that necessarily followed from them.[2]

That is to say, a mathematician with god-like omniscience could deduce the rest of history, for ever and ever, from a mere moment in time. Compared with Hobbes's version of determinism, in which people are automata moving at the insistence of mechanical forces, this is an altogether more constraining straitjacket for the world. Nothing is unknown or uncertain, and there is no deviation from the inevitable play of forces. The fact that no human mind could possibly make such an astronomical calculation is irrelevant: in Boscovich's view, the future was already defined by the present. The eminent French mathematician Pierre-Simon Laplace (1749–1827) made a similar statement in 1814, which, like its author, is far better known. For such an awesome intelligence, said Laplace, 'the future, like the past, would be present before its eyes.'[3]

Mechanism, it seemed, had banished free will.

DISSIPATION AND DEATH

The implications of a mechanical universe were not just philosophical. With the Industrial Revolution in full swing at the dawn of the nineteenth century, there were pressing practical matters for scientists to address. In a short life terminated prematurely by cholera, the French scientist Nicholas Léonard Sadi Carnot (1796–1832) busied himself with one of the most important of these problems: how to optimize the fuel efficiency of a steam engine.

What was true of power generation in Carnot's time is largely true today: extracting work from an engine means generating heat and letting it flow. Think of a coal-fired gas turbine. Heat produced by burning fuel is transferred from the burner to the gas. The hot gas expands, the pressure rises, and a jet is released which drives the rotating blades of the turbine. The rotation turns an electromagnet, creating electricity in the coils. The steam engine, workhorse of the Industrial Revolution, likewise used the expansion of a hot gas: water vapour.

But what, exactly, is heat? In the late eighteenth century many eminent scientists agreed that it was a physical substance called 'caloric' which flows from hot to cold. The American scientist Benjamin Thompson (1753–1814) thought otherwise.[*] Heat, he suggested, is the random motion of atoms in collision. It is not the *product* of such motions – the frictional heating caused by the atoms' surfaces rubbing together. No, it must be identified with these motions themselves. A substance heats up when its atoms are made to jiggle more furiously, for example as a result of atomic collisions when the substance comes into physical contact with another material in which the motions are already very lively.[†] Carnot agreed with this proposal: 'Heat is then the result of a motion',[4] he wrote in 1824. The mechanical world of atoms had rationalized an old mystery.

Engineers needed to capture some of this microscopic motion and turn it into the motion of railway carriages, factory machinery and pumps. Carnot realized that this is contingent on getting heat to *flow* from a hot body to a cooler one. He deduced a general theory for calculating how much of this heat flow could be converted to useful work (the conversion is never perfect because some heat is inevitably squandered) and how this depends on the difference in temperature between the hot heat source and cold heat sink. To develop his argument, Carnot considered an engine in which heat flow allowed a gas to expand (when heated) and contract (when cooled), driving a piston in a cyclic process now known as the Carnot cycle. His analysis

[*] Thompson, who later became Count Rumford, founded London's Royal Institution in 1799.

[†] This was not an entirely unprecedented notion, for a mechanical theory of heat was first proposed by Robert Boyle in 1675.

laid the cornerstone of a new discipline called *thermodynamics* – literally, 'heat movement'.

Most people who have encountered thermodynamics blanch at its mention, because it is an awesomely tedious discipline both to learn theoretically and to investigate experimentally. This is a shame, because it is also one of the most astonishing theories in science. Think of it: here is a field of study initiated to help nineteenth-century engineers make better engines, and it turns out to produce some of the grandest and most fundamental statements about the way the entire universe works.* Thermodynamics is the science of change – and without change there is nothing to be said.

Thermodynamics, like Newton's theory of motion, has three laws. The third is hardly worth knowing unless you are a physicist; the first two should be engraved in the mind of anyone who wants to understand science.

The First Law is the easiest: energy is never destroyed but only transformed. Photovoltaic panels gobble up the energy of sunlight, and turn it – some of it, never all – into electrical energy. (Most of this solar energy is, alas, wasted as heat.) In a turbine, heat is transformed into the energy of motion (called kinetic energy) of the turbine blades, and then into electrical energy. Thus the universe conserves its energy. Only once heat had been identified as the movement (kinetic energy) of atoms could this law be properly formulated.

The Second Law is more remarkable, and some scientists believe we still don't fully understand it. A testament to its importance is C. P. Snow's famous (albeit perhaps overblown) complaint in his

* Physicist Erwin Schrödinger is probably right to point out that thermodynamics owes more to steam engines than steam engines do to thermodynamics. The theory has practical implications, without doubt; but it soon leads us into discussions verging on the metaphysical.

book *The Two Cultures*:

A good many times I have been present at gatherings of people who, by the standards of the traditional culture, are thought highly educated and who have with considerable gusto been expressing their incredulity at the illiteracy of scientists. Once or twice I have been provoked and have asked the company how many of them could describe the Second Law of Thermodynamics. The response was cold: it was also negative. Yet I was asking something which is about the scientific equivalent of: *Have you read a work of Shakespeare's?*[5]

There are several ways of expressing this law. When the German physicist Rudolph Clausius (1822–88) first did so in 1850, he said something along the lines that heat always flows from hot to cold. As anticlimaxes go, this is a pretty damp squib. But what he really meant was that there exist processes which go only one way, which are *irreversible*. Water does not flow uphill, and neither, figuratively speaking, does heat.

This seemingly innocuous statement is really the secret of all change. If there are irreversible processes, then time has an arrow – a singular direction defined by such processes. The Second Law connects to our perception that we are always moving forward in time, never back.

But Clausius did not let it go at that. He contrived the concept that enabled a mathematical theory of change and irreversibility: *entropy*. Entropy crops up in thermodynamics as a rather abstract quantity, but one can in fact measure it just as one can measure the heat released during a chemical reaction. Crudely speaking, entropy is a measure of the amount of disorder in a system (I shall sharpen this definition shortly). The Second Law reduces to the statement that in all processes of spontaneous change (such as heat flowing from hot to cold), entropy increases.

In 1852 William Thomson (1824–1907), later Lord Kelvin, noticed something peculiar about the way energy gets transformed. There is, he said, 'a universal tendency in nature to the dissipation of mechanical energy'.[6] What he meant was that some energy is always 'wasted' as heat (that is, random atomic motion). Think of that rotating turbine, in which friction warms up the bearings. It is hard to win back any useful energy* from this heat, which leaks away into the surroundings. In 1854 the German physicist Hermann von Helmholtz (1821–94) perceived the consequences of this inevitable dissipation: the universe would end up as a uniform, tepid reservoir of heat. No further change would then be possible because there was nowhere colder for the heat to flow. Thus, he said, the universe would ultimately die a 'heat death'. In the behaviour of steam engines we can read the fate of all creation.

THE DANCE OF PROBABILITY

Right from the inception of thermodynamics, scientists wanted to know where its rules came from. If all the world is just atoms in motion, each of them obeying Newton's laws, should it not be possible to deduce the laws of thermodynamics just by considering all those invisible collisions?

Daniel Bernoulli began that quest with his explanation of gas pressure. An Englishman named John Herapath (1790–1869) wondered what manner of motions would be required to account for the pressure a gas exerted, and he calculated that the gas particles (atoms, or molecules – small clusters of atoms) would have to be travelling at speeds of something like two kilometres per second.

The pressure of a gas can be altered by changing its temperature.

* That is, energy that can be used to conduct some mechanical task, such as lifting a weight or moving a wheel. Scientists call useful energy 'work'.

If you heat up a gas in a sealed vessel – that is, in a fixed volume of space – its pressure increases. This is why aerosol cans explode if thrown into a fire. If, on the other hand, the volume isn't fixed – if the walls of the vessel are movable – then a hot gas expands. This is what drives the piston in Carnot's cycle. In other words, the three characteristics of a gas – its temperature, pressure and volume – are rather like the notorious trio one encounters in engineering or business: cost, speed and quality. That is to say, if you specify any two of them, you have no say over the third: it is decided for you. We can arrange for a gas to have a particular temperature and pressure, but then the volume (or equivalently the density – the number of molecules in a given volume of space) is preordained. Another way of putting this is that, if we keep one of the trio constant, there is a mathematical relationship between the other two. At fixed volume, for instance, the pressure of a gas is proportional to its temperature.

These relationships between the temperature, pressure and volume of a gas – the so-called gas laws – were studied in the seventeenth century by Robert Boyle. Nearly a century later, Boyle's investigations were refined by the Frenchmen Jacques Charles (who made the first hydrogen-balloon flight in 1783) and Joseph Louis Gay-Lussac.

The challenge was to see whether the gas laws could be derived from a mechanical model in which atoms are like billiard balls, moving in straight lines until they collide with one another. Rudolph Clausius laid much of the groundwork for this so-called kinetic theory of gases in the 1850s, but it was brought to fruition mostly by one man, the Scottish physicist James Clerk Maxwell (1831–79) (Figure 2.1).

When a snooker player strikes a ball, it is not difficult to calculate what its motion will do to the other balls on the table. But in a single thimbleful of air there are about ten billion billion atoms. We cannot possibly know how they are all moving at any instant; and even if we did, the task of calculating how the motion would be altered by

Figure 2.1 James Clerk Maxwell, whose introduction of statistical ideas into the atomic theory of gases was just one of his major contributions to science. He also clarified the nature of colour, pioneered colour photography, and unified all electromagnetic phenomena in a single theory.

collisions in the next instant, and the next, is imponderable. So how can we hope to account for everyday behaviour, as described by the gas laws, starting from 'first principles' – from atomic motions?

Maxwell's key insight was that we do not need to know all the details. What is important is not the precise trajectory of every gas particle but their *average* behaviour. He pictured a swarm of bees, all buzzing about furiously in the air while the swarm itself hovers as a stationary mass, because on average the bees are no more likely to be flying in one direction than in any other.

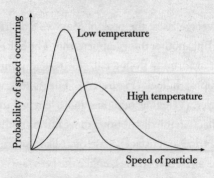

Figure 2.2 Maxwell's probability distribution for the speeds of particles in a gas. As the gas gets hotter, the peak shifts to higher velocities and gets broader and flatter.

All that matters about the motions of the gas particles, said Maxwell, are two things: how fast, on average, each particle is moving – which determines their average motional (kinetic) energy – and how broad is the spread in speeds either side of that average. Maxwell intuited that the distribution of speeds resembles the kind of bell-shaped curve you see in statistical surveys of, for example, the spread of wages. We shall see in the next chapter that this intuition owed a great deal to a nascent science of society.

Maxwell's curve, indicating how many gas particles are moving at each speed, rises smoothly from low speeds, hits a peak at the average, and then tails off smoothly towards high speeds (Figure 2.2). This distribution shows that rather few particles have speeds *much* higher than the average. As Welsh physical chemist Emyr Alun Moelwyn-Hughes once prophetically put it, 'Energy among molecules is like money among men. The rich are few, the poor numerous.'[7]

The average speed of the particles depends on how much kinetic energy the gas as a whole contains. Pump in more energy – heat the gas up – and the average rises: the peak of Maxwell's distribution shifts to higher speeds. But another thing happens too. The bell curve gets flatter and broader, transforming from a tall, steep-sided pinnacle to a lower, more gently sloping hillock (Figure 2.2). That is to say, the spread of speeds gets wider. (Whether pumping more 'energy' into an economy has the same effect on the distribution of wealth is another matter.)

Maxwell's gas does not in fact behave quite like a swarm of bees staying stationary in the air. The particles, unlike the bees, are constantly colliding. This means that their direction of motion is constantly changing, essentially at random. Yet even though each particle moves at random and there is no overall preference for movement in any direction, this does not mean that the particles stay clumped in a swarm. Particles moving at random do actually get somewhere, rather than forever meandering about a fixed position. Their erratic paths take them gradually farther from the starting point, but in a random direction. This is called a random walk, and physicists like to compare it to the path of a drunken man staggering uncontrollably towards no particular destination (Figure 2.3). A particle moving this way is said to be diffusing.

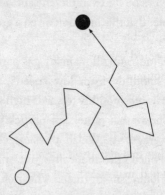

Figure 2.3 A particle bouncing between collisions in a gas executes what is called a random walk, drifting gradually farther from where it began.

Because of diffusion, a cluster of particles released into the air will gradually expand outwards in all directions, rather like an ink droplet dispersing in a glass of water. In the same way, two different gases filling adjacent compartments in a box will gradually mix if the partition between them is removed. Maxwell's mathematical analysis allows one to calculate how rapidly a particle moves by diffusion. This is considerably slower than the particle's actual speed because it takes a highly circuitous route to get from A to B.

It was by observing random walks that scientists finally reached a consensus on the existence of atoms. Maxwell's theory was predicated on the idea that gases are composed of atoms and molecules; he even used it, in 1873, to calculate the sizes of molecules, predicting that a hydrogen molecule is 0.000 000 6 millimetres across. (His estimate was out by only a factor of about 3.) But no one had actually *seen* atoms, and even at the end of the nineteenth century some scientists still refused to countenance them. Ernst Mach, a highly influential German physicist, regarded it as poor science to accept the existence of anything inaccessible to direct experience, and so preferred to withhold judgement on the atomistic theory. But in 1905 Albert Einstein published a seminal paper in which he showed that, by assuming gases to be composed of invisible particles (atoms or molecules) executing random walks, one could explain the hitherto mysterious phenomenon of Brownian motion.

Robert Brown was a great botanist, but he had no intention of pronouncing on physical theory. When in 1828 he first saw pollen grains dancing wildly under the microscope as they sat suspended in water, he thought that this activity revealed the fundamental 'active force' of life, embodied in the old hypothesis of vitalism. He later discovered that unambiguously 'dead' grains, including (bizarrely) fragments of the Egyptian sphinx, behaved the same way, and the various explanations for the movement proposed in the nineteenth century dispensed with these vitalistic notions. Einstein's theory was, however, the first to account convincingly for Brown's observations. Einstein supposed that the tiny pollen grains were small enough to be deflected by collisions with individual molecules of water, even though the grains were microscopically visible and the molecules were not.* Einstein's paper was the first thorough treatment of diffusion, and it made several predictions about Brownian motion that the physicist Jean Perrin verified in a series of extremely precise experiments in 1908. Perrin won the Nobel Prize for Physics in 1926, and for validating a theory over two millennia old that seems fair reward.

FAITH IN NUMBERS

Maxwell's contribution to the kinetic theory of gases is central to the kind of physics with which most of this book will be concerned. He had made physics *statistical*, saying that what matters when we are dealing with huge numbers of virtually identical moving objects is not the detailed behaviour of individuals but the average motions, as well as the extent of deviation from those averages. Anyone interested in broad-scale human behaviour will be familiar with this notion. Demographers do not need to know that Eric Baggins was born on 6 March 1969, but merely what the annual birth rate was. Traffic planners don't care that Mary Parker drove to Safeway in Camberwell on Tuesday morning, but ask simply how many cars use the Walworth Road in the course of a typical day. These statistics become more reliable as the size of the census increases. If we are asking about, say, the behaviour of a lump of matter you can hold in your hand, we are

---

* More strictly, the pollen grains experience an imbalance in the number of collisions from different directions – the deviations of their trajectories happen when more molecules strike them from one side than from another.

typically dealing with billions of billions of molecules, and the statistical behaviour is utterly reproducible from one experiment to another. In other words, for example, for two jars of identical gas at the same temperature the Maxwellian distribution of velocities is absolutely identical.

Along with the introduction of statistics comes the notion of *probability*. Maxwell's distribution tells us nothing exact about the speed of any particular gas molecule. Instead, it tells us the probability that a particle selected at random will have a particular speed. The most probable speed is the average speed; there is a low probability that it will be much faster or much slower than this.* It is indeed extremely convenient that statistics are enough for us to account for the behaviour of gases, for even with modern instruments we could not gather detailed information on every gas particle.

Maxwell evinced a certain uneasiness about his kinetic theory, acknowledging that it broke with the mechanistic tradition of using Newton's laws of motion to deduce the exact trajectories of a system's components, as one does for example to explain planetary motions. This was, in other words, a new way of doing science. Maxwell realized that the theory had profound philosophical implications, and as we shall see later, he may not have risked publishing it had there not already been good precedent.

Maxwell's 'probability distribution' of the speeds of gas particles was a seminal contribution to the kinetic theory, but the truth was that he deduced it using a strong dose of informed guesswork rather than exact mathematics. The job was done more rigorously by a troubled Austrian physicist, Ludwig Boltzmann (1844–1906).

---

* In fact the *mean* and *most probable* speeds in the distribution are not identical – they differ by a small factor. The peak in the distribution is the most probable speed; the mean speed is slightly greater than this, because the curve tails off more slowly than it rises.

As someone employed regularly to scan the scientific literature for news stories, I have come to appreciate that any paper with a title that begins 'Further researches on . . .' should be passed over with alacrity. It tends to be science-speak for 'The odds and ends we did not think worth pursuing in our last paper'. So it is humbling to be reminded that had I taken this attitude in 1872, I would have missed one of the most explosive papers of the century. In 'Further researches on the thermal equilibrium of gas molecules', Boltzmann not only made Maxwell's case watertight but also proved that there truly exist irreversible processes, as the Second Law of Thermodynamics stipulates – and showed why.

Maxwell proved that gas particles, once they achieve a Maxwellian distribution of speeds, will stay that way. But he did not show how they get to that state in the first place. This is what Boltzmann did, by developing a way to calculate how probability distributions change over time. He demonstrated that, for particles moving at random, 'whatever the initial distribution of kinetic energy may have been, it must always necessarily approach the Maxwellian form after a very long time has elapsed.'[8]

Thus Boltzmann put *change* under the lens of the kinetic theory, which at once brought the Second Law into focus. Clausius had proposed that entropy always increases during an irreversible process; Boltzmann clarified what this meant for the probabilities of molecular motions. He showed that entropy can be equated with the number of different arrangements of molecules that, at the everyday scale, look identical.

Picture a child's balloon on the end of a string. It is full of gas molecules moving at random, and the collisions of these particles with the elastic wall create the pressure that keeps the balloon inflated. At any instant, each molecule is moving on a particular path with a particular speed. If we had a camera so sophisticated that it could take snapshots showing all the particles, then two snapshots taken an hour – or a minute, or a second – apart would show very different arrangements. Because of the huge number of particles, we could take a billion snapshots and never see the same picture twice. But on the scale at which we typically make laboratory measurements, the gas is just the same in every case: it still has the same temperature, pressure and volume.

The number of possible arrangements of the molecules here is truly astronomical. But nevertheless it is finite. We can imagine arrangements that would *not* be equivalent – for example, with all the particles in one half of the balloon. In that case, the empty half would deflate. Because the particles are moving at random, there is absolutely nothing in the laws of physics to prevent this arrangement arising by pure chance. But the likelihood that all the particles would suddenly happen to acquire velocities that took them into the same half of the balloon is so tiny that it is hard to distinguish it from zero. The same is true for just about any arrangement other than ones in which all the particles are distributed evenly throughout the balloon.

Thus the balloon stays fully inflated, not because Newton's laws of motion say it must but because the arrangements of particles that ensure this are *overwhelmingly* more probable than any others, simply because there are many, many more of them than there are of any other non-equivalent arrangement. By equating the entropy of a state with the number of equivalent molecular arrangements to which it corresponds, Boltzmann was thus saying that the fully inflated state of the balloon has the highest entropy. He deemed the mathematical equation relating entropy to the number of 'microstates' of a system to be the apogee of his life's work, and this recondite formula, $S = k \log W$, is engraved on Boltzmann's tombstone.

When change happens in any system, entropy increases because the new arrangement of the constituent particles is more probable than the old. To put it another way, the direction of change – the arrow of time – is determined by probabilities. An ink drop diffuses and disperses in water because it is vastly more probable that the random motions of the ink particles will carry them away from the original droplet in all directions than that they will all conspire to make the droplet move coherently sideways, say, or shrink.

The crucial point about this explanation of the Second Law is that it shows how the irreversibility of time can come about through the operation of mechanical laws which have no preferred direction in time. Picture a movie of two billiard balls coming together, colliding, and moving apart. Played in reverse, the movie would not look at all odd: the reverse collision also obeys Newton's laws.* But the coalescence of a droplet of ink within a glass of initially pale blue water would obviously be time-reversed footage, even though each of the individual collisions between particles that 'created' this arrangement would look like those balls hitting in reverse. This is simply due to the effect of very large numbers on the probability of certain processes happening. Entropy does not *have* to increase by cosmic decree – it simply does so because that is overwhelmingly probable.

The theory of Maxwell and Boltzmann was derived from nothing more than the application of Newton's laws of motion to vast numbers of moving molecules – from so-called classical mechanics. It marks the beginnings of the field of *statistical mechanics*. This is the field that provides modern physics with its central organizing framework. By connecting thermodynamics with the properties of atoms in motion, statistical mechanics describes the behaviour of matter from the bottom up.

* This assumes that no kinetic energy is dissipated in the collision as frictional heat or sound. In the real world these dissipative processes do occur, of course: even the striking of billiard balls is an irreversible process that results in an increase in entropy.

The shift from Newtonian determinism to statistical science is what makes a physics of society possible. It was not a smooth ride; but as we shall now see, it may have been bumpier still if scientists and philosophers had not already begun to appreciate that society itself is fundamentally a statistical phenomenon.

3

The Law of Large Numbers

Regularities from randomness

It can be stated without exaggeration that more psychology can be learned from statistical averages than from all philosophies, except Aristotle.

   Wilhelm Wundt (1862)

Taken in the mass, and in reference both to the physical and moral laws of his existence, the boasted freedom of man disappears; and hardly an action of his life can be named which usages, conventions, and the stern necessities of his being, do not appear to enjoin on him as inevitable, rather than to leave to the free determination of his choice.

   John Herschel (1850)

[I]f there is some precision, there is some science.

   Herbert Spencer (1880)

*

Ludwig Boltzmann was seldom a happy man. His poem *Beethoven in Heaven*, written five or six years before he died, expresses an anguish that is all too evidently the physicist's own, as well as a presentiment of what was to come:

With torment that I'd rather not recall
My soul at last escaped my mortal body.
Ascent through space! What happy floating
For one who suffered such distress and pain.[1]

His scientific achievements perhaps brought him more 'distress and pain' than satisfaction and joy, for his ideas were vigorously attacked by several of his contemporaries. Although today most scientists concur with his explanation for the arrow of time, it does leave open some important questions, into which his opponents sunk their teeth. Boltzmann responded robustly to these attacks, but they disheartened him greatly.

Boltzmann was by nature a self-doubting and hesitant man, and his gloom deepened in 1889 when his eldest son died from appendicitis. He grew restless, unable to settle comfortably at any of the several Austrian and German universities he joined. On 5 September 1906, while on holiday with his family in Duino, near Trieste, he hanged himself at the age of sixty-two.

Theoretical physics scarcely sounds like a life-threatening activity. But Boltzmann's self-inflicted death was echoed by that of his brilliant successor, the Viennese Paul Ehrenfest, who shot himself in 1933. (After recounting these wretched episodes, physicist David Goodstein tartly informs the reader of his modern textbook that 'Now it is our turn to study statistical mechanics.'[2])

It would be unwise to draw any conclusions about the psychological hazards of early-twentieth-century physics before asking whether the suicide rate among physicists was any greater than it was in the population as a whole. Vienna at the *fin de siècle* was a furnace of intellectual debate, fired by the likes of Sigmund Freud, Arnold Schoenberg, Ludwig Wittgenstein and Robert Musil; but its citizens were, as Musil observed, nevertheless a soulless, tight-lipped crowd in thrall to convention. 'The notion', he says in *The Man Without Qualities*, 'that people who live like that could ever get together for the rationally planned navigation of their spiritual life and destiny was simply unrealistic; it was preposterous.'[3]

In this rigid and materialistic society, suicides were disturbingly widespread. They claimed the lives of three of Wittgenstein's brothers, Gustav Mahler's brother, and in 1889, the Crown Prince Rudolf of Austria (who killed his mistress first). Boltzmann's sad death does not speak to any broader context until it is seen in the light of the relevant demographic statistics.

To us this seems obvious, but before the nineteenth century hardly anyone would have thought so. Assessing individual events in the context of their average rate of occurrence is a relatively modern practice. Without it, the world is ripe for magic, superstition, miracles and conspiracy theories. A few chance events can become evidence for supernatural influence. Even now the relevance of statistics is routinely overlooked in subjective assessments of risk and coincidence. When the 'psychic' Uri Geller apparently stopped a few watches among his TV audience in the 1970s, no mention was made of the likelihood of such a thing happening by chance given the very large number of viewers.

Whenever one is trying to make sense of mass behaviour, whether it be of atoms or of people, statistics are indispensable. This now seems so beyond question that it is hard to comprehend the urgency of the philosophical arguments that surrounded the use of statistics in nineteenth-century science. At that time, it seemed that God and human free will were being held hostage to numbers. The roots of a physics of society are enmeshed in this debate, so that we shall find some of the moral issues raised by the new discoveries described in this book prefigured by soul-searching from over a hundred years ago.

The history of statistical mechanics outlined in the previous chapter is the orthodox one that physicists tell routinely. Very rarely is any hint given of the way it really began – not just among the insensate gases of the laboratory but in the behaviour of people in society. Speaking of a physics of society perhaps sounds a very postmodern thing to do, but truly there is nothing new under the sun.

MEASURING SOCIETY

In *Leviathan*, Thomas Hobbes was arguably taking to its logical conclusion the analogy drawn by his mentor Francis Bacon between the 'Body Naturall' and the 'Body Politick'. This notion implied that politics might be a kind of natural science, with an anatomy waiting to be dissected by the scalpel of systematic and rational enquiry. In attempting to create such a scientific political theory, Hobbes chose mechanical physics as his framework.

Today we think of physics as a supremely quantitative, not to say mathematical, science. Physicists measure the fundamental numbers of nature down to the tenth decimal place. Their formal literature is dense with symbols, equations and graphs. Things were not quite like this in Hobbes's day, yet still it is striking that *Leviathan* is wholly discursive – there is not a number or an equation in sight. Hobbes liked to make use of physical analogies but he had no intention of making political science mathematical.

That had inevitably to happen, however, if the endeavour was not just to borrow the ideas but to share the demonstrative force of natural science. William Petty, a disciple of Hobbes, seemed to recognize as much when he called for a 'political arithmetic'. 'To practice upon the Politick,' said Petty, 'without knowing the Symmetry, Fabrick, and Proportion of it, is as casual as the practice of Old-women and Empyricks.'[4]

What numbers was this arithmetic to manipulate? Why, naturally, those that measured society. In the 1660s John Graunt (1620–74), a London haberdasher and a friend of Petty, introduced the study of 'social numbers' as a means to guide political policy. Chief among the numbers with which he concerned himself were death rates. In *Observations upon the Bills of Mortality* (1662) he drew up tables of mortality figures 'whereby all men may both correct my Positions, and raise others of their own'.[5] How could one reasonably legislate and govern the population, he asked, without knowing the numbers in which they come and go?

Graunt's statistics were hardly a model of methodological finesse. As he freely admitted, those humble souls responsible for recording deaths were all too easily induced 'after the mist of a Cup of Ale, and the bribe of a two-groat fee, instead of one', to list the cause of death as something anodyne (consumption, say) when the truth was more scandalous (such as syphilis). Yet his tables of causes and ages of death were seen as a bountiful resource for those seeking to understand the flux of society. Graunt, although a mere businessman, was elected a Fellow of the Royal Society, and Charles II averred that 'if they found any more such Tradesmen, they should be sure to admit them without any more ado'.[6]

William Petty continued to revise the *Observations* after Graunt's death. He was the first to study political economy by means of such social statistics, which he argued could provide a rational basis for formulating policy. In this respect he was an empiricist, working with observations of social aggregates rather than trying to derive theories based on assumptions about the fundamental psychology of individuals in the manner of Hobbes. Petty enjoyed the favour of Charles I, Charles II and James II (and managed, pragmatically, to serve Cromwell too) and he was a founding member of the Royal Society. Yet his policy recommendations were largely ignored, and frankly this was often just as well. Petty often exemplifies the dangers of a hyper-rational, analytical approach to social policy that takes no account of its human costs.

Population measures – birth and death rates – were the major pre-occupation of early quantifiers of society. It was thought to be of paramount importance for a nation to multiply its subjects – an injunction that was, after all, sanctified in the Bible. The power and glory of a country were believed to be reflected in the size of its population, so much so that some savants proposed that wars of conquest were driven largely by a desire to increase it. In the mid-eighteenth century, Johann Peter Süssmilch (1707–67), a German army chaplain, argued that war could be avoided by removing all checks to the growth of population, obviating the need for kings to gather new subjects from outside their realm.

A focus on mortality was understandable in an age that knew so much of it. Masses died in noisome cities, the 'Places of the Waste and Destruction of Mankind' according to Thomas Short in 1767.[7] Famine and starvation were endemic in the countryside. Few wars were quite as devastating as the Thirty Years' War, but war still seemed to be an ever-present part of human affairs and a steady source of attrition in the population. Procreation was the only remedy. Ironically from today's perspective, Protestants in England and Germany denounced Catholicism because its advocacy of celibacy compromised population growth.

By 1826, when the English economist Thomas Malthus (1766–1834) published his *Essay on the Principle of Population* – a compelling critique of unchecked population growth which had a profound influence on both Darwin and Marx – governments in Europe and the United States had begun to appreciate the wisdom of counting their citizens. Censuses in fact date back to the Norman efforts to record in the *Domesday Book* the population of England in the eleventh century, although this was not so much an exercise in quantification as the establishment of a bureaucratic basis for the exploitation of a conquered population. By the eighteenth century such social numbers were considered to encode insights into how society functioned. Süssmilch, for example, argued that the differences in birth and death rates of boys and girls balanced perfectly so as to provide marriage prospects for them all. In other words, from the chaos of human life arose a kind of law of the masses that stabilized society.

Süssmilch's observations helped to establish the idea that society observed rules that were ordained by no government yet could be revealed by counting. This led Immanuel Kant in 1784 to speak of 'universal laws' which,

[h]owever obscure their causes, [permit] us to hope that if we attend to the play of freedom of human will in the large, we may be able to discern a regular movement in it, and that what seems complex and chaotic in the single individual may be seen from the standpoint of the human race as a whole to be a steady and progressive though slow evolution of its original endowment.[8]

On the one hand, this belief in 'laws' of society that lay beyond the reach of governments was a product of the Enlightenment faith in the orderliness of the universe. On the other, it is not hard to see within it the spectre of the Industrial Revolution with its faceless masses of toiling humanity like so many swarming insects. Before the nineteenth century, the laws that applied to Graunt's 'social numbers' were regarded as evidence of divine wisdom and planning. To later commentators they were the preconditions for catastrophe and revolution.

This study of social numbers needed a name. In 1749 the German scholar Gottfried Achenwall suggested that since this 'science' dealt with the natural 'states' of society, it should be called *Statistik*. John Sinclair, a Scottish Presbyterian minister, liked the term well enough to introduce it into the English language in his epic *Statistical Account of Scotland*, the first of the 21 volumes of which appeared in 1791. The purveyors of this discipline were not mathematicians, however, nor barely 'scientists' either; they were tabulators of numbers, and they called themselves 'statists'.

THE CHURCH OF NEWTON

Those who collected statistical data soon realized that these data tell us something not just about what has occurred but about the general *probability* of things happening. Statistics thus began to attract the interest of mathematicians who concerned themselves with one of the most philosophically recondite and sometimes counter-intuitive branches of their subject: probability theory.

This discipline has its origins not in social numbers but in gambling. If you play a game of chance, you are wise to know the odds beforehand. Anyone who does not appreciate that betting on black or betting on number 15 in roulette are quite different things, with different odds, is not destined to spend long in the casino. Games involving dice and other agents of chance have a long history, and in the eighteenth century mathematicians began to investigate the rules that govern them. In this seemingly frivolous topic the French mathematician Marie-Jean-Antoine-Nicolas Caritat de Condorcet (1743–94) found the tools he needed to build one of the most optimistic of all science-based utopias.

Condorcet described his vision in a book written shortly before the architects of the French Revolution condemned him to the guillotine. His *Esquisse d'un tableau historique des progrès de l'esprit humain*

('Sketch for a Historical Picture of the Progress of the Human Mind'), written in 1793, is a hymn to a kind of rationality that Condorcet must have found in short supply in the Reign of Terror. Society, he believed, must be founded upon reason and guided by its great formalization: science. The eighteenth century was a truly radical time, when belief in liberty and equality (Robespierre's Terror notwithstanding) was more than a matter of good intentions. Many Enlightenment philosophers genuinely trusted that these principles, allied to reason, would usher in a glorious age of freedom. Condorcet argued for the equality of women, and in 1792 he proposed that all patents of nobility be burnt – including his own (he was a marquis). He became a friend of Voltaire, whose utopian writings were, however, rather more cynical.

Condorcet was a precocious mathematical genius, and his early efforts came to the attention of the eminent French Academician Jean Le Rond d'Alembert. Under d'Alembert's influence, he turned from the pure mathematics of probability to consider social and economic issues. In his study of democratic decision-making, *Essai sur l'application de l'analyse à la probabilité des décisions rendues à la plurité des voix* ('Essay on the Application of Analysis to the Probability of Majority Decisions', 1785), he concurred with his fellow statists that if (as he maintained) there is indeed a science of human affairs with its own axioms and laws, then it must be a statistical science. To read these laws, one must gather enough numerical data: 'All that is necessary to reduce the whole of nature to laws similar to those which Newton discovered with the aid of calculus, is to have a sufficient number of observations and a mathematics that is complex enough.'[9]

Condorcet foresaw that the scientist so equipped would be able to predict the outcome of democratic decision-making, so that history itself could become a true science. Then we would indeed be on the threshold of a true utopia, as he indicated in his *Esquisse*:

How consoling for the philosopher who laments the errors, the crimes, the injustices which still pollute the earth and of which he is often the victim is this view of the human race, emancipated from its shackles, released from the empire of fate and from that of the enemies of its progress, advancing with a firm and sure step along the path of truth, virtue and happiness![10]

It is not hard to read in this passage an attempt by the author to console himself in the face of a bleak future. The *Esquisse* was written hurriedly in hiding while Robespierre's agents hunted for its author. His was a dramatic fall from favour, and tells us much about the nature of revolutions.

In 1792 Condorcet's intellectual reputation and his support for the Republican cause earned him a seat on the Committee of Nine that was charged to draft the new French Constitution. Among Condorcet's colleagues was Thomas Paine, a French citizen after his exile from Britain following the publication of his book *The Rights of Man*. The draft Constitution was blocked by Robespierre, who resented having been excluded from the Committee. When a new version, hastily redrafted by another makeshift committee and full of loopholes, was accepted, Condorcet published an anonymous letter urging the people to reject it. His authorship did not stay secret for long, and he was convicted of treason.

The *Esquise*, penned in a safe house in Paris run by one Madame Vernet, is, particularly under the circumstances, strikingly optimistic. Condorcet regards mankind as having 'evolved' from the level of beasts to a state of higher intelligence whereby people acquire an innate altruism. He sees no reason why this evolution (anticipating Darwin) cannot continue until people are 'perfected' – an idea in stark contrast to Jean-Jacques Rousseau's view that civilized man is corrupted. In the future utopia, says Condorcet, medical science will conquer all disease, and people will be too enlightened to go to war. Education will abolish social inequality, and all people will speak the same language. 'Are we not arrived at the point', he asked,

where there is no longer any thing to fear, either from new errors, or the return of old ones? . . . Everything tells us that we are approaching the era of one of the grand revolutions of the human race . . . The present state of knowledge assures us that it will be happy.[11]

This soaring vision was not that of a worldly man. Although Condorcet eluded his captors as they came to arrest him, his refined manner aroused suspicion in the country inn to which he fled, and he was swiftly apprehended. He was taken to prison at Bourg-la-Reine, near Paris. With the guillotine his likely fate, he seems to have committed suicide by poisoning himself in his cell. Had he remained hidden for just several months more, he would have escaped his persecutor for ever: Robespierre himself went to his death in July 1794.

The *Esquisse* became posthumously celebrated. Malthus read it but did not share its rosy outlook. Condorcet was aware that population growth could eventually overwhelm available resources and threaten the stability of civilization, and he had a simple remedy – birth control. Malthus did not think it would be so easy. He reckoned that the 'passions of mankind' put population outside the control of governments, whether they sought either to encourage or to limit its rise. It was, he believed, a 'law of nature' that the populace would multiply exponentially, while society could not increase the means of feeding itself at the same rate. Thus nations must succumb sooner or later to overcrowding, misery, poor health and social unrest – which would bring with them the stark choice between repression and revolution. To escape this fate, people would do well to accept that government alone, however good, cannot steer them clear of catastrophe. Rather, said Malthus in his profoundly influential *Essay* of 1826, one needed to know the irrevocable laws, the 'internal structure of human society'.

Others, while less pessimistic, concurred with Malthus's view that there *was* an internal structure, a set of laws, that dictated the way society behaved and evolved. These laws were deemed to stand in relation to society as Newton's mechanics stood in relation to the motion of bodies. The idea was particularly popular in France; the Baron de Montesquieu (Charles Louis Secondat de la Brède, 1689–1755) adduced it in *The Spirit of the Laws* (1748), which preceded Condorcet by several decades. Claude-Henri de Rouvroy, the Comte de Saint-Simon (1760–1825) shared Condorcet's dream of a society governed by scientific reason, and he imagined that it might lead to the founding of a 'Religion of Newton'. The vision was particularly explicit in Jean Théophile Desaguliers's *The Newtonian System of the World: The Best Model of Government, an Allegorical Poem* (1728), in which he wrote that the notion of a force of attraction 'is now as universal in the political, as the philosophical world'.[12] The Scottish philosopher David Hume (1711–76) expressed a desire in his *Treatise of Human Nature* (1739–40) to become the Newton of the moral sciences by reducing human nature to first principles through empiricism rather than Cartesian *a priorism*. When in 1741 Hume proposed 'that politics may be reduced to a science',[13] the idea had already become so commonplace as to be the subject of satire: Jonathan Swift's Gulliver berates the Brobdingnagians for 'not having hitherto reduced *Politicks* into a *Science*, as the more acute Wits of Europe have done'.[14]

It was Hume who in the 1760s introduced Adam Smith, on a European Grand Tour with his pupil the young Duke of Buccleuch, to François Quesnay (1694–1774), physician to Louis XV at Versailles. In his sixties, Quesnay had begun to take an interest in economics and was collecting facts and figures in the hope of discerning among them laws and 'social forces' akin to the physical forces of Newton. Quesnay's *Tableau économique* (1758) was one of the first works of economic theory, and indeed his followers were the first to be classified as *les économistes*. His work leaves its mark clearly on Smith's *Wealth of Nations* (1776) (see page 221), whose author would have dedicated it to Quesnay had the Frenchman not died two years before it was published.

Belief in a 'scientific' political theory has tended to flourish mostly at the liberal end of the political spectrum. (It is partly for this reason that Thomas Hobbes can rightly be regarded as a liberal.) When Thomas Jefferson wrote the American Declaration of Independence, his vision of a free and happy nation was that of a man in love with Newtonian mechanics and the ideals of the Enlightenment, one who believed that humans are compelled towards happiness just as apples are pulled by gravity towards the earth. Jefferson considered that happiness could be measured as quantitatively as matter could be weighed. The Irishman Edmund Burke (1729–97), often considered to be the 'Father of Conservatism', abhorred this sort of thing, holding that a state's laws and institutions cannot be deduced from first principles but can only emerge empirically from particular historical processes. For that reason, one should regard those laws which existed as tried and tested, honed by the whetstone of experience and tradition and not to be meddled with on account of abstract, 'rational' theories. After all, Burke argued, people are too complex to permit of any 'scientific' analysis of the histories they produce:

In the gross and complicated mass of human passions and concerns, the primitive rights of men undergo such a variety of refractions and reflections, that it becomes absurd to talk of them as if they continued in the simplicity of their original direction.[15]

Their 'trajectories', in other words, become randomized. But how telling that even Burke felt compelled to phrase his argument in the terminology of Newtonian mechanics and optics!

The spirit of the Enlightenment position was, however, brought to its zenith by the French positivist philosopher Auguste Comte (1798–1857), who grounded his thinking in a rational religion motivated by the advancement and well-being of humanity. Like Adam Smith, he believed that these ends could be attained by uncovering the natural laws of society, rather than by political interference. Although he did not share the statists' enthusiasm for quantification, Comte coined a term that encapsulated his desire for and faith in a science of civilization: *physique sociale*, social physics. In his *Cours de philosophie positive* (1830–42), he argued that this would complete the scientific description of the world that Galileo, Newton and others had begun:

Now that the human mind has grasped celestial and terrestrial physics, mechanical and chemical, organic physics, both vegetable and animal, there remains one science, to fill up the series of sciences of observation – social physics. This is what men have now most need of; and this it is the principal aim of the present work to establish.[16]

ORDER FROM CHAOS

No one did more to propagate the notion of a scientific understanding of society than the Belgian astronomer Adolphe Quetelet (1796–1874) (Figure 3.1). In Quetelet's work, all the developing strands of social enquiry – Hobbes's mechanistic politics, the value of statistical quantification, and the belief in natural laws of society – came together, and for a heady half century or so there seemed to be no demarcations between physics, mathematics, economics, politics and sociology.

Figure 3.1 Adolphe Quetelet, whose *mécanique sociale* was an attempt to find 'laws of society' analogous to Newton's laws, which governed the motions of inanimate bodies.

Like Hobbes, Quetelet had a vested interest in proving that a scientific view of society could promote stability. He pursued his career at a time of great political upheaval in Belgium. Most of the country had become part of France in the late eighteenth century, but the southern provinces were absorbed into the Netherlands. In 1830 the Belgians revolted in a bid for independence, and the resulting conflict made academic study all but impossible. Scientists left their posts to enlist in the army, and the universities and colleges were disrupted. The Royal Observatory in Brussels, where Quetelet was Director and which he had helped to construct, was occupied by soldiers and, in his own despairing words, 'converted into a fortress . . . surrounded with ditches and ramparts'.[17] Just months after the revolution, Quetelet published his first paper on *mécanique sociale*, a discipline allied explicitly with Comte's social physics. By making direct analogies between the organizing forces of a solar system and those of an orderly social system, he aimed to show that society was as rule-based as astronomy.

There was good precedent for such comparisons. Indeed, Quetelet felt that astronomers were to thank for introducing statistics to social studies. The first table of mortality figures, he pointed out, was drawn up by Newton's contemporary and friend, the astronomer Edmond Halley, and published in 1693. Quetelet argued that it was natural that astronomers should perceive order in the social sphere:

The laws that concern man, and those that govern social development, have always had a special attraction for the philosopher, and perhaps most especially for those who have directed their attention to the system of the universe. Accustomed to considering the laws of the material world, and struck with the admirable harmony that reigns there, they can not be persuaded that similar laws do not exist in the animate world.[18]

All this became evident to Quetelet when in 1823 he was sent to the French Royal Observatory in Paris to expand his astronomical knowledge in preparation for his directorship at the planned observatory in Brussels. Quetelet indeed discovered much about celestial science, but he also found that the stars of the French astronomical firmament had a broad interest in statistics.

THE SHAPE OF ERROR

The dominating figure of French astronomy at the time was Pierre-Simon Laplace, who extended Newton's mechanics and used it to discover new aspects of planetary motion. Laplace was a formidable mathematician, and he and his contemporaries knew that their measurements of celestial motions rarely reflected an exact adherence to the mathematical regularity that Newton's laws demanded. All measurements incur errors that can cause apparent deviations from the smooth relationships predicted by the laws of mechanics.

The French astronomers developed methods for dealing with these errors which involved finding the smooth curves or lines that best fitted a scattering of data points. Laplace and his pupil Siméon-Denis Poisson (1781–1840) assumed that measurement errors were essentially random. They could be of any magnitude, but not with equal probability. Rather, the likelihood of an error of a particular size – a deviation of that magnitude from the 'true law' governing the observations – decreased as its size increased. In the same way, if you measure the length of your foot using a ruler, you are more likely to be out by a millimetre than by a centimetre. The error will not be the same for every measurement, even if it is made using the same instruments and techniques. If you measure the feet of all the members of your family, you might sometimes be out by just half a millimetre, sometimes by two millimetres. It depends not only on how accurate your ruler is, but on how careful you are each time. The error is largely a matter of chance. This, the astronomers realized, connected error rates with probability theory.

To know the likelihood of a particular error, we need to know how *often*, in a large and representative sample of measurements, it occurs. We need to collect statistics about errors. The French scientists found that errors were always distributed in the same way. Not only were there always more small errors than large ones, but this decrease in number with increasing size was predictable. Plotted on a graph, the statistics of errors fitted onto a particular curve which became known as the 'error curve' (Figure 3.2). Its ubiquity also gained it the name *normal distribution*. The German mathematician Carl Friedrich

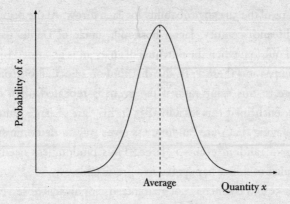

Figure 3.2 The error curve. This bell-shaped curve describes the statistics of all random processes. (Strictly speaking, mathematicians call these *stochastic* processes, meaning that each outcome or observation is independent of the others.)

Gauss (1777–1855) bequeathed another name to this bell-shaped curve when he analysed its properties in 1807: measurements that fall on this curve are now most commonly said, by physicists at least, to obey *gaussian* statistics. Whatever one calls it, the error curve is basically the probability distribution of the outcomes of a random process.

This curve was already known to mathematicians interested in probability, for in 1733 Abraham De Moivre showed that it described the distribution of outcomes from tossing a coin. There is an equal probability of the coin falling head up or tail up. The result depends (if the toss is fair) on pure chance. Nevertheless, for a large number of throws there will be roughly as many heads as tails.

That this predictable result arises from a series of chance events is not terribly surprising, for it simply reflects the cancelling out of equally probable deviations in either direction. A series of several heads in a row is likely to be balanced eventually by a temporary, chance prevalence of tails. The equal numbers that result merely remind us of the 50 : 50 probability for each throw. At the beginning of the eighteenth century, Jacob Bernoulli, uncle of Daniel (page 41), pointed out that when the outcome of an event is governed by a fixed probability ratio (here 1 : 1), the distribution of actual outcomes will converge to this same ratio if the event is repeated often enough. Poisson enshrined this idea in 1835 in his 'law of large numbers', a way of saying that pure randomness gives way to determinism if the number of random events is large. Thus randomness need not, in itself, rule out things happening in a predictable way.

But a 50 : 50 balance is not guaranteed, nor always observed. If we toss a coin 10 times, a ratio of 4 heads to 6 tails should not surprise us – chance has shifted the balance in favour of tails so that there is a 20 per cent deviation from equal numbers. The more tosses we make, the more closely the numbers converge on a 50 : 50 distribution. For 100 tosses, 49 heads and 51 tails is not an unlikely outcome – again, a difference of 2, but this time the deviation is a smaller proportion (2 per cent) of the whole. Another round of 100 tosses might produce a different outcome – say, 52 heads and 48 tails. De Moivre showed that for many rounds of a fixed number of tosses, the bell-shaped error curve always provides a good fit to the distribution of outcomes.

For coin tossing, it is possible in theory to calculate the shape of the error curve – that is, to write an exact equation describing the probability of each outcome in a single round of tosses. But this calculation is very laborious, especially with eighteenth-century techniques. De Moivre showed instead that the curve could be accurately approximated by a simpler mathematical equation.

We could regard the deviations from a 1 : 1 ratio of heads and tails as 'errors' which shift the 'measurement' away from its 'true' value. It might seem perverse to do so, since if the coin tossing is done fairly and the counting is accurate, there is no 'error' as such but just the operation of chance. However, in the 1770s Laplace realized that errors in measurement were also the outcome of factors which, being too complex to quantify, caused random divergences from true values. So he began to use De Moivre's approximate equation to quantify errors in astronomical measurements.

In the early part of the nineteenth century the French mathematician Joseph Fourier (1768–1830) began to apply the error curve more widely. As director of the Bureau de Statistique of the Département de la Seine, Fourier published several papers on demographic statistics, and he helped to introduce the curve into social science. Laplace too had sought to put De Moivre's equation to work in social statistics. In 1781 he argued that the near-equality of male and female births in Paris, which others had previously read as a sign of divine providence, was merely the expected result of a random process with two equally probable outcomes, whose variations were consistent with the error curve.

Quetelet was impressed by these examples when he encountered Laplace's work in Paris, and he began to suspect that the error curve was the fundamental leitmotif of human demography. In 1844 he showed that human dimensions – height and girth – were distributed with this same hump-backed profile. To Quetelet this was a sign of regularity and order in nature. Look at a crowded street and you will see people of many sizes. It may well seem at first that (within obvious limits) there is no predictability about the dimensions of a human being, but collect enough data and the bell-shaped curve will emerge.

ORDERLY BEHAVIOUR

Quetelet acquired from the French scientists the idea that variation is linked with error. Instead of regarding height differences as a characteristic feature of nature, he saw them as departures from an ideal form. These 'errors' become less prominent as greater numbers of people are taken into account, just as Poisson's law of large numbers says they should. This is true, Quetelet decided, not only for physical characteristics but for *behaviour*, since the foibles of the individual temperament average out among the tendencies of the mass. Quetelet wrote in 1832 that

whatever concerns the human species, considered en masse, belongs to the domain of physical facts; the greater the number of individuals, the more the individual will is submerged beneath the series of general facts which depend on the general causes according to which society exists and is conserved.[19]

Since it was clearly a desirable thing that society should 'exist and be conserved', this implied that average behaviour was the 'right' behaviour. And so Quetelet's social physics became founded on the concept of the 'average man' (*l'homme moyen*), whose dimensions and physical features and also moral and aesthetic attributes represented a perfect mean to which all should aspire. To be great was to be average:

an individual who epitomized in himself, at a given time, all the qualities of the average man, would represent at once all the greatness, beauty and goodness of that being.[20]

This disturbing worship of uniformity has as its corollary an abhorrence of all variation:

Deviations more or less great from the mean have constituted . . . ugliness in body as well as vice in morals and a state of sickness with regard to the constitution.[21]

The idea of a physical and moral perfection of humankind which is reflected in the conformity to a mathematical ideal dates back to the

Renaissance; but now there existed the tools to quantify what perfection was. In retrospect it is easy to read into Quetelet's theory of the average man overtones of racial purity and stifling social conformity; but in an age that believed in physiognomy, such views might have been deemed perfectly rational, although that does not excuse their sinister aspect.

In any event, the French Government was soon to discover that *l'homme moyen* had his uses. In 1844 Quetelet compared the distribution of men's heights in the general population with that of 100,000 conscripts drafted into the French army, and found a discrepancy which led him to conclude that about 2,000 men had lied about their height so as to fall below the minimum and evade conscription.

Many of Quetelet's contemporaries were deeply impressed by and enthusiastic about the regularities that he perceived in the statistics of human affairs. Reviewing Quetelet's work in 1850, the eminent English astronomer John Herschel wrote:

No one has exerted himself to better effect in the collection and scientific combination of physical data in those departments which depend for their progress on the accumulation of such data in vast and voluminous masses, spreading over many succeeding years, and gathered from extensive geographical districts.[22]

Florence Nightingale urged that Quetelet's social physics, which she deemed to be an indicator of God's design, be taught at Oxford.* Karl

---

* Longfellow's romanticized image of Florence Nightingale as the 'Lady of the Lamp' does her achievements few favours. She was more formidable than angelic, and would not otherwise have had such an impact on medical conditions in the Crimean War. Less well known is her love of mathematics, in which as a girl she had begged her parents to let her be tutored. In the Crimea she collected statistics on mortality rates and devised a way of displaying them as wedge-shaped graphs. Quetelet's *Social Physics* was said to be the devout Nightingale's second Bible, and her copy was covered with her annotations.

Marx used Quetelet's statistical laws in developing his labour theory of value. And John Stuart Mill, the utilitarian successor of Jeremy Bentham, felt that Quetelet's work lent support to his conviction that society and history were bound by laws as absolute (if harder to discern) as those of the natural sciences. In *A System of Logic* (1862), Mill had the universal error curve in mind when he wrote that the

very events which in their own nature appear most capricious and uncertain, and which in any individual case no attainable degree of knowledge would enable us to foresee, occur, when considerable numbers are taken into the account, with a degree of regularity approaching to mathematical.[23]

THE SCIENCE OF HISTORY

Awareness of Quetelet's work was promoted in Britain by one of the most avid proponents of a law-bound social physics, Henry Thomas Buckle (1821–62). Like many adherents of Comte's positivist philosophy, Buckle wanted to fortify the world of human affairs against the meddling influence of governments. In Quetelet's view, governments had little effect on his statistical laws, which transcended human intervention. But Buckle maintained that one should not even try to tamper with them: like Adam Smith, he argued actively for the principle of laissez-faire and for the need for people to be allowed to govern themselves. Left to their own devices, he believed, societies automatically produced 'order, symmetry, and law', while 'lawgivers are nearly always the obstructors of society, instead of its helpers.'[24]

Metaphysical philosophers, in Buckle's view, had in the past pursued the futile goal of trying to unravel the way society works by worrying about what makes individuals tick. The empirical science of social statistics avoided such imponderables by discovering laws within the numbers. To make his case, Buckle felt he needed to show the play of these laws throughout history. Traditionally, history had been a narrative about kings and queens. Buckle's history would be different: it would be a science.

Kant anticipated this quest for 'historical laws' in alluding to the way that collective behaviour smoothes out individual unpredictability. In his essay 'Idea of a universal history from a cosmopolitan point of view' (1784), he said:

Individual men, and even whole nations, little think, while they are pursuing their own purposes . . . that they are advancing unconsciously under the guidance of a purpose of nature which is unknown to them.[25]

Buckle agreed; for him, history was ruled by a 'great truth' – that

the actions of men . . . are in reality never inconsistent, but however capricious they may appear only form part of one vast system of universal order . . . the underlying regularity of the modern world.[26]

Buckle set out the case for this universal order in his book *History of Civilization in England*. The first two volumes of this ambitious work were published between 1857 and 1861, but, exhausted by his efforts, he died before he could complete it. As a consequence, the *History* says more about the rest of the world than about the author's homeland. Buckle had initially planned a world history and could not resist dispensing the fruits of this great vision in the initial volumes – even though he thought that England showed the operation of the laws of history more clearly than did any other country.

In this great compilation of regularities in social statistics, Buckle drew on many of the same examples as had Quetelet: rates of birth and death, crime, suicide and marriage. The *History of Civilization in England* helped to shape the British intellectual climate of the mid-nineteenth century around the idea of a liberal laissez-faire that held government to be largely unnecessary and consequently unwelcome. According to William Newmarch of the Statistical Society of London, writing in 1860,

men are gradually finding out that all attempts at making or administering laws which do not rest upon an accurate view of the social circumstance of the case, are neither more nor less than imposture in one of its most gigantic and perilous forms . . . Every topic from the greatest to the least which the old legislators dealt with according to . . . caprice . . . have all been found to have laws of their own, complete and irrefragable.[27]

In the same year, the British economist Nassau Senior summarized the Zeitgeist thus: 'the human will obeys laws nearly as certain as those which regulate matter'.[28]

To some observers, this new way of looking at human behaviour was decidedly strange. In an 1850 issue of *Household Words*, a periodical edited by Charles Dickens, Frederick Hunt's amusement seems to sugar-coat a degree of scepticism:

the savants are superseding the astrologers of old days, and the gipsies and wise women of modern ones, by finding out and revealing the hitherto hidden laws which rule that charming mystery of mysteries – that lode star of young maidens and gay bachelors – matrimony.[29]

Ralph Waldo Emerson was also wary of the supposed certainty of statistical laws. In his 1860 essay 'Fate', he explained the central claim of 'the new science of Statistics': 'that the most casual and extra-ordinary events – if the basis of population is broad enough – become matter of fixed calculation.' He went on to mock what he saw as the rigidity of such an idea:

In a large city, the most casual things, and things whose beauty lies in their casualty, are produced as punctually and to order as the baker's muffin for breakfast. *Punch* makes exactly one capital joke per week; and the journals contrive to furnish one good piece of news every day.[30]

Mark Twain, meanwhile (in a remark he attributed to Benjamin Disraeli), scarred statistics for life by claiming that, 'There are three kinds of lies: lies, damned lies, and statistics.'[31] And Friedrich Nietzsche, whose belief in the shaping of history by a few 'great men' was second to none, was characteristically acerbic: 'so far as there are laws in history, laws are worth nothing and history is worth nothing.'[32]

FROM PEOPLE TO ATOMS

Others found inspiration in these manifestations of regularity within apparent randomness. The reliability of the statistical laws of society invited scientists to use them as analogies for similarly random processes in the natural world. Among them was James Clerk Maxwell.

A few months after the publication of Buckle's great work, Maxwell wrote to his friend Lewis Campbell:

One night I read 160 pages of Buckle's *History of Civilization* – a bumptious book, strong positivism, emancipation from exploded notions and that style of thing, but a great deal of actually original matter, the result of fertile study, and not mere brainspinning.[33]

When Maxwell came to study the problem of gases in which the constituent particles were constantly engaging in collisions that none could hope to follow, he recognized this as a problem of the same class as those that Buckle had pondered in society, in which the immediate causes of individual behaviour must forever be inscrutable:

the smallest portion of matter which we can subject to experiment consists of millions of molecules, not one of which ever becomes individually sensible to us. We cannot, therefore, ascertain the actual motions of any one of these molecules; so that we are obliged to abandon the strict historical [Newtonian] method, and to adopt the statistical method of dealing with large groups of molecules . . . In studying the relations between quantities of this kind, we meet with a new kind of regularity, the regularity of averages, which we can depend upon quite sufficiently for all practical purposes.[34]

As he indicated in 1873, the experiences of social statisticians lent him confidence that this statistical approach could extract order from the microscopic chaos:

those uniformities which we observe in our experiments with quantities of matter containing millions of millions of molecules are uniformities of the same kind as those explained by Laplace and wondered at by Buckle arising from the slumping together of multitudes of causes each of which is by no means uniform with the others.[35]

Would Maxwell have dared abandon the 'strict historical method', the obligation to explain everything in terms of the Newtonian mechanics of individual particles, if studies of society had not revealed laws at work even in complex systems where the direct causes of behaviour were obscure? How otherwise might he have found the faith to look for laws in the face of woefully incomplete knowledge about motions?

Maxwell began his work on the kinetic theory of gases shortly after reading Buckle. But in his early work he also drew on the more analytical studies of Quetelet, whose wide application of the error curve came to his attention through John Herschel's description of the Belgian scientist's work in 1850. There, Herschel had already alluded to connections between social physics and the early kinetic theory of gases.

Maxwell knew that Rudolf Clausius had used probability laws in 1857 to deduce the role of molecular collisions in the pressure exerted by a gas on the walls that confined it. But Clausius was interested only in the average velocity of the particles. Maxwell wanted to know how these velocities were distributed around this average. If the error curve worked so well for describing variations from the average in social physics, why then, it would do for him too. In 1859 he proceeded on the assumption that particle motions could be described by Quetelet's error curve, and was able to show what this implied for the measurable properties of a gas.

As we saw in the previous chapter, Maxwell's velocity distribution was merely an assumption until Ludwig Boltzmann showed in 1872 that any group of moving particles in a gas *must* converge on this distribution. Boltzmann too knew of Buckle's work and was not slow to draw analogies between his particles and the individuals in the social censuses that furnished Buckle's statistics:

The molecules are like to many individuals, having the most various states of motion, and the properties of gases only remain unaltered because the number of these molecules which on the average have a given state of motion is constant.[36]

Boltzmann likened the gas laws, a statement of the invariance of statistical averages, to the uniform profits of insurance companies. In

1886 Maxwell's friend Peter Guthrie Tait compared the statistical approach of the kinetic theory with

the extraordinary steadiness with which the numbers of such totally unpredictable, though not uncommon phenomena as suicides, twin or triple births, dead letters,[*] &c., in any populous country, are maintained year after year.[37]

Today, physicists regard the application of statistical mechanics to social phenomena as a new and risky venture. Few, it seems, recall how the process originated the other way round, in the days when physical science and social science were the twin siblings of a mechanistic philosophy, and when it was not in the least disreputable to invoke the habits of people to explain the habits of inanimate particles.

The limitations, not to say the dangers, of reducing human affairs to statistical laws were, however, amply illustrated in other spheres. When Charles Darwin apparently turned humans into highly evolved apes, first in *The Origin of Species* (1859) and then more explicitly in *The Descent of Man* (1871), he appealed to chance and randomness as the engine of nature's variation. The analogy with the kinetic theory of gases was quickly appreciated. In 1877 Charles Peirce wrote:

Mr Darwin proposed to apply the statistical method to biology. The same thing has been done in a widely different branch of science, the theory of gases. Though unable to say what the movements of any particular molecule of gas would be, Clausius and Maxwell were yet able . . . by the application of the doctrine of probabilities . . . to deduce certain properties of gases, especially in regard to their heat-relations. In a like manner, Darwin, though unable to say what the operation of variation and natural selection in any

---

[*] That is, those which remain in the postal system because they are badly addressed. Laplace had commented on how this was a constant fraction of the total turnover of the postal service.

individual case may be, demonstrates that in the long run they will, or would, adapt animals to their circumstances.[38]

Darwin's cousin Francis Galton saw that, as natural selection was basically a statistical theory, natural variation within a species could be tamed by Quetelet's error law. Galton's investigations of the statistical distributions of human features and behaviour led him to conclude that there was 'better' and there was 'worse' – that such a distribution implied that men are not 'of equal value, as social units, equally capable of voting, and the rest'.[39] It was then but a short step to the idea of selective breeding to improve the distribution, as he argued in *Hereditary Genius* (1869). Galton's insistence on the need for statistics in studies of inheritance led him to establish the central mathematical basis of biometrics, the measurement of biological variation; but he is also notorious now as the progenitor of eugenics, an enthusiasm of the socialist left until its acceptance by the fascist right alerted the world to its real implications.

Herbert Spencer, whose infamous phrase 'the survival of the fittest' has created much confusion about Darwin's theory, regarded Quetelet's findings as justifying an evolutionary approach to sociology. He also took from Quetelet's concept of *l'homme moyen* the dubious idea that natural selection in nature weeded out flawed variations and resulted in the survival of one 'perfect specimen'. More generally, the statistical aspect of Darwinism led many in the late nineteenth century to regard it as a theory as 'mechanical' as Maxwell's, with effects arising from definite, if unknown, 'forces'. That was how Boltzmann, for one, saw it.

WILL AND DESTINY

Before the 1850s, 'statists' were generally people who collected data about social habits and trends, and statistics itself was best regarded as the empirical arm of political economics. After that time, statistics became less a form of social science and more a *method*, a means of handling quantification in all scientific disciplines. Condorcet and Laplace were among those who foresaw this broad role of the mathematics of probability, but as the economist Antoine Augustin Cournot admitted in his book on probability theory in 1843, many (himself not included) had more limited visions: 'Statistics . . . is principally understood as the collection of facts to which the aggregations of men in political society give rise.'[40]

It was the Englishman J. J. Fox who, in a paper presented to the Statistical Society of London in 1860, first clearly enunciated the idea that statistics was not a scientific discipline but a technique. Statistics, he said,

has no facts of its own; in so far as it is a science, it belongs to the domain of Mathematics. Its great and inestimable value is, that it is a 'method' for the prosecution of the other sciences. It is a 'method of investigation' founded upon the laws of abstract science; founded on the mathematical theory of probabilities; founded upon that which had been happily termed the 'logic of large numbers'.[41]

This might make statistics seem a most modest thing, an instrument like Euclid's geometry or the calculus of Newton and Leibniz. Are tools not there simply to be taken up and used? That the truth was so much otherwise reminds us not only how deeply nineteenth-century scientists examined the philosophical basis of their work but also how much they remained influenced by religious thinking. What statistical thought held up to the looking-glass was nothing less than the concept of human free will.

From the outset, statistical approaches to social science were controversial. As statistics looked ever more likely to reveal the supposed natural laws of society, the question of what that implied for individual human behaviour became impossible to ignore. This may quite reasonably be the first question in the minds of those encountering the new 'physics of society' for the first time. The debate that raged (and it did rage) in the nineteenth century can usefully inform us about the arguments.

First, there is the issue of cause and effect: the conclusions we might draw by looking *back* to derive causes from their effects. Many statists believed, reasonably enough, that there was little point in assembling numbers if one was not going to interpret them. But interpretations immediately become politically charged. One of the central concerns of statists of the early nineteenth century (and it still preoccupies sociologists today) was crime. Could statistics tell one how to reduce it, and thus to achieve the great goal of the times, social improvement? For that, one needed to deduce the causes of crime.

The warning that a correlation between numbers does not necessarily reveal cause and effect – 'correlation is not causation' – is now almost a mantra among statisticians (though it is not always heeded). That is to say, if two sets of statistics show the same trend, it need not follow that one is caused by the other. Yet in the early days of statistics, many had no qualms about leaping to cause-and-effect conclusions that confirmed their prejudices. When the Frenchman A. Taillandier found in 1828 that 67 per cent of prisoners were illiterate, the conclusion seemed clear to him: 'What stronger proof could there be that ignorance, like idleness, is the mother of all vices?'[42] (Taillandier did not even bother to state, and presumably did not know, the illiteracy rate in the population as a whole.)

In the face of such abuses, the council of the Statistical Society of London, whose cofounders in 1834 included such luminaries as

Malthus, Charles Babbage and William Whewell, tried to keep its practitioners within the proper bounds by announcing that:

The Science of Statistics differs from Political Economy, because, although it has the same end in view, it does not discuss causes nor reason upon probable effects; it seeks only to collect, arrange, and compare.[43]

William Farr of the British General Register Office echoed the exhortations of the Society to 'exclude all opinions', telling Florence Nightingale in 1861 that 'The statistician has nothing to do with causation . . . Statistics should be the dryest of all reading.'[44] This was, all the same, frequently honoured only in the breach. As Alphonse De Candolle remarked in 1830, in the hands of policy-makers statistics could become 'an inexhaustible arsenal of double-edged weapons'.[45]

Second, there is the matter of what statistics has to say about future probabilities: the conclusions we might draw by looking *forward*.

So long as social numbers stay in their tables – a mere record of events – they are uncontroversial. But statistics took on an entirely new significance as it became appreciated that these numbers held within them the potential for prediction. If, say, 6 in every 100 people in England who were alive at the beginning of 1790 had died by the end of the year, and if the same was true for 1791 and 1792, then would we not be justified in expecting, as 1793 dawned, that 6 per cent of the current population would be dead by Christmas? That seems eminently reasonable; and yet making the extrapolation proved to be furiously contentious.

It is one thing to know what *has* happened; it is quite another to claim to know what *will* happen. Of course, death rates are not rigidly constant – it may be that 5 people in 100 died in 1791, and 7 in 100 in 1792. All the same, it would seem valid to suppose that the deaths in 1793 will be close to 6 in 100, as opposed to 20 or 50. Yet such a suggestion would have excited outraged opposition in 1793. How can one possibly know what the death rate will be? What if some fatal and infectious disease ravaged Europe that year? Or if another war broke out? Or even if it was simply an exceptionally healthy year and only 1 in 100 died? No one, surely, could be certain that these things would not happen.

The distinction here is between *statistics*, which are numbers that represent facts, and *probabilities*. One deals in certainties, as long as good counting methods are employed; the other deals in the unknown. To some philosophers and scientists, these were chalk and cheese and should not be conflated. To do so was not just mathematically or logically bad form – it was heresy, promoting fatalism and undermining free will.

Kant had already recognized in 1784 that the regularities evident in Johann Peter Süssmilch's tables of births and deaths seemed to confront belief in free will with a kind of determinism:

Whatever concept one may hold, from a metaphysical point of view, concerning the freedom of the will, certainly its appearances, which are human actions, like every other natural event are determined by universal laws.[46]

Statistical regularities were, however, not difficult to explain away in theological terms: they were evidence of God's wisdom. By engineering, for example, roughly the same number of baby boys as girls, He ensured that in principle there were marriage prospects for all and thereby maintained the stability of society.

But other constancies of statistics could hardly be rationalized so blithely. For what kind of God would arrange for the number of suicides or of murders and other crimes to remain constant year after year? Moreover, these records were enough to trouble the unbeliever too. An atheistic biologist might not need to look far for reasons why the proportions of the sexes should be roughly equal. But acts of suicide and crime were utterly volitional and not obviously explicable in terms of any natural 'mechanism'.

To Adolphe Quetelet, statistical regularities in wilful acts such as crimes placed these acts outside the responsibility of the individual: Quetelet was the first to suggest that crime is 'caused' not by wickedness but by society. It should, he said, be attributed 'not to the will of individuals, but to the customs of that concrete being that we call the people, and that we regard as endowed with its own will and customs, from which it is difficult to make it depart.'[47] In other words, 'Society made me do it, m'lud' – or in Quetelet's words, crime is a 'budget that is paid with frightening regularity'.[48] In a mechanistic age, this implied not so much that the conditions of society tended to create a constant proportion of criminals as that there was some 'force' that compelled people to break the law until the quota was fulfilled. Within such a philosophy the world is wholly deterministic and free will is nowhere to be found.

Wasn't this a convenient excuse for criminals? That argument was neatly dispatched with the satirical response that, by the same token, a deterministic force compelled judges to sentence criminals in order to meet the year's quota of prisoners. But there were many, in the 1860s and 1870s, who were deeply disturbed by the fatalism that the statistics of Quetelet and Buckle seemed to invite. William Cyples lamented in the *Cornhill Magazine* that humankind was threatened with 'a fate expressive in decimal fractions, falling upon us, not personally, but in averages'.[49] In 1860 Prince Albert told the International Statistical Congress in London that some regarded statistics as leading 'necessarily to Pantheism and the destruction of true religion, as it deprives, in man's estimation the Almighty of his power of free self-determination, making His word a mere machine'.[50]

Fyodor Dostoevsky's alter ego in *Letters from the Underworld* rages against the determinism that statistics threatened:

As a matter of fact, if ever there shall be discovered a formula which shall exactly express our wills and whims; if there ever shall be discovered a formula which shall make it absolutely clear what those wills depend upon, and what laws they are governed by, and what means of diffusion they possess, and what tendencies they follow under given circumstances; if ever there shall be discovered a formula which shall be mathematical in its precision, well, gentlemen, whenever such a formula shall be found, man will have ceased to have a will of his own – he will have ceased even to exist.[51]

If that were possible, as Buckle insisted, then the future was bleak indeed:

All human acts will then be mathematically computed according to nature's laws, and entered in tables of logarithms which extend to about the 108,000th degree, and can be combined into a calendar . . . in a flash all possible questions will come to an end, for the reason that to all possible questions there will have been compiled a store of all possible answers . . . man will become, not a human being at all, but an organ-handle, or something of the kind.[52]

To avoid such a rationalistic, mathematical fate, says Dostoevsky's reclusive scribbler, men will always strive to exert their volition, even to the extent of making themselves act irrationally or insanely:

He might act thus for the shallowest of reasons; for a reason which is not worth mentioning; for the reason that, always, and everywhere, and no matter what his station, man loves to act as he *likes*, and not necessarily as reason and self-interest would have him do.[53]

In his more measured way, Leo Tolstoy struggled in *War and Peace* with the questions posed by Buckle's deterministic view of history, with what Tolstoy called the relation of free will to necessity. He suspected that this 'new conception of history' might answer the fundamental question of international affairs: 'What is the force that moves nations?'[54] But even conceding the existence of such forces seemed to throw into doubt the notion that we can choose our fate: 'A particle of matter cannot tell us that it is unconscious of the laws of attraction and repulsion and that the law is not true; but man, who is the subject of history, says bluntly: I am free, and am therefore not subject to laws.'[55] This defiance, Tolstoy concluded, depends on there being a residue of ignorance about the causes of events: 'Free will is for history only an expression connoting what we do not know about the laws of human life.'[56]

The writer and critic Maurice Evan Hare was more whimsical in 1905:

> There once was a man who said 'Damn!
> It is borne in upon me I am
> An engine that moves
> In predestinate grooves
> I'm not even a bus, I'm a tram.'[57]

Defenders of free will argued (with justification) that because statistical laws are not true laws in the sense of describing cause and effect, as Newton's law of gravitation does, they cannot be applied to individuals and so say nothing about how any one person might act. Conversely, the nineteenth-century faith in the 'naturalness' of rationality allowed others to regard statistical regularities as a *demonstration* of free will – for had not Kant argued that free will itself tended to guide men towards orderly behaviour? In this view, free will was to be equated not with mere caprice but with its opposite.

From a modern perspective we can adduce several other considerations. Identifying randomness with maximal unpredictability seems intuitively sound, but is not necessarily so (as we shall see in

Chapter 10 in particular). Looked at from a distance, randomness becomes total uniformity: the randomly moving particles of a gas push equally in all directions and inflate a balloon into a sphere. Those phenomena that often strike us as the most complex are, in contrast, not random. The random, white-flecked 'static' of an untuned television screen presents a chaotic image that in the end looks monotonous and unchanging, while in a movie we can never be sure (this is perhaps a generous assessment) what will happen next.

And there are, as we shall see, many examples of social behaviour where a kind of regularity and order comes not from any predestination in the fates of the participants but from the very limited range of their viable choices. When we walk down a corridor, our options in principle include taking a zigzag path from wall to wall, and progressing in Ptolemaic epicycles, but no one whose faculties are intact chooses these over a more direct route.

Moreover, many of the phenomena we shall encounter in this book involve not steady behaviour but abrupt changes between certain alternative modes of behaviour. In the eighteenth and nineteenth centuries there was a strong belief in the equilibrium of society, its maintaining a steady, unchanging state. Here we shall often be more concerned with the making of choices and the sudden shifts this can provoke.

THE WILFUL DEMON

If social statistics challenged free will at the level of the individual, the Second Law of Thermodynamics as formulated by Clausius posed a deeper kind of determinism that was equally troubling. As William Thomson and Hermann von Helmholtz argued in the 1850s, the inexorable rise of entropy – heat passing forever from hot to cold –

implied a cosmic 'heat death' in which all of creation was reduced to a vague buzz of faint and useless heat, unable to bring about the organization that life requires. Not with a bang but with a whimper, indeed. And the Second Law insists that, before this dreary end, all change has a preferred direction, in which entropy increases. Does this not suggest that, just as a ball released at the top of a hill has to roll down it, so humans composed of so many dancing atoms have to behave in a certain way? Free will implies that we can do one thing or another; the Second Law seemed to be saying that, wherever change is possible, it has to happen in a certain way.

Maxwell was as deterministic as Laplace when it came to the operation of Newton's laws of motion, but unlike Laplace he had a very strong need for the hypothesis of God. He was a devout Christian and could not accept a universe in which God deprived man of free will. How, though, could free will operate without contravening thermodynamics? Alfred Tennyson sought escape from the fatalism of atoms with the poetic licence a scientist could not afford. Maxwell recounted Tennyson's description of a dream in which the poet

> Saw the flaring atom streams
> And running torrents of her myriad universe
> Running along the illimitable inane,
> Fly on to clash together again, and make
> Another and another frame of things,
> For ever.[58]

He noted how Tennyson 'attempted to burst the bonds of Fate by making his atoms deviate from their courses at quite uncertain times and places, thus attributing to them a kind of irrational free will.'[59]

Maxwell's answer was more sophisticated. He appreciated that the Second Law is a statistical law – a 'law of large numbers'. In 1867 he saw a way to 'pick a hole', as he put it, in this cosmic edict. The statistical inevitability, he said, is just the result of our ignorance: we do not and cannot know the motions of all the atoms even in a tiny scrap of matter. But they were not unknowable *in principle*. Suppose, said Maxwell, a 'very observant and neat-figured being'[60] were to exist at a scale that allowed him to watch the atoms in flight, one by one. Such a being could subvert the Second Law by selectively picking out for special treatment atoms moving in a certain direction: by exercising his free will in conjunction with his superhuman knowledge.

Say, for instance, this perspicacious being operated a trapdoor in a wall dividing a gas-filled vessel into two compartments. By opening the trapdoor only to gas particles passing in one direction, he could cause the number of particles, and thus the pressure, in one compartment to increase at the expense of the other. This would contravene the Second Law, as it produces a *less* probable configuration of particles than the one we started with. If we humans were to make a hole in the dividing wall, we would never bring about anything but an equalization of pressures – that is by far the most 'probable' outcome.

William Thomson dubbed Maxwell's being a 'demon' – much to the devout Maxwell's disapproval. Despite the apparent success of his argument, Maxwell feared that his demon alone might not be enough to rescue free will. Through the 1870s he continued to look for other loopholes in physical law that would permit free will to operate without violating the principle of energy conservation (the First Law of Thermodynamics). But this quest was doomed. Decades after thermodynamics had been connected to the theory of information devised by telecommunications engineer Claude Shannon in the 1940s, scientists uncovered the flaw in Maxwell's argument: he had neglected to take into account the thermodynamics involved in the processing of information that the demon must conduct. That is, the demon cannot make a choice about whether or not to open his trapdoor without generating at least as much entropy as is 'saved' by letting a particle through. So even Maxwell's demon is not immune to the Second Law.

It is often said that quantum mechanics destroyed the deterministic world of Newtonian mechanics by introducing probability into the very heart of matter. For sure, there is a big difference between uncertainty about the motion of objects in practice, as Maxwell and Boltzmann accepted in developing statistical mechanics, and uncertainty about such motions in principle, as embodied in Erwin Schrödinger's wave mechanics and, in particular, Werner Heisenberg's Uncertainty Principle in 1927. Quantum mechanics says that there are some things that we not only do not know but cannot know.

Yet the path towards the probabilistic physics of quantum mechanics was surely cleared by the introduction of statistics into 'classical' physical science in the late nineteenth century. By 1918 the Polish physicist Marian Smoluchowski considered probability to be central to modern physics:

From this trend, only Lorentz's equations, electron theory, the energy law, and the principle of relativity have remained unaffected, but it is quite possible that in the course of time exact laws may even here be replaced by statistical regularities.[61]

The way to statistical science would have been more tortuous if nineteenth-century experience with social statistics had not given scientists the confidence to believe that large-scale order and regularity in nature can arise even when we do not know, or cannot even meaningfully propose, a determining cause for each event. In such situations, we must trust that there are laws within large numbers.

4

The Grand Ah-Whoom

Why some things happen all at once

Nature has her own best mode of doing each thing, and she has somewhere told it plainly, if we will keep our eyes and ears open.

Ralph Waldo Emerson (1860)

In the statistical method of investigation, we do not follow the system during its motion, but we fix our attention on a particular phase, and ascertain whether the system is in that phase or not, and also when it enters the phase and when it leaves it.

James Clerk Maxwell (1878)

The Attractions of Gravity, Magnetism, and Electricity, reach to very sensible distances, and so have been observed by vulgar Eyes, and there may be others which reach to so small distances as hitherto escape Observation; and perhaps electrical Attraction may reach to such small distances.

Isaac Newton (1704)

*

Towards the end of Kurt Vonnegut's book *Cat's Cradle*, something happens that is inevitable from the outset, yet it is no less chilling when it comes:

There was a sound like that of the gentle closing of a portal as big as the sky, the great door of heaven being closed softly. It was a grand AH-WHOOM.[1]

That was the sound of the sea freezing to ice. Not ordinary ice, which one would expect to creep down slowly from the poles if the entire Earth were to drop below freezing point. The sea in Vonnegut's book freezes to *ice-nine*, a hypothetical form of ice that is stable up to a 100°C. When a block of ice-nine falls into the waves, it transforms the water of the oceans instantly into this more stable solid.

This is the nature of freezing: it is abrupt. Either a substance is liquid and mobile (above its melting point), or it is solid and rigid (below this temperature). There is nothing in between – water does not become sluggish and treacly before it turns to ice. There is always a Grand Ah-Whoom. The same is true when water boils: it is either bubbling in the pan or escaping in wisps of steamy vapour.* There is always a surface at which liquid water changes unambiguously to steam.

Such transformations between solid, liquid and gas are called *phase transitions*, and they provide the missing link between the kinetic theory of gases and a molecular-scale understanding of the other forms of matter. In ice, water and steam, the particles are all identical – they are molecules of water, the atomic trio of H_2O.† But they are organized differently: dashing about in isolated, random frenzy in the gas,

* The wispy stuff you see above boiling water is, however, liquid – tiny droplets of water that have condensed from the vapour as it rises into the cooler air. Water vapour is as invisible as air itself.

† Strictly speaking this is not quite true, since a few molecules fall apart into charged fragments called ions.

regimented and immobile in the solid, and jostling each other like a dense, unruly crowd in the liquid.

Maxwell and Boltzmann uncovered the microscopic underpinnings of the behaviour of gases. But what could statistical mechanics say about liquids and solids? There was no way to derive the Grand Ah-Whoom from the Maxwell–Boltzmann theory of gases. It was plain enough to every nineteenth-century scientist that the way to make a liquid from a gas is to cool it down. Cooling even further gives you a solid. But if you lower the temperature in the kinetic theory of gases, all that happens is that the molecules on average move more slowly. The bell-shaped curve gets sharper and shifts its peak to lower speeds. Only by cooling to the lowest temperature possible – absolute zero, or minus 273°C – do you end up with particles that have no energy and do not move. But there is no prescription in the theory to make them line up regularly as they do in crystalline solids like ice. And in any event, real solids appear well before absolute zero is reached. And what about the liquid state which precedes them?

This chapter is about phase transitions, which turn out to be one of the central concepts underlying the physics of society. Others have invoked phase transitions, often unknowingly or in a metaphorical way, to explain processes of sudden change in social contexts, for example in René Thom's catastrophe theory or Malcolm Gladwell's concept of a 'tipping point' in trends, norms and fashions[2] (page 403). Postmodern architect Charles Jencks talks of 'the creativity and surprise of a universe that evolves in phase changes – sudden jumps in organization' to justify an architecture based on surprise, 'broken symmetry' and apparent instability.[3] Thomas Kuhn's 'paradigm shifts' in the evolution of scientific thought[4] have also been touted rather vaguely as an analogue of phase transitions. But we shall see that phase transitions are not merely a convenient allegory for abrupt shifts in modes of behaviour or thought. They really do appear to happen in society, and the physical theory developed to understand them is to some extent directly transferable to descriptions of social behaviour.

In appreciating the role that phase transitions play in statistical physics, you can get a long way simply by knowing that they exist, that they are abrupt, and that they connect states of matter in which particles are arranged in different ways. But there is a crucial aspect of the new physics of society, to understand which requires us to know something of *why* and *how* phase transitions occur. That aspect has to do with the curious abruptness of these changes. For how, if each particle is happily going its own way, do they all suddenly conspire to create a Grand Ah-Whoom?

CONTINUITY PROBLEMS

At face value, it is precisely this abruptness that might seem to make the lack of an explanation of phase transitions in the kinetic theory of gases a less pressing concern. For does it not imply that something special is going on at this point where steam condenses to water, something which does not apply when steam is content to be steam? The theory's failure to account for the transition meant that it was missing something, but as long as it was used only to explain the particles' behaviour above the boiling point of water, wasn't that all right? Nineteenth-century physicists knew that, sadly, it was not right at all. They knew that it was possible to transform a gas to a liquid *without* going through a phase transition. In the jargon of the time, they knew that there was *continuity* between the two states.

A French nobleman named Charles, Baron Cagniard de la Tour was the first to show that this was so. In 1822 he found that if ether, alcohol or water is heated in a sealed tube, the substance can apparently be interconverted smoothly between liquid and gas, without the transitions of evaporation or condensation. 'There can be little doubt', John Herschel concluded in 1830,

that the solid, liquid and aeriform states of bodies are merely stages in a progress of gradual transition from one extreme to the other, and that, however strongly marked the distinctions between them appear, they will ultimately turn out to be separated by no sudden or violent lines of demarcation, but shade into each other by insensible gradations.[5]

From his careful studies of the 'Cagniard de la Tour state' in the 1860s, Thomas Andrews, a chemist at Queen's College in Belfast, concluded that, indeed:

The ordinary gaseous and liquid states are, in short, only widely separated forms of the same condition of matter and can be made to pass into one another by a series of gradations so gentle that the passage shall nowhere present any interruption or breach of continuity.[6]

This was possible only above a certain temperature, which varied from one fluid to another. Below this temperature an abrupt phase transition always intervened. We call this changeover point the *critical temperature* or, more loosely, the critical point.*

The Dutch physicist Johannes Diderik van der Waals (1837–1923) achieved the tremendous feat of explaining all of this while connecting the kinetic theory of gases to the existence of liquids. He did so in his doctoral thesis of 1873 – on which he had embarked without this aim

---

* For water this critical temperature is 374°C. In other words, below 374°C you cannot change water from liquid to gas without an abrupt phase transition; above 374°C you can make water progress smoothly from a liquid-like to a gas-like density (by altering the pressure) without any sudden jump.

in mind at all. He had the more modest ambition of explaining certain aspects of the theory of capillarity. Derived largely by Pierre-Simon Laplace in the early nineteenth century, this theory describes how liquids behave close to solid surfaces. But van der Waals dug into ground that, by the 1870s, was fertile enough to yield a more abundant harvest.

The kinetic theory explains how the pressure of a gas results from the motion of its molecules. Van der Waals knew that a liquid also exerts a pressure on a surface with which it is in contact. It was recognized that this pressure was actually huge, perhaps in the region of several thousand atmospheres – for liquids are much denser than gases and so there are many more molecules bouncing off a given area of surface. But no one knew how to calculate the pressure of a liquid from its known physical characteristics. That is what van der Waals decided to do. The only theory that enabled a pressure to be derived from the microscopic motions of the individual particles responsible for it was the kinetic theory of gases, so this was van der Waals' starting point.

Like gases, liquids are made up of molecules in rapid, disorderly motion. But liquids, unlike gases, have a degree of cohesion: a raindrop holds together as it falls. Laplace supposed in 1806 that this cohesion was due to the tendency of all the particles to attract one another. This attractive force explains why liquids have a surface tension: the mutual tug of molecules at the surface forms a kind of skin.

The kinetic theory of gases did not need to take account of such an attractive force. This is not because the molecules in a gas do not attract one another as they do in the liquid. Rather, it is because the attraction has a very short range – typically it becomes negligible once two molecules are separated by a distance of more than their own width. In the gas, the average distance between molecules is much larger than this, so they are simply not close enough to 'feel the force'. Molecules in a gas do approach one another and collide occasionally –

but the attractive force does not cause them to stick together, for they are moving too fast. So it is an effective simplification to ignore attractive forces altogether in a gas. To describe the pressure of a liquid, van der Waals thought of it as a kind of sticky gas: a gas whose behaviour is modified by attractive forces.

He also included another factor in his theory. Molecules are very small, but not infinitely so. Yet the standard kinetic theory treats them as infinitesimal 'points' which possess mass but no size. Van der Waals supposed that in a liquid, which is much denser than a gas, the space occupied by the molecules is not insignificant. The volume of space in which each molecule can wander is then not the same as the volume of the container in which they are confined, but is less than this by an amount equal to the total volume of all the other molecules.

None of this was exactly new. In the eighteenth century Daniel Bernoulli had pointed out the need to account for molecular size, and in 1863 the French scientist Gustave-Adolphe Hirn had considered how both this size and the existence of attractive forces might modify the way a gas behaves. It was already known that many gases don't behave quite as the gas laws (and the kinetic theory, which accounts for them) say they should, and it was generally assumed that factors such as the finite size and mutual attractions of particles probably explain this deviance.

In van der Waals' thesis, 'On the continuity of the gaseous and liquid states', however, all these notions came together to unify the understanding of liquids and gases. By making certain assumptions about the character of the attractive forces and of particle-size effects, the Dutchman showed that within a certain range of temperatures and pressures a fluid can exist in either of two different densities. He realized that the denser state must correspond to the liquid and the more rarefied state to the gas. Moreover, he could see that, at some point while a gas was being compressed or cooled, it would become unstable and switch to the liquid state. He predicted that there would be a phase transition.*

The crucial thing about this switch is its abruptness. You cool a gas, and go on cooling it, and it is still a gas, until suddenly – *whoom!* – there is a liquid. Van der Waals showed that the density of a fluid is not arbitrary: it may take a small value (gas-like) or a fairly large value (liquid-like), but not something in between. These are the only two stable states of the collection of particles. That is the key fact we must bear in mind. You might look at the particles and say, 'Oh, I'd like to arrange them in *this* way' – but even supposing you could perform such a feat of manipulation, you would find that 'this way' isn't actually a stable state of the system, and the particles quickly revert to one of the two allowed configurations.

In other words, when it comes to collective behaviour, not all things are possible.

What about the *continuity* of liquids and gases? After all, this was the topic that van der Waals' thesis announced in its title. His theory predicted that, as the temperature gets higher, the difference between the density of the liquid and gaseous states gets smaller. At a certain temperature this density difference vanishes, and the fluid can then exist in only one state: neither liquid nor gas, but something in between. Thus he predicted the critical point.

The big deal about van der Waals' theory is not immediately obvious, but it is vital to what follows. When the fluid switches from a gas to a liquid or vice versa, the individual particles do not change. They are still behaving like hard little balls, each with its own attractive, short-ranged force field. But their *collective* state is decidedly (and abruptly) different: one moment a tenuous gas, the

---

* Van der Waals could not, however, predict at what point this phase transition would take place as the gas's temperature or pressure were changed. Maxwell, who gave the theory his cautious approval, showed how to do this a few years later.

next a viscous liquid. No prescription for condensation or evaporation is 'built into' the particles – these things just happen, as if all the particles make up their minds to do the same thing at the same time. Except that particles have no minds. So we cannot inspect a single particle and say, 'Oh yes, this is obviously a particle that likes to form liquids.' Only when a whole mass of particles is considered together can we know what they will do.

Van der Waals identified the ingredients necessary for this behaviour. At face value, he seemed to be saying that an attractive interaction between particles was enough to promote liquefaction of a gas. But in fact there is a subtle balance of attraction and repulsion. Van der Waals thought that molecules experience a kind of repulsion due to heat – due, in other words, to the fact that they are in rapid motion. But the truth was that he had introduced the notion of repulsion as soon as he took account of molecular size: for this implies that, once particles are touching one another, they cannot approach any closer. That is to say, there is a repulsion that kicks in once particles come into contact. This seems obvious, but if we neglect size and treat the particles as infinitesimal dots, there is nothing to stop them from coming infinitely close.

So a phase transition arises out of compromises. A balance of attraction and repulsion makes for a stable liquid. If the force of disorder (that is, heat) gets too great, the gas becomes more stable instead. Moreover, the tension between these factors causes not gradual but catastrophic change: a landslide victory for one influence or the other. That landslide is the Grand Ah-Whoom.

A UNIFYING PRINCIPLE

Van der Waals' theory helped to bring the liquid state of matter within the province of statistical mechanics. It garnered widespread praise: Maxwell himself averred that 'there can be no doubt that the name of van der Waals will soon be among the foremost in molecular science.'[7] And so it was, for he was awarded the Nobel Prize for Physics in 1910. But the theory did not seem at first to be saying anything more general about the way matter behaves. It could not describe the phase transitions of freezing and melting that separate a liquid from a solid. Still less did it appear to have any broader relevance for physics.

Yet phase transitions crop up in quite different situations. It has been known for centuries that magnets lose their magnetism if they are heated (and regain it when cooled). For magnetic iron these changes happen at about 770°C, a temperature well within the reach of a medieval blacksmith's forge. In the same year that van der Waals presented his thesis at Leiden, William Barrett proposed that heat-induced 'demagnetization' is not a gradual process but happens suddenly at a particular temperature: it is a phase transition.

In 1889, John Hopkinson of King's College in London came to the remarkable conclusion that the loss of magnetism, when put in quantitative terms, shares mathematical similarities with the way that a fluid loses its distinct liquid and gas states at the critical point. This discovery is often mistakenly attributed to the Frenchman Pierre Curie, husband of Marie, who came to the same conclusion in 1895. This is because the French physicist Pierre Weiss, who developed a theory of the magnetic transition in 1907, named a magnet's critical temperature the 'Curie point' in memory of his compatriot, who had died in a road accident the year before.*

The atoms in a solid block of iron are stacked like eggs in an eggbox: immobile and ordered, not at all like the chaos of particles in liquids and gases. And yet Weiss's theory uses some of the same concepts that

---

* Hopkinson had equal claim to such a sentimental memorial too: he died tragically in a climbing accident in 1898, along with three of his children.

van der Waals had invoked to explain phase transitions and the critical point in fluids. How could that be?

To explain this, we will be helped by considering a model of magnetism devised in 1920, after Weiss's work, by the German physicist Wilhelm Lenz. A lump of iron is magnetic because each of its atoms acts individually like a tiny magnet. We can think of these atoms as little compasses, with magnetized needles that have a tendency to line up with one another. Physicists call these 'atomic needles' *spins* (although this does not mean that they are literally spinning). In iron the orientation adopted by each spin is determined by the magnetic fields created by all the surrounding spins. Thus each atom's spin influences the spins of all its near neighbours in the lattice-like array. Typically, these magnetic interactions encourage adjacent spins to align with one another. So the most stable state of the array is that in which all 'needles' point in the same direction. In this configuration the little magnetic fields of each atom add up to create one big magnetic field, and the iron is a magnet.

But just as heat's disruptive influence opposes the attractive forces between atoms in a liquid, so too does it tend to undermine the alignment of magnetic spins. The effect of heat is like shaking each of the atomic compasses, subverting their ability to line up with one another. Even though the atoms themselves stay in their serried ranks, sufficient heat can shatter the orderly alignment of spins and leave them pointing in random directions. Then their little magnetic fields cancel one another out on average, and the iron becomes non-magnetic.

So is the demagnetization transition like the evaporation of a liquid? Not quite. The temperature at which it occurs – the Curie point – is sharp and well defined, but the magnet does not switch abruptly here from being a strong magnet to being non-magnetic. Instead, the strength of the magnetism (called the magnetization) falls steadily towards zero as the Curie point is approached. Thus the Curie point resembles the liquid–gas critical point, at which the distinction between liquid and gas dwindles steadily to nothing.

This is a genuine phase transition – there is a clear-cut change between a non-magnet above the Curie point and a magnet below it. But it is a different kind of phase transition from evaporation, condensation, melting or freezing. Phase transitions that happen as a system passes through a critical point are called, naturally enough, critical phase transitions (or sometimes, for technical reasons, second-order phase transitions). Ones that involve sudden jumps in some property of the system, such as density, are called first-order phase transitions (Figure 4.1).

Lenz decided to describe all of this with the simplest model he could think of. Instead of regarding iron atoms as compasses with

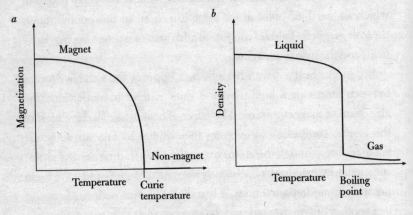

Figure 4.1 Two different kinds of phase transition. In a *critical* (or second-order) transition, some property that is characteristic of the entire system (such as the magnetization of a magnet) falls gradually to zero as a 'control parameter' (such as temperature) is changed (*a*). In a *first-order* transition (such as freezing or boiling), such a property (here the density of a fluid) changes abruptly at the transition point (*b*). Both types of phase transition are encountered in the physics of society.

smoothly swinging needles, he assumed that their spins could point in only one of two opposing directions – for argument's sake, 'up' and 'down'. Thus, neighbouring spins could either be aligned or opposed, but nothing in between. Some magnetic metals do indeed behave like this (but not iron, as it happens). Lenz also assumed that each atom, which occupies a position on a regular grid, feels the magnetic fields only of its immediate neighbours.

In 1925 Lenz's student Ernst Ising at Hamburg figured out how this model behaved in the simplest possible case. A real magnet contains a three-dimensional lattice of atoms stacked in ordered layers. This exposes each atom to the influence of an awful lot of neighbours. A simpler case is a two-dimensional magnet, with atoms sitting on a flat grid like a checkerboard: then the neighbours are fewer. But Ising opted for an even simpler model: a one-dimensional magnet, in which the atoms are lined up in a row, each one flanked by just two others. This of course bears very little resemblance to a lump of iron, and so it was perhaps not surprising that Ising obtained a rather disappointing result. His one-dimensional magnet exhibited no phase transition until it was cooled to absolute zero, minus 273°C. Only then did all the atomic compasses become aligned. The slightest heating enabled the forces of disruption to overwhelm those of order, and the row of atoms lost their magnetization.

Ising never achieved anything else of note in science – he became a schoolteacher, fled the anti-Semitism of Hitler's Germany in 1938, and ended up teaching physics in the USA, where he died at the grand old age of ninety-eight. But he has been granted some kind of immortality, for models based on a lattice of magnetic atoms are still widely used by statistical physicists and have become known, not after Lenz, but as Ising models. In the Ising model, short-range 'order-promoting' interactions between individual particles compete with the generalized disrupting effect of heat. Yet to extract from it anything vaguely

a

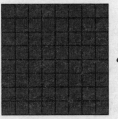
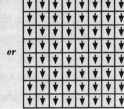

Above critical transition
temperature

Below critical transition temperature

b

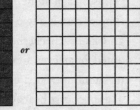

Above critical transition
temperature

Below critical transition temperature

Figure 4.2 The Ising model of magnetism (*a*) assumes that the 'magnetic needle', or spin, of each atom can point in only one of two opposed directions. The atoms lie on a regular lattice in one, two (as shown here) or three dimensions. In the magnetic state, all the spins are aligned. Above a critical transition temperature, the spins get randomized by heat and there is no overall magnetization. The same model can also be used to describe the transitions between a liquid and gas, and their merging into a single fluid state at the critical point (*b*). In this case there are two types of grid site in the model: occupied by a particle (corresponding to the dense liquid state) and unoccupied (gas). Below the critical point, the system is either a liquid or a gas. Above the critical point, it is (on average) a single fluid of intermediate density.

resembling real-world behaviour, one has to advance from one to two dimensions. This is the hard part, and it took nearly twenty years more before any progress was made.

The Norwegian physicist Lars Onsager (1903–76) virtually had to invent a new kind of mathematics before he could deduce how the 2D Ising model behaves, which he finally accomplished in 1942.[*] Unlike the 1D model, this checkerboard magnet (Figure 4.2a) does undergo a phase transition from a magnetic to a non-magnetic state at some temperature above absolute zero, and it is a critical phase transition like the one that happens at the Curie point of real magnets. But if solving the 2D model was difficult, doing it for the real-world three-dimensional case seems to be impossible. No one has yet succeeded, and there is some evidence that it is an intractable problem.

When scientists are now confronted with intractability, rather than throw up their hands in despair they switch on their computers. Although it seems impossible to solve the algebraic equations that describe the 3D Ising model, it is quite straightforward to simulate how the model should behave on a computer. It is rather like predicting the weather: it is impossible to solve all the equations that describe the patterns of air movement, but we can still run a computer simulation to see what comes out. Not surprisingly, the 3D Ising model turns out to have a critical phase transition too.

I have probably not made a very strong case for why we should expect any analogy between critical phase transitions in liquids and gases and those in magnets. Sure, there are some vague similarities in the atomic-scale picture of interacting particles – but there are some pretty substantial differences too. In 1947 a Japanese scientist named S. Ono suggested that the Ising model could be regarded as a crude model of fluids, because both have a 'binary' nature. Instead of an

[*] The German-British scientist Rudolf Peierls predicted this behaviour in 1936, but was unable to prove it rigorously.

array of cells containing atomic spins that point 'up' or 'down', one can imagine an array of cells either occupied or unoccupied by a particle, corresponding to (dense) liquid-like and (tenuous) gas-like states (Figure 4.2b). But this too might seem to be stretching a point.

Yet there is in fact an astonishing and deep connection between these different transitions. I shall consider it in more detail in Chapter 10; here I shall sketch only the briefest of supporting arguments. I have implied that the closer you get to the critical point of a liquid and gas, the smaller becomes the density difference between the two states. One can put numbers to this. At 99 per cent of the critical temperature the density difference might be, say, a factor of 2 (the liquid is twice as dense as the gas). At 99.5 per cent it might fall to 1.5, and so on. There is a characteristic *rate* at which the fluid approaches its critical point. The same is true of a magnet near its Curie point, where it is the magnetization rather than the density difference that is falling to zero. And the extraordinary thing is that, expressed in terms of percentages, these rates are identical for fluids and for a certain class of magnets. Experimentally, the magnetization of these magnets and the density difference in a fluid fall to zero at precisely the same rate. What is more, a computer simulation of a 3D Ising model shows that it too exhibits the same rate of 'critical onset' – even though the Ising model is only a very crude model of either a fluid or a magnet. The details differ dramatically in these three cases, yet their 'critical behaviour' follows the same trajectory.

Physicists call this *universality* – a term which aims to convey that there are some processes in the world for which the details simply do not matter. It is surprising enough that (as we shall see below) two different fluids, such as carbon dioxide and methane, which have quite different critical temperatures, should approach their critical points at the same rate in relative (that is, percentage) terms. It is baffling that two wholly different *kinds* of system – a fluid and a magnet – also display this universality. What this suggests is that phase transitions are generic phenomena: they happen in the same way for a wide range of apparently different systems. A physicist can stand up and talk for an hour about a 'first-order phase transition', and no one will stop him to ask, 'Wait, are you speaking here about freezing, or evaporation, or what?' The audience knows that what is being said applies to all such cases. In the same way, a town planner can speak about traffic jams without anyone protesting, 'Do you mean the snarl-up last Thursday at Parson's Corner?' Every traffic jam involves a different set of vehicles and circumstances, but there are features common to them all. This still sounds, perhaps, like a very broad generalization to be drawing from some correspondences in the mathematical details of the way magnets and fluids behave. But there is more to phase transitions than this.

NEAR ZERO

Van der Waals' theory described how gases and liquids *in general* are connected by a phase transition below their critical point. What some scientists wanted from such a theory was the ability to predict how and when this change of state would happen for particular substances. All fluids are clearly not the same: water is liquid under conditions in which carbon dioxide is a gas; and carbon dioxide liquefies at a higher temperature than nitrogen. This is no mystery: different molecules have different sizes and exert different forces of attraction. But van der Waals realized that the critical point is the point of reference that reveals all these specifics.

If you were to survey the heights of all schoolgirls between the ages of twelve and fourteen in a London school, and plotted the results on a graph, you would obtain something like the familiar bell-shaped curve of Figure 3.2 (page 75). The number of pupils of each specific height would rise to a maximum and then tail off. If you repeated the survey for all the schools in London and plotted the results on the same graph, the curve would have the same shape but would be much taller – because the absolute numbers (plotted on the vertical axis) are bigger. In a survey of the whole country the peak would be even higher. But the *proportions* would be the same in each case: the fraction of girls 4 cm shorter than average would be more or less identical in every survey. So you could collapse all the bell-shaped curves onto a single 'master curve' by plotting not absolute numbers, but relative proportions: the fraction of the total that are 130 cm tall, 140 cm, and so on.

If you did the same for boys in the same age range, the shapes of the curves would be the same but they would be slightly shifted on the horizontal axis: the commonest height would be slightly greater.* Yet even the two master curves for boys and girls can be superimposed onto a single curve – by making the heights relative instead of absolute. That is, you plot the fraction of boys or girls that deviate from the mean by 2 per cent, 5 per cent, and so on.

Van der Waals did much the same 'rescaling' of curves for liquids and gases. He saw that the relationships between pressure, temperature and density for different substances all fall onto the same master curve if expressed in relative terms by reference to the values of these quantities at the critical point. The relative temperature, for instance, is the temperature divided by the critical temperature. This so-called 'principle of corresponding states' shows that all liquids and gases are (to a first approximation) the same 'master fluid' rescaled by some factor related to the critical point. And the rescaling is determined by the properties of the individual particles (molecules) from which the

* It is actually only around the age of twelve to thirteen that boys start to attain a greater average height than girls.

substance is constituted, because the critical temperature, pressure and density can be calculated, in van der Waals' theory, from these single-particle characteristics: their size and the range and strength of their attractive forces.

The director of the Physical Laboratory at Leiden, where van der Waals was working, was Heike Kamerlingh Onnes (1853–1926). He became interested in the principle of corresponding states for very practical reasons. It allows one to predict how fluids will behave far below the critical point purely from a knowledge of the critical properties (the critical temperature, density and pressure). By scaling the 'master curve' according to these quantities, you can deduce the relationships between pressure, temperature and density for the entire range at which the liquid state is stable. Kamerlingh Onnes was interested in finding out how cold some gases had to be before they would liquefy.

Helium was especially resistant to liquefaction, and at the turn of the century the existing cryogenic techniques could not produce temperatures low enough to condense it to a liquid. By fitting the experimentally observed behaviour of helium at temperatures above its critical point to van der Waals' equations, Kamerlingh Onnes predicted that he would need to reach just 5 or 6°C above absolute zero before he could make it turn to liquid. This was a reasonable estimate: in fact helium does not condense until 4.2°C above absolute zero at atmospheric pressure, a temperature that Kamerlingh Onnes finally achieved in 1908.

Having liquid helium at his disposal, Kamerlingh Onnes was able to use it as a cooling fluid to investigate how other substances behave at such frigid extremes. Physicists at that time suspected that, because the heat-induced vibrations of atoms in metals disturb the flow of electrical current through them, they might become much better conductors at very low temperatures, approaching 'perfect' (resistance-free) conductivity at absolute zero. Yet when Kamerlingh Onnes put this to the test with mercury in 1911, he got a surprise. The electrical resistance of mercury did not decline smoothly as the temperature dropped: it fell abruptly to zero at around the boiling point of helium. At this point mercury becomes a superconductor, able to carry a current that is totally unimpeded by electrical resistance.

It was soon found that other metals display the same behaviour above (but close to) absolute zero. Lead, for example, superconducts at 7.2°C above zero. The change to a superconducting state has all the characteristics of a critical phase transition: the resistance plunges swiftly to zero as the superconducting transition temperature is approached, just as the magnetization of iron dips to zero close to its Curie point. But that was not all. In 1937 the Soviet physicist Pyotr Kapitsa in Moscow discovered that if liquid helium is cooled even further below its boiling point, at a little over 2°C above absolute zero it develops extremely bizarre properties. It loses all viscosity, and once it begins to flow it never stops. This form of liquid helium will even crawl up the side and out of a container. It becomes a so-called superfluid.

A theoretical explanation for superfluidity came in the late 1930s. The superconductivity of metals was harder to understand, and was not satisfactorily explained until 1957. Both of these exotic phenomena happen because at very low temperatures the laws of quantum mechanics take precedence over the laws of 'classical' physics, which dominate at higher temperatures. Yet both come about via genuine phase transitions. They are both manifestations of collective behaviour arising from the interactions between the atomic-scale components of the materials. Thus phase transitions are not restricted to particles behaving in accord with the good old-fashioned Newtonian laws of motion. It is often said that quantum mechanics overturned all of the 'classical' physics that went before. But it certainly did not over-turn the physics of phase transitions: quantum mechanics merely lends a different flavour to it (one which becomes evident only at low temperatures).

Indeed, phase transitions seem to persist however much one reshapes the fundamentals of physics. To understand the internal structure of subatomic particles such as the protons and neutrons that make up the nuclei of atoms, the 'traditional' quantum theory of the 1920s is not enough. Physicists in the 1970s had to develop a new theory called quantum chromodynamics. This uses many of the conceptual tools of statistical mechanics, such as lattice models like the Ising model, and finds abrupt phase transitions taking place among its subatomic constituents: sudden jumps between different stable arrangements of these particles. And many cosmologists believe that a phase transition reshaped the entire universe during an absurdly brief instant a tiny fraction of a second after the Big Bang. They propose that during this 'inflationary era' the universe experienced a cosmic phase transition in which it expanded from a size much smaller than a proton to a size comparable to its present gargantuan dimensions. As phase transitions go, this has to be the most dramatic Ah-Whoom ever.

TIME FOR A CHANGE

You might now be imagining that all abrupt change is due to phase transitions. That is not so. If I turn on a light by flicking a switch, there is suddenly light where before the room was dark – but no phase transition has made this happen. I have opened a valve, that's all, and an electrical current has flowed. The key notion about a phase transition is that it happens everywhere throughout a system (that is, globally) all at once, and it does so because of a conspiracy of countless constituents.

A phase transition is a sudden, global change in behaviour arising from the interactions of many constituent particles. Typically these interactions are short-range, *local* ones: each particle takes heed only of its immediate neighbours, and neither knows nor cares what is happening farther afield. The phase transition happens when some global influence which acts upon the particles reaches a certain threshold value. One moment the particles are behaving 'normally', as if nothing were amiss; then without warning (or almost so, as we shall see) they switch to some entirely different mode of behaviour.

The statistical mechanics of Maxwell and Boltzmann, humbly conceived to explain how gases behave, has now mutated into the discipline called, less archaically, *statistical physics*. Statistical physicists have traditionally been interested in the complexities of how inanimate matter behaves, particularly when it undergoes phase transitions, and especially critical transitions. This science is statistical because it generally deals with systems made up of huge numbers of particles or separate components, all interacting with one another so that the average behaviour is often all that counts.

Statistical physicists are now starting to glimpse vistas far more enchanting and seductive than gases in pots. To reach this vantage point, they have first had to labour furiously to reshape the tools developed by the great scientists of a century ago to new uses for which they were never originally designed. Traditional statistical mechanics deals with some extraordinary things, but it struggles to say much about the dynamic world in which we go about our business. Nineteenth-century thermodynamics dealt with equilibrium states, to which nothing was ever added or taken away and in which, on average, nothing ever changed. In the following chapter we shall see how statistical physics today is more like life: full of the processes of growth and decay.

5

On Growth and Form

The emergence of shape and organization

The reasonings about the wonderful and intricate operations of Nature are so full of uncertainty, that, as the Wise-man truly observes, hardly do we guess aright at the things that are upon earth, and with labour do we find the things that are before us.

   Stephen Hales (1727)

Are the attributes of a society, considered apart from its living units, in any way like those of a not-living body? or are they in any way like those of a living body? or are they entirely unlike those of both?

   Herbert Spencer (1876)

Order is the handmaiden of all virtues! But what leads to order?

   Georg Christoph Lichtenberg

*

A phase transition brings magic to high places. Walking in wintry woods in 1856, Henry David Thoreau eulogized about the tiny ice crystals that lay sparkling all around him:

How full of the creative genius is the air in which these are generated! I should hardly admire them more if real stars fell and lodged on my coat. Nature is full of genius, full of the divinity; so that not a snowflake escapes its fashioning hand.[1]

Snowflakes form in the atmosphere when water vapour freezes into ice – a transformation directly from gas to solid. What can statistical mechanics tell us about these six-pointed crystals, which scientists have been studying since the seventeenth century? Until rather recently, the answer was not a lot. If you had asked a statistical mechanician of the 1940s to tell you about snowflakes, he might have shuffled in an embarrassed way and said, 'Well, you see, the theory wasn't designed for that.' Wasn't designed to tell us about one of the commonest phase transitions of one of the most plentiful substances in the world? What kind of theory is it, then?

Traditional statistical mechanics arguably owes its current form to one man above all others: the American scientist Josiah Willard Gibbs (1839–1903). In his book *Elementary Principles in Statistical Mechanics* (1902), this Yale professor brought all the threads together and wove them into a robust and elegant tapestry. Building on the work of Clausius, Maxwell, Boltzmann and van der Waals among others, Gibbs made thermodynamics thoroughly coherent and established exactly how its laws arise from the underlying microscopic description of a system.

Gibbs showed that the key consideration in processes of change is *minimization*. Simply put, this is akin to water flowing downhill and gathering in depressions as ponds and lakes. The higher the water is to start with, the more gravitational or 'potential' energy it has. If unimpeded, it will minimize this energy by moving to as low a position as possible. Although it is not immediately obvious, this is a direct corollary of the Second Law of Thermodynamics – the stipulation that entropy increases in all spontaneous change.

When water collects in a lake, it reaches a kind of equilibrium: it stops moving downwards and just sits there. Thermodynamics is all about the drive to find equilibrium states like this. It describes processes of change from one equilibrium state to another, even more stable state – like the opening of a sluice gate in a mountain reservoir to let water flow from a higher basin to a lower one. Gibbs stated that, for a system to be in an equilibrium state (a stable state with no immediate propensity to change), 'it is necessary and sufficient that in all possible variations of the state of the system which do not alter its energy, the variation of its entropy shall either vanish or be negative.'[2] Which is a way of saying that when a system is in equilibrium, you can push, pull and prod it all you like, and it will simply return to how it was to begin with. This is the paradox of 'classical' thermodynamics. It seeks to account for change, but it can't actually say anything about the process of change itself. It can only provide prescriptions for the starting point and the end point, and falls silent on the question of what happens in between.

Now think of the snowflake. It began as water vapour and ended as ice: two equilibrium states, separated by a phase transition. But in one case the snowflake grows into a six-pointed star with branching, Christmas-tree arms. Another snowflake has arms ending in clover-like clusters. A third is made of blocky hexagons; a fourth is like a bulbous six-petalled flower. You can look through the classic snow-flake portrait album *Snow Crystals* (1931) by Wilson A. Bentley and W.J. Humphreys and see two thousand crystals, none of them identical. Each has a unique history of growth: it clearly matters very much *how* the water vapour became solid.

While snowflakes show nature at its most inventive, they are by no means peculiar in their great variety. Does one tree ever resemble another in exact detail? Is one day's global weather pattern ever like any other? In a river flowing incessantly around a rocky outcrop, is the pattern of water flow at one instant ever exactly like that at another? As the Greek philosopher Heraclitus pointed out in the sixth century BC, you can never step in the same river twice.

The processes of change that go on around us are very seldom the kind of leaps between platonic equilibrium states that thermodynamics is designed to handle. Many are ongoing processes – as if that stream of water were forever meandering in the hills, looking for a nice stable basin to fill, while rain forever replenishes its source. They are, in short, processes that are not in equilibrium and never will be, or at least not in our lifetimes.

Even when there do seem to be stable starting and end points to a transition, like vapour and crystal, the form that results from the change can be complex and impossible to predict. This is because, in the case of a snowflake, the growth process takes place far away from any equilibrium states. It is as though, rather than opening a sluice gate to transfer water smoothly downhill, the entire bank of a reservoir is suddenly destroyed so that the water gushes forth on a course that no one can anticipate.

This is not to imply that thermodynamics is theoretically neat but practically useless. On the contrary, it is immensely useful. It enables us, for instance, to understand and predict the direction of a change, and to know under what conditions the change will occur – to explain, for example, why water freezes at 0°C at sea level. The thermodynamics of equilibrium states can be used to understand energy generation in living cells and in power stations, the formation of minerals in the Earth's crust and heat dissipation in a computer. But it is a limited theory. If we are dealing with processes taking place far from any stable, unchanging equilibrium, and in particular if we are dealing with processes of growth and form, something else is needed.

HISTORY MATTERS

I have taken my chapter title from one of science's truly classic texts, fit to sit on the bookshelf alongside *On the Origin of Species*, *Principia mathematica* and *Traité élémentaire de chimie*. But although Darwin's book initiated modern biology, Newton's modern physics and Lavoisier's modern chemistry, the Scotsman D'Arcy Wentworth Thompson's (1860–1948) idiosyncratic and eclectic book *On Growth and Form* spawned no definable field of science. Instead, since its publication in 1917 it has awoken generations of scientists to the realization that we live in a world of profound beauty in which there is still much we do not understand.

Thompson's book was an attempt to challenge the tendency, prevalent in his time (and far from absent today), to answer all questions of biological form by pulling Darwin out of a hat and proclaiming the magic word 'Adaptation!' Thompson argued that much in nature could be explained purely on the basis of geometry, mathematics, physics and engineering, without the need to invoke natural selection.

In many ways *On Growth and Form* was limited by its being ahead of its time, for it laid out many instances of natural form resulting from what we would now identify as *non-equilibrium* growth processes. In 1917 the concept of change away from equilibrium was barely grasped even by the founders of thermodynamics, let alone by a Scottish zoologist, however erudite he was in Greek, Latin and geometry. When considering ink droplets dispersing in water, cracks spreading in mud or liquids circulating convectively in a pan, Thompson was forced to be descriptive rather than analytical. He noted that some systems undergoing change reach 'a "steady state", if not a stable equilibrium',[3] but could say no more about the nature of these steady states. They were clearly not, however, Gibbs's placid equilibrium states.

Lars Onsager (page 113) knew this. He was one of the first to face up to the shortcomings of standard thermodynamics, and in the 1930s he began to do something about it.

By restricting itself to equilibrium states, thermodynamics might appear to have painted itself into a corner. How can change happen, except by disrupting that tranquillity? Once you have opened the sluice gates of our mountain reservoir, the water it contains is no longer in equilibrium. So until all the water has emptied into the lower reservoir and stillness is restored, non-equilibrium reigns. The same is true of water freezing to ice. It cannot do so infinitely fast, and while it is freezing thermodynamics can no longer describe what is going on.

Classical thermodynamics gets around this with ingenious legerdemain. It treats the process as if it were happening very, very slowly – strictly speaking, infinitely so. That makes it possible to look at the system at any moment and observe, 'Why, nothing is happening: it is still in equilibrium.' It is like drilling the tiniest of tiny holes into the bank of the reservoir so that the trickle it releases is barely perceptible. True, the lower reservoir then fills up drip by drip; but at any one moment both the higher and lower reservoirs appear to be in stable equilibrium states.

This might seem to be at odds with my earlier suggestion that phase transitions are abrupt changes. But by 'abrupt' I do not necessarily mean 'instantaneous', although I must confess that invoking Kurt Vonnegut's Grand Ah-Whoom rather implies that interpretation. I mean that next to no change in the prevailing conditions is required to induce the transition. To cross the boundary from liquid water to ice, you need only alter the temperature by an infinitesimally small amount. At one hundredth of a degree above 0°C, the equilibrium state is that in which all the water is liquid. At one hundredth of a degree below zero, all is ice.

Rather than the kind of pseudo-equilibrium process that thermodynamics can handle, many transformations in the real world happen when the 'driving force' is much greater than the minimum necessary – for example, when the temperature of water drops suddenly far below zero. Snowflakes form in this way, at temperatures several degrees below water's normal freezing point.

Onsager considered that the best starting point for developing a thermodynamics of non-equilibrium states was one in which the states depart only slightly from equilibrium, impelled by relatively small driving forces. We don't open the sluice gates, but simply make the hole rather larger than a pinprick.

Not even non-equilibrium processes escape the Second Law of Thermodynamics; they too proceed by increasing the total entropy of the universe. But whereas equilibrium thermodynamics tells us only that the entropy will be greater *after* than *before*, an unfolding, dynamic non-equilibrium process prompts us to consider how entropy is changing over time. Onsager began wondering about the *rate* at which such processes generate entropy.

For equilibrium states the criterion for determining how a system will behave is the Gibbs condition: the system will configure its components so as to minimize the energy.* Onsager looked for a rule equivalent to the Gibbs condition that determines the character of a non-equilibrium 'steady state'. Even away from equilibrium, systems are capable of adopting states that in some sense remain the same. A

---

* By a 'system' I mean any collection of components that we choose to define, such as the water molecules that comprise a snowflake. We could equally well consider as a system a mountain lake and the surrounding slopes and valleys, or the planets orbiting the Sun. What matters is that we be able to define, to a sufficient degree of accuracy, the relevant components of the system and how they interact with one another. A further point is that 'energy' is actually too vague a term here; Gibbs gave a more precise definition in terms of a quantity called the 'free energy', which is a well-defined portion of the system's total energy.

river is not at equilibrium, since its waters are constantly flowing from high ground to low; but it generally stays in a steady state in which the waters are confined between its banks at a roughly constant level. Living cells are also in such a 'dynamic' steady state, maintaining their integrity and their function while constantly burning up energy and churning out wastes. Non-equilibrium steady states are ubiquitous: a whirlpool's vortex, a cruising automobile, the ebb and flow of the tides.

Onsager did not, in truth, get very far with elucidating the criteria that make specific non-equilibrium steady states preferred over other possible states of a system. He showed that, close to equilibrium, there are general rules relating the driving force to the rate of entropy production in a system, which in itself was a phenomenal achievement in virgin territory, deemed worthy of a Nobel prize in 1968. But he found no universal Gibbsian principle for non-equilibrium thermo-dynamics.

There is a good reason for this: it is almost certain that no such principle exists. Ilya Prigogine, a chemist born in Russia and working in Brussels, believed he had found the magic formula in the 1940s. He claimed that the most favourable non-equilibrium steady state, at least for only small departures from equilibrium, is that which minimizes the rate of entropy production. One still sees this criterion quoted as fact, but sadly it is not universally true. Does this mean, then, that away from equilibrium anything goes – that all things are possible? Clearly not. Non-equilibrium processes do seem to make choices which, while not always predictable in advance, are nevertheless consistent and reproducible. And what is truly extraordinary is that many of these choices correspond to states that are not chaotic or messy, but instead display a high degree of orderliness.

A classic example of an ordered, non-equilibrium steady state is the one discovered by the French scientist Henri Bénard in

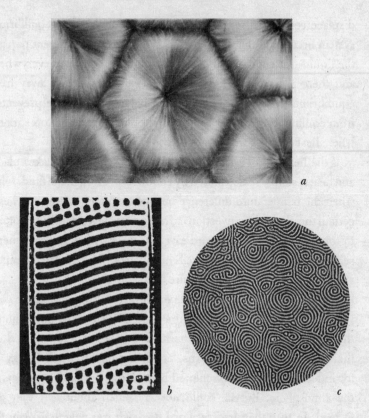

Figure 5.1 Henri Bénard found that a convecting fluid can become patterned into circulation cells defined by rising and sinking streams of fluid. In *a* these cells, visualized by adding metallic flakes to a liquid, form an orderly hexagonal arrangement. In *b* and *c*, for a different fluid and a different visualization method, the cells are roll-like.

1900.[*] He warmed a thin layer of liquid in a copper dish, which set up convection currents: warmer, less dense liquid at the bottom rises to

[*] Although these patterned states are associated with Bénard's name, they were actually first observed thirty years earlier by the German physicist Georg Hermann Quincke.

displace cooler, denser liquid at the top. This is a non-equilibrium system because of the differences in temperature in different parts of the liquid – at equilibrium, the temperature is equalized everywhere and there is no convective flow. The system is driven away from equilibrium by the heat coming from below, and will be prevented from equilibrating as long as heat is supplied (since the top surface, where heat escapes, will always be cooler).

If the heating is very gentle, however, no convection takes place, and heat is simply redistributed by conduction through the fluid. Only when the temperature difference between top and bottom reaches a certain threshold do convection currents start to circulate from bottom to top and back again. Bénard saw that the currents organize themselves into roughly hexagonal cells in which the fluid rises in the centre and sinks around the edges (Figure 5.1a).

Under different experimental conditions, a variety of other convection patterns can be seen (Figure 5.1b,c). D'Arcy Thompson noted this diversity in convection patterns in smoke, and commented on their similarity to cloud patterns, 'as of a dappled or mackerel sky'[4] – for indeed clouds can be sculpted by orderly convection currents in the atmosphere. But he could not say much about *why* they are formed. In 1916 the physicist Lord Rayleigh was able to explain the roll-like convection state shown in Figure 5.1b using the theory of fluid mechanics; a better theoretical understanding of the other convection patterns emerged in the 1960s. Yet there is still no theory that can reliably predict which pattern will appear in any given experiment under a given set of conditions – there is no Gibbs-like criterion for non-equilibrium pattern formation.

Searching for a priori predictability is in fact futile because Rayleigh–Bénard convection patterns (as they are now called) can differ even under seemingly identical conditions, if their means of preparation are different. That is to say, if you take two different experimental routes to reach the same end point – heating at different rates, say, or stirring and not stirring initially – then the resulting patterns may differ. These non-equilibrium steady states depend on their *histories*.

Rayleigh–Bénard convection patterns are examples of *dissipative structures*: organized arrangements in non-equilibrium systems which are dissipating energy (the convection patterns are maintained by the constant supply of heat, for example) and thereby generating entropy. In the 1950s and 1960s Prigogine and his collaborators suggested that dissipative structures are reached when a non-equilibrium system is driven to a crisis point, called a *bifurcation*. Close to equilibrium, the system might not do anything very remarkable – the fluid in a Bénard dish, for example, just sits there conducting heat, apparently at rest. At the bifurcation point, however, the system is abruptly compelled to change its state dramatically.

As the term implies, there are generally two choices on offer. In the roll-cell Rayleigh–Bénard state (Figure 5.1*b*), adjacent rolls rotate in opposite directions like the rollers of a mangle. But any one roll could rotate in either of two senses – clockwise or anticlockwise – provided the directions of all the others were reversed too. So there are two possible roll-cell states, equivalent in every respect except the direction of rotation. What determines the choice? Pure chance – or more properly, fluctuations, which physicists often call *noise*.

Noise is everywhere. At any temperature above absolute zero, atoms are shaking with thermal energy. This sets up a random background 'buzz' that pervades all matter. The buzz gets 'louder' as the temperature rises: the forces of disorder exert themselves relentlessly. Because of this haphazard aspect of atomic motions, all processes incur small random variations, or fluctuations. If one could measure the pressure on a very small area of a balloon's rubber wall with tremendous sensitivity, one would find tiny variations caused by the moment-to-moment differences in the number of gas molecules striking

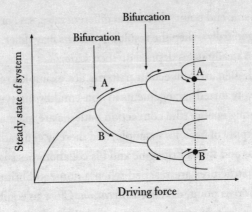

Figure 5.2 Ilya Prigogine predicted that non-equilibrium steady states are reached via a cascade of bifurcations, at each of which the system must choose between two alternatives. Because this choice is determined by random fluctuations, two initially equivalent systems (here A and B) driven out of equilibrium can end up on different 'branches' – in different steady states – while experiencing the same driving force, because they have a different history of choices.

it. Indeed, scientists now routinely make highly precise measurements of quantities such as pressure and temperature at very small scales, and invariably find that these quantities fluctuate about some average value.

Normally these fluctuations have a negligible effect, in keeping with their tiny magnitude. But at a bifurcation point a non-equilibrium system is poised on a razor's edge. Will it take this path, or that one? The smallest chance fluctuation can tip the balance and irrevocably determine the future fate of the system. Prigogine explained that 'in the neighbourhood of the bifurcation points fluctuations play an essential role and determine the "branch" that the system will follow.'[5] Increasing the driving force of a non-equilibrium process even further beyond a bifurcation point can produce a switch to another steady state: a different pattern, attained by a second bifurcation. In general, said Prigogine, there is a cascade of bifurcations through which a non-equilibrium process might progress as it is driven further from equilibrium (Figure 5.2).

At each branching point the options are well-defined, but the choice is arbitrary. So two systems that are wholly identical at the outset might end up on quite different branches while experiencing the same driving force, simply because they happened to take different paths at each junction. 'Time forks perpetually towards innumerable futures', as Jorge Luis Borges says in his story 'The Garden of Forking Paths'.[6] But whereas the people in the book written by Borges' character Ts'ui Pên may venture simultaneously down both of the routes open to them, real systems can take only one of the possible ways ahead. So, of course, can real people, and a long and countless succession of such choices makes the world what it is and not some other place. In this way, says Prigogine, 'the bifurcation introduces *history* into physics and chemistry, an element that formerly seemed to be reserved for sciences dealing with biological, social, and cultural phenomena.'[7]

Away from equilibrium, then, Gibbs's determinism gives way to historical contingency. Ironically, perhaps, it is this that stymies Prigogine's attempt to find a minimization principle for non-equilibrium thermodynamics – for in *any* steady state out of equilibrium, what matters is not just the prevailing conditions influencing the system's components but how these conditions were arrived at. Nevertheless, there is a striking and important similarity between non-equilibrium bifurcations and equilibrium phase transitions. A bifurcation is a sudden global change to a new steady state. That sounds familiar. Indeed, the bifurcation point bears a strong resemblance to a critical point, like the Curie point of a magnet.

As a metal is cooled through its Curie point, it changes from a non-magnet to a magnet.* In the non-magnetic state all the atomic 'compass

* Note that not all metals are magnetic, and not all magnetic metals align all their spins at a Curie point.

needles' (spins) are randomly oriented, while in the magnetic state they are aligned in an orderly fashion. This critical phase transition therefore corresponds to the onset of an ordering process. Similarly, when a dish of fluid is heated above the threshold for convection to begin, this non-equilibrium bifurcation orders the fluid into ranks of roll cells. In both cases, physicists say that *symmetry-breaking* has taken place.

Why symmetry-*breaking*? Do we not associate order and pattern with symmetry, and randomness with lack of it? Maybe so, but randomness has its own kind of symmetry. A system in which all the components are moving at random is, on average, more highly symmetrical than one in which they are moving in concert in a certain direction. In the random state, any direction is indistinguishable from any other. In a convecting roll-cell state, the direction parallel to the rolls is clearly different to the direction perpendicular to them: the rolls 'point out' a special direction in space. So when a uniform fluid is transformed to one that is circulating in roll cells, some symmetry is lost, or 'broken'. The same is true of a magnet: the aligned spins point out a special direction in space where in the non-magnet there was none.

A bifurcation offers two equivalent choices of steady state – and so does the transition to a magnet. Think of Ising's model, in which each spin may point in just one of two directions, 'up' and 'down'. They can line up in either of these two senses, with equal but opposite magnetization (see Figure 4.2, page 112). And what of the liquid–gas critical point, which we compared with the Curie point? The fluid can adopt either a gas-like or a liquid-like configuration below the critical point. Moreover, these choices at a critical phase transition are, like those at a bifurcation point, at the mercy of fluctuations. There is no reason for the magnet to prefer one direction of alignment over the other, but a chance excess of one type of spin in some part of the system can tip the balance throughout. So systems undergoing critical phase transitions become hypersensitive to fluctuations. We shall see later that this gives rise to very special and remarkable behaviour at critical points.

These correspondences do not imply that there is after all some non-equilibrium statistical mechanics that parallels that of equilibrium systems. No, the truth is both simpler and deeper. The two kinds of transformation – phase transitions and non-equilibrium bifurcations – have features in common because they are both fundamentally of the same ilk: they are both *collective* modes of behaviour arising from the mutual, local interactions of many individual components. There are conditions both in equilibrium and away from it for which these interactions can make one part of a system almost miraculously sensitive to what is happening far away. Every particle is suddenly in touch with all the others via intricate networks of mutual nudges – and all at once, a new steady state emerges.

THE SHAPE OF CULTURES

To develop a physics of society, we must now take a bold step that some might regard as a leap of faith and others as a preposterous idealization. You may have guessed it already: particles will become people. To make that bold step a little easier I shall introduce a stepping stone, which will bring *life* into the picture before we have to worry about such things as free will.

It would be hard to make the case that bacteria, lacking anything resembling a brain or a nervous system, possess free will. Yet they are certainly alive. Possessing only primitive means of communication, bacterial cells can display rich and varied patterns of collective behaviour in non-equilibrium growth processes. The Japanese physicist Mitsugu Matsushita of Chuo University discovered in the

Figure 5.3 Under certain conditions, the bacteria *Bacillus subtilis* grow in colonies that have a complex *fractal* branching shape, resembling the pattern seen in many growth processes in non-living systems.

1980s that these patterns derive from a kind of bacterial physics. Louis Pasteur once said that fortune favours the prepared mind, and Matsushita was prepared for his foray into the life sciences by his background in the statistical physics of non-equilibrium growth. When he saw the complex branching pattern of a growing colony of *Bacillus subtilis* bacteria (Figure 5.3), he knew what he was looking at.

It was a *fractal*: a structure that reiterates at successively smaller scales what it displays at larger ones. Matsushita recognized here the kind of fractal branching pattern that is characteristic of a process called diffusion-limited aggregation (DLA), which occurs, for example, when metal collects at a negatively charged electrode dipped into a solution of a salt: the phenomenon of electrodeposition. What captivated physicists in the early 1980s, when they first began to study fractal shapes like this, produced in non-biological processes like electrodeposition, was that these patterns resemble the 'organic' forms of nature: branched fractal mineral deposits in rocks, for example, are sometimes mistaken for fossilized ferns. Now here in Matsushita's laboratory was a genuinely biological fractal.

The fractal shape of a DLA deposit looks fairly chaotic, but it is not arbitrary. No two deposits are ever identical, but they all have a feature in common, namely the efficiency with which the branches fill up the space available to them. No matter how big the structure grows, there will always be 'fjords' of empty space penetrating deep into the ramified form. There is a mathematical measure of this efficiency of space-filling, called the *fractal dimension*.* All DLA growth patterns have the same fractal dimension. It is a badge of identity which can distinguish the branching growth from some other superficially similar structure.

In 1981, US physicists Tom Witten and Len Sander proposed a theoretical model of the DLA process, motivated by their attempts to describe how dust particles clump together in air. The model invokes particles undergoing a Brownian random walk (that is, diffusing; see page 50) in air and sticking together when they come into contact. They found that this process creates unstable branch tips, which cannot help but split in two at every opportunity. In 1984 Matsushita showed that the clusters that emerge from this DLA model are precisely like those formed in electrodeposition in a flat dish. Matsushita's branching colonies of *B. subtilis* did not just resemble DLA clusters –

* The fractal dimension is not a whole number, implying that fractal objects incompletely occupy the space they inhabit. A block of stone is a three-dimensional object: it fills up a specific three-dimensional volume of space. Likewise, a square drawn on the flat, two-dimensional plane of a piece of paper occupies a specific surface area: it is a two-dimensional object. And a simple line is one-dimensional. But a fractal shape like the one in Figure 5.3 is somewhere between one-dimensional and two-dimensional. This particular shape is around 1.7-dimensional: it has a fractal dimension of 1.7. This means that it 'fills up' more of the two-dimensional space than does a simple line, but not as much as, say, a disc.

they had the same fractal dimension. This suggested (although it did not prove) that the formation processes in both cases shared the same essential features, most notably constant tip-splitting induced by random fluctuations in the advancing edge of the cluster.

DLA is a non-equilibrium process, and fractal patterns are one of the motifs of non-equilibrium growth. The immediate and irreversible sticking of particles wherever they strike the growing cluster leaves them no opportunity to find their most stable equilibrium configuration. The DLA cluster is a map of frozen accidents of history. Matsushita and his colleagues wondered what might happen if they modified the driving force that pushes their fractal bacterial colonies away from equilibrium. So they altered two factors they presumed would affect the growth process. The colonies were cultured on a layer of transparent gel called agar, suffused with the nutrients needed for the cells to grow and multiply. As the amount of water in the gel diminishes and the material becomes more rigid, the cells stick ever more firmly to the surface. So by varying the agar-to-water ratio in the gel, the researchers could control the cells' mobility. And by changing the nutrient content they could alter the 'health' of the colony – its ability to generate new cells.

When they altered these two parameters, Matsushita's team found that the colonies developed growth shapes very different from the branching DLA structure. At high nutrient levels the colony was dense, with fat, finger-like branches only at the perimeter, rather like a blob of lichen on a rock face. This shape resembles that produced in a theoretical growth model devised by the mathematician M. Eden in 1961 to describe the development of cancer tumours. Both the DLA-like and Eden-like colonies appear on hard agar surfaces, on which the cells cannot move. The colony advances as new cells are generated at the edge. But if the gel is softer and the cells can move, other patterns appear: thin, radiating branches at low nutrient levels, and Eden-like

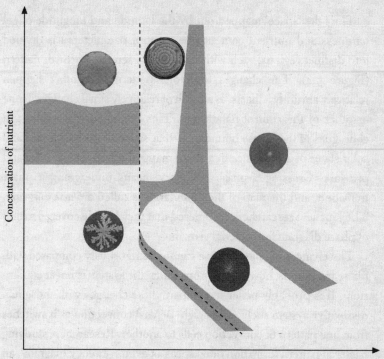

Figure 5.4 Bacterial colonies grown in a Petri dish of agar gel develop patterns that depend on two factors: the amount of nutrient available and the hardness of the gel. These patterns can be organized in a *morphology diagram*, as shown here. The switches between distinctly different types of pattern can be fairly abrupt. The boundaries between morphologies are shown here in grey, while the dashed line indicates the change from immobile (left) to mobile (right) cells as the gel medium becomes softer.

concentric rings if nutrient is abundant. If the bacteria are both fully mobile and abundantly fed, they advance rapidly in a roughly circular front, leaving behind a colony so sparse as to be barely visible.

Thus the 'space' mapped out by the latitude and longitude of gel hardness and nutrient content (the 'control parameters') is divided into distinct regions, each with its own characteristic growth pattern (Figure 5.4). The changes from one pattern to another happen relatively abruptly – that is, as a result of relatively small changes in one or other of the control parameters. This reminded Matsushita and colleagues of the 'phase boundaries' that separate the gas, liquid and solid states of a substance in a space mapped out in temperature and pressure. Crossing a phase boundary means undergoing a phase transition, and the map of the boundaries is called a *phase diagram*. The Japanese researchers considered that they had discovered a kind of phase diagram for bacterial growth.

The changes in colony shape cannot be rigorously compared with phase transitions, however. For one thing, the system is not at equilibrium. It is probably better to compare these changes with those in a dissipative system such as Rayleigh–Bénard convection as it switches from one pattern of convection cells to another. Researchers studying bacterial pattern formation often speak of a 'morphology diagram', an analogue of a phase diagram that classifies the shapes of colonies and the conditions under which they occur (morphology here means nothing more than 'shape').

Some other kinds of bacteria show different, complex growth patterns, and also undergo abrupt changes from one to another as the growth conditions are varied. We shall see in the next chapter that some of these patterns can now be understood by making simple assumptions about what controls the cells' movements, in the same way that the complex DLA cluster can be understood from the model of particle motions and interactions developed by Witten and Sander. From simple 'rules' dictating how individual cells behave, we can deduce the global patterns into which they can form.

ICE FLOWERS

The concept of a morphology diagram for non-equilibrium growth processes has a history. In the 1930s the Japanese scientist Ukichiro Nakaya, working at Hokkaido University, drew a similar picture for snowflakes. Bentley and Humphreys' book is filled with wondrous six-pointed stars – but in the final pages one finds something quite different. Here are snowflakes that look more like architectural designs: columns topped with table-like plates, rectangular blocks and even what appear to be ice sundials (Figure 5.5). These are snowflakes formed under unusual atmospheric conditions, such as extremely low temperatures. Nakaya and his team conducted experiments that mimicked the formation of natural snowflakes. They grew flakes

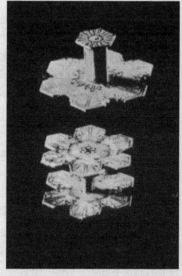

a   b

Figure 5.5 Not all snowflakes are branched. Some have less familiar shapes, such as columns and plates.

attached to a single strand of rabbit's hair placed within a stream of moist air in a very cold room.

The researchers found that as the air temperature changes, the shape of the snowflakes alters at certain thresholds. Hexagonal columns of ice grow below about minus 25°C. The air humidity also affects the shape. Between about minus 5 and minus 22°C the flakes are flat plates if they are formed in relatively dry air, whereas the familiar star shapes appear if the moisture content is higher. So temperature and humidity are the control parameters that map out a 'morphology space' for snowflakes, just as gel hardness and nutrient level do for bacteria.

Even though no two snowflakes, nor any two bacterial colonies, are identical, they can nevertheless be divided into distinct *classes* of growth pattern, separated by what we might call shape transitions. There is, in other words, a kind of order underlying the apparent profusion of shapes: each individual growth pattern may be uniquely embellished, but for a given set of growth conditions there is an inevitability in what we might regard as its platonic form. It is in this sense that an experiment in snowflake growth is repeatable: the details may differ, but the form stays the same.

Yet where does this individuality come from? It illustrates the contingency of non-equilibrium growth: a snowflake arm branches here, and not there, because of some fluctuation during the growth process which sparked off the budding of a new arm at that stage. Snowflakes experience so-called growth instabilities which amplify small, random bulges into new branches. The same happens in the DLA process, resulting in the ramified, fractal patterns of bacterial colonies. But snowflakes are not so random: their alluring beauty lies in their six-pointed symmetry, which was first clearly identified by the German astronomer and mathematician Johannes Kepler in 1610. This hexagonal symmetry is imposed by the regular arrangement of water

Figure 5.6 'Bacterial snowflakes' can be grown by letting a colony develop on a sheet of gel which has been stamped with a hexagonal grid of grooves. This biases the direction in which new branches sprout and extend.

molecules in ice: a crystal lattice that picks out six 'special' directions in space. New branches of crystalline ice sprout preferentially in these directions. In effect, the growth of the crystals is constrained by an underlying hexagonal grid, such that the geometrical orderliness at the molecular scale becomes manifest as a regularity which is evident at the much larger scale of the whole snowflake. This interplay of chance and regularity in snowflake growth is a subtle affair whose details came to be understood only in the 1980s.*

The influence of an underlying symmetry on an otherwise random growth process has been demonstrated strikingly by growing fractal bacterial colonies on a layer of gel stamped with a hexagonal grid of

* I discuss the modern understanding of this process in *The Self-Made Tapestry*.

grooves. The grid channels the branching process so that the resulting colony looks rather like a snowflake (Figure 5.6).

Non-equilibrium growth and the patterns it produces are an active area of research in physics. But while a great deal is now understood, there is no overarching theory comparable to the statistical physics that describes the states and phase transitions of systems at equilibrium. The structures and patterns that arise in systems away from equilibrium are often complex and rather subtle. Nevertheless, as we have seen, many of the tools and ideas developed for equilibrium statistical physics can be applied or adapted to non-equilibrium situations, which are by no means immune to predictability and regularities of behaviour. Since most of the processes that the new social physics seeks to understand are non-equilibrium phenomena, this is a reassuring message to carry forward with us.

6

The March of Reason

*Chance and necessity in
collective motion*

It is not nearly as bad to explain a phenomenon with a little bit of
mechanics and a strong dose of the incomprehensible as to try to
explain it by mechanics alone.

Georg Christoph Lichtenberg

A condition of order at the junction of crowded city thoroughfares
implies primarily an absence of collisions between men or vehicles
that interfere with one another.

Edward A. Ross (1901)

He smiled and said that he would go so far as to assert that, if a
craftsman were willing to construct a mechanical figure to the specifications he had in mind, he could use it to perform a dance that
neither he nor any other accomplished dancer of his time would be
capable of imitating.

Heinrich von Kleist (1810)

*

Thomas Hobbes was an elderly and venerable scholar when Antoni van Leeuwenhoek first saw bacteria through his home-made microscope in the 1670s. A Delft linen merchant by trade, Leeuwenhoek ground magnifying lenses which he used to inspect cloth, and gradually he became immersed in the microworld. Nothing was too dangerous or too distasteful that he would shrink from examining the wonders it held beyond the resolution of the human eye. Among his microscopic specimens were explosive gunpowder and his own excrement.

There was no theory, no natural philosophy, that could have prepared him for the multitude of 'little animals' which seemed to swarm in just about any material derived from or touched by living matter. Marsh water was full of them, a miniature menagerie. 'And the motion of most of these animalcules in the water', he said in his report to London's Royal Society in 1674, 'was so swift, and so various upwards, downwards, and round about, that 'twas wonderful to see.'[1] Here, then, was Leeuwenhoek, peering with a god-like eye into the Commonwealth of the Bacteria and seeing them going avidly about whatever business such creatures might have. Who could guess their purposes?

When Israeli physicist Eshel Ben-Jacob became fascinated by Mitsugu Matsushita's bacterial patterns in the 1990s, those purposes were the key to understanding what was going on in these multicellular colonies. As someone who had helped to decode the mystery of the snowflake's branching shape, Ben-Jacob, like Matsushita before him, recognized in the bacteria's delicate traceries a fingerprint of processes familiar from the inorganic world. Here were living cells behaving like blind and insensate gas particles. Could the rules that governed their motion really be so simple?

Ben-Jacob and his colleagues set out to explore the factors that dictated the growth patterns of Bacillus cells, and soon discovered

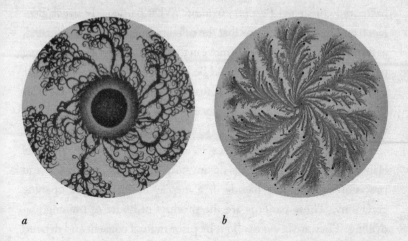

a   *b*

Figure 6.1 (*a*) 'Chiral' growth in a colony of Bacillus bacteria. (*b*) The 'vortex' growth mode. In the blobs at the ends of each branch, the cells swarm in circles.

marvellous, new, uncharted regions on the map of shapes called the morphology diagram. Some colonies expanded by sending out baroque, curling tendrils (Figure 6.1*a*), while others formed whorls whose spiralling arms ended in blobs (Figure 6.1*b*). When he inspected these blobs under a microscope, Ben-Jacob saw something that surely never passed before the eyes of van Leeuwenhoek. The blobs were vortices. Tens of thousands of long, thin cells swarmed in loops like traffic on an immense roundabout. These vortex-forming bacteria, which the researchers called the 'vortex morphotype', were mutants of the original strain in which the propensity for turning in circles seemed to be genetically hard-wired. If a few cells were extracted and used to seed a new colony, that too developed vortical blobs.

This remarkable behaviour was not unprecedented. Something similar had been reported in the 1940s in the movements of a species of bacterium christened *Bacillus circulans*. What particularly struck Ben-Jacob and his colleagues was that the cell movements were coordinated, as though they had all agreed to move in the same manner. The branching patterns seen by Matsushita could be interpreted by assuming that the cells followed a random walk across the agar gel, like gas particles executing their erratic paths in space. But in both the curling and the vortex growth patterns there was a kind of organized, collective motion.

Humans organize themselves to move together in all kinds of elaborate ways. But there is no mystery about the coherence and precision of a military parade or a display at an Olympic opening ceremony. These patterns are the product of hours of training and drilling. They are brought about by prior mutual consent and depend for their execution on each participant paying careful attention to detail. Someone has determined where the marchers will go and what, at each stage, they will do next: there is a guiding intelligence whose instructions are being followed by every individual. Social scientists have tended to assume that any kind of complex behavioural pattern like these requires complex motives and planning.

But bacteria do not have complex motives – after all, they do not have brains. They cannot conspire to decide what to do next. Neither is there some Master Bacterium dictating their movements. And yet the exactness of the spiralling motion is astonishing. If, as this indicates, such patterns of motion can arise without volition, is it conceivable that people might sometimes fall into similar patterns, unguided by any plan or intention?

THE CHOREOGRAPHY OF SWARMS

Bacillus bacteria are not freaks of the microbial world which happen to have an unusual penchant for group activities. Collective behaviour is a common phenomenon in nature, and these feats of cooperation sometimes go well beyond swimming in circles. A microbial Hobbes would have no difficulty pointing to his Leviathan, his 'Multitude united in one Person': he need only cite the case of the slime mould *Dictyostelium discoideum*. These single-celled organisms go their own individual ways when times are good – when food and water are plentiful and the weather is warm. But come drought, famine or bitter cold, the slime cells seek mutual support.

They cluster into aggregates of cells, each heading for a population centre like starving peasants streaming into a city. Once clustered in groups of tens to hundreds of thousands, they behave like a single, multi-celled organism, a 'slug'.* The slug moves as a coherent mass. Eventually it sets down roots and changes shape. It begins to resemble a bizarre plant, with a narrow stalk and a blob-like head called the fruiting body. This body contains cells that have become spores which can survive hardship virtually without nourishment, waiting for better times to come along. It is just as Hobbes pictured the 'Essence of the Common-wealth':

One Person, of whose Acts a great Multitude, by mutuall Covenants one with another, have made themselves every one the Author, to the end he may use the strength and means of them all, as he shall think expedient, for their Peace and Common Defence.[2]

How are these 'mutuall Covenants' agreed between cells which can neither speak nor hear nor see? For Dictyostelium, as for all single-celled organisms that display collective behaviour, communication is effected by a kind of sense of smell. In times of stress, some cells exude a chemical substance which acts as an attractant to other cells, rather as higher

* The clustering process involves some of the most fantastic spontaneous patterning seen in nature, and is discussed in *The Self-Made Tapestry*.

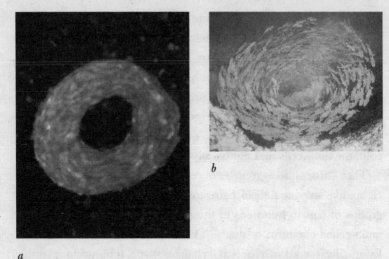

a

b

Figure 6.2 (*a*) Vortex motion in cells of the slime mould *Dictyostelium discoideum*. (*b*) Some fish swim in the same kind of vortex pattern.

animals emit pheromones to attract mates. Certain Dictyostelium cells become 'pacemakers', emitting pulses of the attractant, and nearby cells then follow the trail to the source of these waves. This sort of chemically induced cell motion is called chemotaxis.

At the stage where Dictyostelium cells are converging into their aggregate citadels, they usually organize themselves into circulating vortices, rather like the vortex morphotype of Bacillus (Figure 6.2*a*). This kind of movement is familiar to biologists of a quite different stamp, for it is common in schools of fish (Figure 6.2*b*). Yet fish do not communicate by chemotaxis – they have eyes with which to see one another. And, moreover, they have brains, albeit small ones, which allow them a far greater and more sophisticated range of responses to their environment. Is it mere coincidence, then, that we find the same collective mode of behaviour in slime moulds and in fish?

Many kinds of animals aggregate into co-moving groups, often for specific and readily identifiable reasons. Groups can protect vulnerable young, and there is safety in numbers in the face of predators. A throng of bees regulates the temperature of a hive using body heat; ants forage more effectively en masse. But as biologists Julia Parrish and Leah Edelstein-Keshet have pointed out, 'it is hard to argue the case that all animal aggregations have a functional purpose . . . Pattern and structure can arise . . . through nonlinear interactions whether or not the units are alive.'[3] In other words, seeking a 'biological' (which is to say, an adaptive) explanation for all patterning and collectivity in animal populations runs the risk of invoking a contingent evolutionary explanation for something that is in fact an immediate consequence of the 'physics' of the situation.

Regardless of questions about purpose, there is still the matter of *how* animals coordinate their behaviour and motion. A striking example of collective motion in the animal world is to be seen at dusk above the treetops, as flocks of starlings swoop and dive this way and that in near-miraculous close formation. Again, no individual is calling the shots – there is no leader. Yet every bird seems to make the same decision at the same time. Bees do something similar when they swarm. This is fertile ground for those who wish to believe that science is overlooking some deep and mysterious aspect of nature.

Until the late 1980s there was no satisfactory explanation for the manoeuvring of flocks of birds. One researcher was even forced in the 1930s to propose that birds use a kind of thought transference. In 1984, Wayne Potts of Utah State University suggested that manoeuvres can be instigated by single individuals, whose movements spread like waves through the other members of the flock. Potts claimed to identify this behaviour in film of dunlin flocks swooping over Puget Sound. He asserted that each bird watched the approaching wave and timed its manoeuvre to coincide with the wave's arrival, just as the members of a chorus line synchronize their high kicks.

But it was not really a convincing explanation, for it required each bird to be aware of what was happening a long distance from itself – the waves Potts reported moved too fast for each bird simply to be reacting to its nearest neighbours. The 'chorus-line' theory asks rather too much of avian brains. Something else was needed. The extra ingredient came eventually not from biologists, but from physicists.

NEWTON'S PUPPETS

Scientists tend to get drawn into the tropes of their fields. Zoologists look for explanations of animal behaviour in the characteristics of individual animals; molecular biologists demand a genetic explanation. Physicists, on the other hand, are more accustomed to thinking in terms of interactions *between* things, whether those things be atoms, electrons, quarks or gas molecules. But one thing is clear: in a phenomenon such as flocking, the precise flight path for each day's flying obviously cannot be genetically pre-programmed into every bird.

Craig Reynolds is not exactly a physicist. But as a software engineer with the Symbolics computer company in California in 1987, he was more familiar with the physicist's rule-based, axiomatic mode of thinking than with the biologist's phenomenological approach. So when he used to sit and watch blackbirds flock in the local cemetery, he began to wonder what rules they were following. 'All evidence', he later wrote, 'indicates that flock motion must be merely the aggregate result of the actions of individual animals each acting solely on the basis of its local perception of the world.'[4] This, he felt, was the key: *local perception*. A bird cannot possibly anticipate nor keep track of what all its flock-mates are doing, but it can respond very quickly to what its immediate neighbours do.

So Reynolds wrote a computer program in which particles were commanded to obey a set of three simple rules, each based only on information about the particle's nearby flock-mates. It was hardly fair to call such responsive, 'sentient' agents 'particles', so Reynolds called them *boids*: robot-like virtual birds, a fusion of 'bird' and sci-fi 'droid'. Each boid reacted to all the others within a certain distance of itself – let's call this the local sphere. The rules of motion were these: each boid would try to match its speed to the average speed of others within its local sphere; it would move towards the centre of mass of this group; and it would avoid collisions with others.* The stipulations on speed and direction are cohesive rules: they encourage boids near one another to stick together as they move. But there is nothing in this prescription that dictates the overall behaviour of the flock – nothing that allows boids on one side of the group to directly influence the movements of boids on the other side. The rules contain no built-in tendency to form a coherent flock.

Yet that is just what the boids do. Their motion, simulated with computer graphics, is startlingly reminiscent of the movement of real birds. Reynolds was able to include additional influences on the motion, such as permanent obstacles or a tendency to move towards a certain position (as if the birds were seeking a new roost or a food source). In this way he could mimic a wide range of the flocking movements seen in the real world. So convincing was his software that it was adapted by Hollywood movie-makers for animated special effects such as the swarms of bats in *Batman Returns* and the herd of stampeding wildebeest in *The Lion King*.

A boid is a form of automaton: a robot-like creature programmed to follow a set of rules which dictate its response to its immediate environment. The behaviour of an automaton is entirely determin-istic: it takes a look at its surroundings, performs some 'mental'

---

* The centre of mass is like the centre of balance of a pirouetting dancer, located in that case somewhere in the belly.

computation and applies to the result a set of criteria that prescribe its next act or move.

The concept of the particle-like automaton stems from the work of Hungarian mathematician John von Neumann (1903–57) in the 1930s, when he was formulating early ideas about computers. He was interested in the possibility of thinking machines that could reproduce and increase their complexity. The Polish mathematician Stanislaw Ulam suggested to him a tractable simple model of this process: a chessboard universe with an automaton in every square (or cell). Each cellular automaton can exist in one of several different states, and which of them it chooses is determined by the states of the automata in neighbouring cells. In effect one can think of each automaton as a kind of memory cell holding information, just as each memory element in a computer can exist in one of two binary states, 1 and 0, like a switch that is on or off. Von Neumann and Ulam looked at how certain patterns of information might be able to duplicate themselves on the grid.

One can devise all kinds of games to play with cellular automata, each with different rules specifying how a cell responds to its neighbours. Yet it was difficult to investigate the dynamic world of cellular automata until real digital computers were invented. In the late 1960s John Horton Conway, a mathematician at Cambridge University, devised a kind of chessboard cellular automaton game which he provocatively called the Game of Life. It was a crude model of how living cells or organisms proliferate. Alone they die; in a community they thrive and multiply. But if the community gets too overcrowded, they expire through lack of food and resources. In the Game of Life, each cell can be alive or dead. A living cell stays alive if it has two or three living neighbours adjoining it; if there are fewer or more, it dies. A dead cell is revived (one should really think of it as an empty space that is reoccupied) if it has exactly three living neighbours.

Conway's Game of Life is the prototype for research on artificial life. The sheer diversity in the shape and behaviour of the living cell clusters the Game of Life generates is legendary; there are several web groups devoted to exploring them. Some clusters propagate across the grid, winding like snakes or gliding like birds. Some clusters eat others; some spit out a stream of new clusters. It is a strange world, full of richness and surprise – all arising from a few simple rules about local interactions between cells.

Craig Reynolds' boids are automata which can move under their own steam, and are not confined to a grid – but they are automata all the same, rigidly bound by rules. They are arguably the first of such artificial-life games to reproduce the kind of complex phenomena we see in the living world. They are one of the star exhibits in the catalogue of research commonly now grouped under the banner of 'complexity theory', whose key concept is *emergence*. Flocking emerges spontaneously from the boids. It is not programmed in: the rules specify only the motion of individuals. Yet something about the way those individuals interact produces a kind of coherent group behaviour. Emergent properties show that the whole can be more than the sum of its parts.

In this context, 'complexity' has become a word that can mean whatever you want it to. Often it has little to do with physics as such. Most of the work conducted under this heading is a kind of empirical computer science – devising games for automata and playing them to see what happens. Much of this, fascinating though it is, comes up in the end against a brick wall, for it has no underpinning concepts. What it yields is a description, or rather a prescription, not a deeper theoretical understanding. As Harvard biologist E. O. Wilson puts it, 'By itself, emergence can be no explanation at all if you don't have any insight into the mechanics of the system.'[5]

That is why boids and all of the other marvellous computer models of collective animal behaviour, from ant colonies to herd animals on the prairie, would remain little more than hi-tech parlour games, had physicists not begun to see that they are really a form of non-equilibrium statistical physics.

THE PHYSICS OF GROUP MOTION

Tamás Vicsek had never heard of boids when he and Eshel Ben-Jacob teamed up in the early 1990s to figure out why mutant Bacillus bacteria were going round in circles. Like Ben-Jacob, Vicsek had established himself as an expert on the growth and form of non-equilibrium systems, such as the branched clusters formed by diffusion-limited aggregation (see page 136). He recognized that, whatever rules the vortex bacteria were following, they must be simple ones.

But bacteria are not like gas particles. Most importantly in this context, they have their own form of propulsion. Bacteria burn up nutrients and use the energy released to whirl their propeller-like flagellae – whip-like strands of protein – to drive themselves along. When gas particles collide, they obey Newton's laws and conserve momentum (more or less); bacteria can flagrantly disregard such strictures, stopping or speeding up at will. This self-propulsion is what makes a bacterial colony a non-equilibrium system: as long as the cells are burning up fuel, they are driven away from an equilibrium state.

In 1994 Vicsek and his student András Czirók devised a model to describe bacterial motion. They treated each bacterial cell as a self-propelled particle – like one of Maxwell's dancing gas particles, but with its own source of propulsive power and with a simple 'program' dictating its motion. This program specified only that the cells all main-tain the same speed and that each one travels in the average direction of motion of all the others that lie within a certain distance. These rules are similar but not identical to those governing Reynolds' boids. Vicsek and

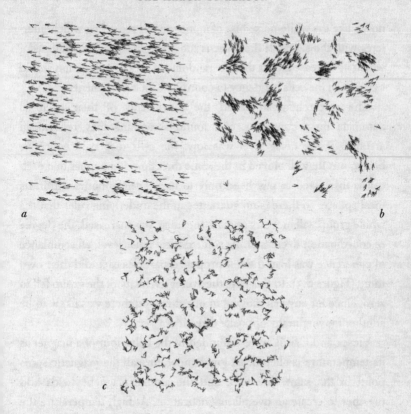

Figure 6.3 A phase transition in self-propelled particles. If the noise is low, the particles align their motions (*a*). For a higher noise level, the motion breaks up into smaller, co-moving clusters which tend to circulate like vortices (*b*). If the noise exceeds some critical value, the motions become random (*c*). The direction of a particle's motion is shown here by an arrowhead, attached to a short line which shows its most recent trajectory.

Czirók also assumed that there would be an element of randomness in the bacterial motions – the rules would not be obeyed perfectly, for real life is never so neat and tidy. This randomness is a kind of background noise, like the white-noise hiss of a poor sound recording which interferes with the signal. If the noise is too great, the signal is swamped – which in our case means that the random element of cell motion would overwhelm the cells' tendency to coordinate their movements.

The researchers simulated the movements of their bacterial automata on a computer, and found something they recognized at once. If the noise level was low, the cells displayed collective behaviour: they all moved in the same direction. Remember that each cell is instructed to pay heed only to its near-neighbours within its local sphere, so there is no guarantee in the underlying rules that the whole group will move as one. As the noise was increased, the degree of coordination decreased. At some critical noise level, all semblance of coherence was lost: the self-propelled particles each did their own thing (Figure 6.3). In this case the average velocity of the group fell to zero, since for any cell moving in one direction there was likely to be another moving in the opposite direction.

Vicsek and Czirók were reminded of the behaviour of a magnet as its temperature is altered. At low temperature all the magnetic spins point in the same direction, and the atomic magnetic fields add together to create an overall magnetization. At high temperature the needles point in random directions and the fields cancel out, averaging to zero. Between the two states – magnetic and non-magnetic – there is a phase transition at a critical temperature. The self-propelled particles were undergoing an analogous phase transition from an aligned to a non-aligned state, the average velocity playing the role of the magnetization and the noise playing the role of temperature. Just as the liquid–gas critical point can be 'mapped' onto the phase transition of a magnet, so were the researchers able to map their model of bacterial motion onto this phenomenon.

It is more than simply drawing an analogy. Despite some important differences connected to the fact that the ordering of self-propelled particles is a *non-equilibrium* transition, one can demonstrate a formal, mathematical equivalence between the two processes. They both share 'universal' features.

MOB RULES

The cooperative behaviour of social animals was not lost on Thomas Hobbes, and he anticipated that it might provoke objections to his bleak prediction about humanity's savage State of Nature:

It is true, that certain living creatures, as Bees, and Ants, live sociably one with another, (which are therefore by *Aristotle* numbered amongst Politicall creatures;) and yet have no other direction, than their particular judgements and appetites . . . therefore some man may perhaps desire to know, why Mankind cannot do the same.[6]

Hobbes's answer was that people are different: their thoughts, desires and sensations are more complex, and provoke them into conflict. For example,

men are continually in competition for Honour and Dignity, which these creatures are not . . . [and] these creatures, having not (as man) the use of reason, do not see, nor think they see any fault, in the administration of their common business.[7]

This answer will ring true for many social scientists, who have long assumed that people are just too complicated to yield to any mathematical model of behaviour. We are each moved by a thousand impulses in a blend unique to every one of us. So what is the point of making idealizations of human activity?

The objective of any physical model of the kind I have been discussing so far is to get out more than you put in. If I were to devise a model of a crowd in which every person had a pre-programmed and complicated trajectory, replete with sudden stops to check direction signs or slowing down to peruse a shop window, that would not be much of a model. Any information I got out would be pretty much what I had put in.

In 1971, L. F. Henderson of the University of Sydney saw that, beyond this miasma of individuality, there might lie some quantifiable *statistical* characteristics of what humans do in moving groups. In a sense this is obvious, but it is worth stating. Football crowds move into stadiums before a game, spread out over the terraces, and flow out again after the final whistle. The spectators are not each so lost in their internal worlds that these collective movements are rendered impossible. Shoppers in a busy high street move along the narrow pavements mostly in one direction or its opposite. People do not travel on foot at 30 kilometres per hour; they do not walk with their eyes closed, colliding with all in their path. There are some general rules, some constraints, some trends and averages.

Henderson's decision about how to search for these statistical properties is telling: he wondered whether they might fit the Maxwell–Boltzmann kinetic theory of gases. That is to say, he suspected that the distribution of velocities of people walking along a pavement would trace out the bell-shaped curve used by Maxwell and verified by Boltzmann. He tested this idea by observing various moving crowds: students walking along a footpath at the University of Sydney campus, people using a pedestrian crossing in a busy street, children moving in all directions in a playground. In all cases, the velocities fitted the Maxwell–Boltzmann curve quite closely – but with one curious difference. Each of the crowd curves had two distinct peaks, giving the appearance of two intermingled populations of Maxwell–Boltzmann particles with slightly different average velocities. These, Henderson considered, were males and females, who seem to move in the same general manner but at different rates.

Henderson suspected that perturbations to the crowd motion might bring about changes in the 'global' state of the throng:

At one end of the passage, the crowd may be forced to slow down, probably by the need to surrender a ticket at a barrier, when it may be expected that people will bunch into a densely packed crowd which moves along with a slow elbow-to-elbow shuffle. This transition will be called a phase transformation – the loosely packed phase is a crowd gas and the densely packed phase is a crowd liquid.[8]

In other words, he foresaw phase transitions in crowds, brought about in a manner analogous to liquefying a gas by compressing it. Yet he did not in fact report any observations of transitions of this sort.

Although it was valuable in establishing the potential connection to statistical physics, Henderson's study did not really contain anything very surprising. We'd expect people in a crowd to have a more or less bell-shaped spread of velocities – it's hard to imagine anything else. Not all bell-shaped curves are mathematically identical to a Maxwell–Boltzmann curve, but Henderson's data on just a few hundred moving individuals was unlikely to reveal subtle distinctions between them. And in any case, hadn't Adolphe Quetelet long ago established the bell-shaped error curve as the characteristic fingerprint of social statistics?

When Dirk Helbing began to think about models of crowd motion in the late 1980s, when he was at the Georg-August University at Göttingen in Germany, he thought Henderson's 'gas-kinetic' description would be a good starting point. But he realized that it said nothing about what was actually motivating the walkers: what their intentions were, and how they responded to their surroundings. Only by including such factors could a model hope to capture the complex patterns of movement that develop in human crowds. Henderson had gone on to propose that pedestrian motion might be treated like the flow of a fluid. But if so, it was a fluid with a mind of its own, a fluid that could defy Newton's laws by stopping dead or breaking into a run.

Helbing considered that the motion of any individual was conditioned by two influences: internal, or 'personal aims and interests', and external, or 'perception of the situation and environment'. Often these will conflict. I am forced to slow my habitual pace because there is a slow-moving group in front of me. I want to cross the pavement diagonally to enter a shop, but there is a stream of people in the way. In a particularly dense crowd, I am prevented from coming to a stop by the crush of people behind me.

External influences on behaviour are caused by interactions with others. In 1945 the psychologist Karen Horney identified three ways in which people interact: 'moving towards people', 'moving away from people' and 'moving against people'. Motion here is to be understood largely as a metaphor, but it could also be interpreted literally. People walking with friends, partners or as members of a group aim to stick together; famous people at cocktail parties attract a coterie of admirers and sycophants through their own personal (or manufactured) magnetism. But in most crowds people do not know one another, and so have no propensity to stick together. Rather, they try to keep their distance from strangers. Such interactions resemble forces of attraction and repulsion. (Horney's 'moving against' is a rather special case: she had in mind deliberate obstruction, motivated by aggression and conflict. It happens in some social dynamics, but is rarely to be seen in simple walking motions in crowds.)

American social psychologist Kurt Lewin saw that the kinds of attractive and repulsive interactions sketched by Horney could be broadly applied. He postulated in the 1950s that there might be an analogy between the forces of electromagnetism that act on particles with electrical charge and the social pressures that determine the behaviour of people. According to Lewin, individuals could be considered to 'move' in an abstract field of ideas, beliefs, habits and notions. The field is conditioned for each person by the behaviour he or she sees in others, and it pushes and pulls the person towards certain dispositions.

This is very much the modern equivalent of Hobbes's mechanistic vision of humanity. Motivated by Lewin's concept of 'social forces', Helbing and his colleague Péter Molnár at the University of Stuttgart developed a physics-based mathematical model of pedestrian movement. Of course, it is one thing to speculate about 'forces' between people, but quite another to describe them by mathematical equations like those one can write for electromagnetism or gravity. Yet that is what Helbing and Molnár did.

They assumed that walkers simply want to move in a particular direction at a certain preferred speed. This impulse is tempered by external factors, foremost of which is the wish to avoid collisions – indeed, to avoid coming close enough to others to compromise their own 'personal space'. Thus, people in crowds behave as if there is a force of repulsion between them which increases in strength as the distance between two people gets smaller. Modern theories of liquids invoke a similar 'soft' repulsion between particles, which starts small but rises steeply as the particles approach one another. Recall that van der Waals' theory of fluids implicitly included a 'hard' repulsive force that was felt only when two particles collide.

Van der Waals' theory also took account of attractive forces between particles. Helbing and Molnár could include in their model attractive forces between people if they wanted to investigate certain special situations such as group cohesiveness, but otherwise they gave

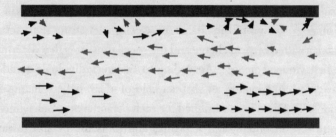

Figure 6.4 Pedestrians moving along a corridor in the Helbing–Molnár model become organized into counter-flowing streams. Each line here shows the path of a 'peoploid' particle over several time steps. Black and grey particles are moving in opposite directions down the corridor. The simulations can be seen at http://www.helbing.org/Pedestrians/Corridor.html

their walkers a shared aversion to closeness. These walkers are not like boids or Vicsek's self-propelled bacteria, because they do not possess any tendency to align their motions and thus to form coherent flocks.* Instead, each individual – which, borrowing from David Bowie's *Diamond Dogs*, we might call a peoploid – ploughs his or her own furrow, subject to the constraints that the other pedestrians impose.

In this sense the model is one of self-centred people with no real

---

* Of course – and as we shall see – people sometimes do exhibit herd-like collective behaviour. One can see this manifested at football matches in the form of the Mexican wave, or *La Ola*, which became a fad in the 1986 World Cup in Mexico. Spectators would leap to their feet with their arms in the air and then sit down again, in a wave that passed swiftly round the crowd. Helbing, Vicsek and their colleague Illés Farkas have modelled these waves by assuming that spectators switch between three states: excitable (those ready to leap up as the wave approaches), active (those who are leaping) and passive (those who've done their bit and sat down again). These simulations are available at http://angel.elte.hu/wave. The waves generated in this model are closely analogous to the concerted ripples of cardiac tissue that make the heart beat.

social graces or courtesies. And yet when Helbing and Molnár ran their computer simulations, they found that certain types of group dynamic did emerge spontaneously. Some of these even looked like good behaviour. For example, peoploids walking in opposite directions down a corridor tended to organize themselves into counterflowing streams, which reduces the need for collision-avoiding manoeuvres (Figure 6.4). This happens frequently in real life. Barriers placed along the centre of a thoroughfare, such as pillars and trees, can encourage this stream-formation, even when it is not specified on which side people should pass. The lanes become organized spontaneously, but it is a matter of pure chance to which direction each of them becomes designated.*

When two groups of peoploids try to pass in opposite directions through a single doorway, alternating bursts pass first in one direction and then in the other (Figure 6.5). Each group temporarily 'captures' the doorway – a trailblazer passes through and several others follow in its wake. The other group looks almost as if it stands back to let this stream through, though this apparent courtesy is the result of nothing more than the wish to avoid close contact.

This model can be used to improve pedestrian walkways so as to reduce discomfort, congestion and inconvenience. Dividing a

* This may not be quite true in practice – people generally have a preferred passing direction around obstacles (including other walkers). This preference is culturally determined: in Germany, for example, it is to the right. The bias seems to be learned, since children (and tourists) tend to lack it. The learning is subconscious, but is beneficial to all individuals since it reduces the risk of collisions from two approaching walkers who choose to swerve the 'same' way: one to the left, the other to the right. In adults, lane directions in a divided corridor will also no doubt be influenced by instincts acquired from road driving.

Helbing and his co-worker Kai Bolay have been able to simulate this learning process, finding that if walkers are allowed to adapt their behaviour to improve collision avoidance, they develop a preference for passing on a particular (but arbitrarily determined) side.

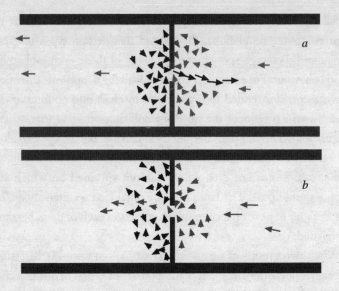

Figure 6.5 People pass through a doorway in alternating bursts which allow passage first in one direction (*a*) and then in the other (*b*). The arrows show the direction of motion; their length is proportional to the person's speed. The simulations can be seen at http://www.helbing.org/Pedestrians/Door.html

corridor by installing pillars is one way to improve the flow. To ease the bottlenecks caused by doorways, one might imagine that making the door wider would help. But it is not that easy – with a wider door, the alternations in direction of flow simply become more frequent. A better solution is to introduce two doors. Even if it is not specified which is to be used in each direction, a crowd will automatically organize itself into two counterflowing streams which pass through one door or the other (Figure 6.6). Two doors are thus more efficient than a single door of their combined width.

Intersections are tricky. In the simulations, the flow at intersections never settles into a stable state in which collisions or congestion are

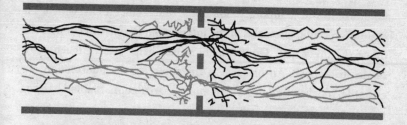

Figure 6.6 If two doorways are available, each will find itself assigned to pedestrians moving in a particular direction – even though this is not specified at the outset. Here I show not instantaneous snapshots, as in Figures 6.4 and 6.5, but the trajectories of black and grey walkers over many consecutive time steps.

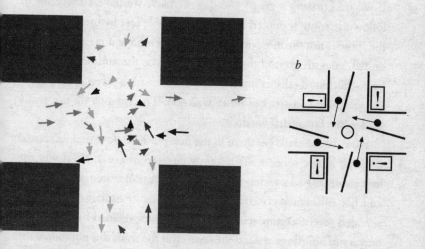

Figure 6.7 At intersections, the motions of pedestrians often settle temporarily into a circulating mode which reduces the chances of collision and obstruction (*a*). Here the arrows are shaded differently for different approach directions; each person is trying to cross the intersection to the opposite side. Intersections can be designed to promote the efficient circulating mode by using barriers and signs (marked '!') designed to draw the walkers towards them (*b*).

minimized. But temporary solutions often emerge in which the pedestrian traffic circulates at the intersection in one direction or the other (Figure 6.7a). Good planning can help these efficient flow modes to appear. An obstacle at the centre of the intersection, for example, blocks off routes which impede circulation. Slanted barriers give further guidance, stabilizing circulation in a given direction – especially if helped by posters or displays which aim to draw walkers to one side of the corridor (Figure 6.7b).

In 1997 Helbing and Molnár, together with computer scientist Joachim Keltsch in Tübingen, used this pedestrian model to deduce how trails evolve organically in open spaces over which people walk. If we are crossing a grassy meadow or park, we have a tendency to follow the route trampled flat by innumerable feet before us – even if the route is not the most direct one. Why? Perhaps it is easier walking on flat, smooth ground than on grass. Maybe the authoritative voice from childhood tells us not to wander off the path (even if the path is the arbitrary construct of other walkers). The reasons do not really matter; the fact is that we do it.

How do the trails get there in the first place? Before people started crossing it, the space was uniformly covered in grass. No doubt the first pioneers had a variety of destinations. A trail system that Helbing and his colleagues observed on the campus of Stuttgart University revealed several common entry and exit points, defined by the various university buildings around the lawn. But the trails did not follow the shortest path between any of these points (Figure 6.8).

The trails build up from the passing footprints of every walker who crosses the space. The researchers emulated this in their model by assuming that each peoploid crossing a space wears away the grass to a small degree. But the grass regrows at a steady rate, so that trails which remain unused for a long time will disappear. The more worn a trail becomes, the more visible and tempting it is to subsequent

Figure 6.8 Trail patterns formed spontaneously on an open space at Stuttgart University. Notice the lack of sharp intersections. None of the paths represents the most direct route between any two points of entry or exit.

walkers, who, in the model, are attracted towards a particular trail segment with a pull that depends on its distance from them and its visibility.

Helbing and his team let loose a steady stream of peoploids across an open space, all coming from and going to a few destinations at the periphery. They found that the trails that evolved depended on how strongly peoploids were attracted to existing paths. At first, people simply trod down fairly direct routes between the various destinations (Figure 6.9a). If the attraction is small, these direct routes persist and become well-trodden trails. But if the attraction is appreciable, the direct routes evolve into something else: a trail system that represents a compromise between directness and the tendency to follow existing paths (Figure 6.9b). In this case, the diagonal routes between opposite

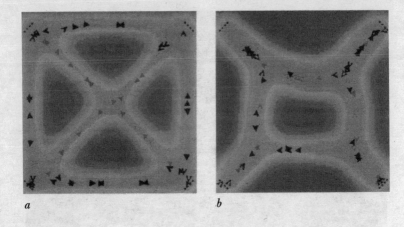

a   *b*

Figure 6.9 The model developed by Helbing and colleagues simulates the way that trails develop. In the early stages (*a*), walkers take fairly direct routes. The final state (*b*) represents a compromise between the various 'most direct' trails.

corners and 'edge' routes between adjacent corners merge into gently curved trails which take all walkers slightly out of their way. No one passes through the centre: the trails intersect to leave a central island, just as they tend to do in real life (Figure 6.8).

Park and garden planners often incline to linear thinking: they create paths which are straight lines, intersecting at sharp angles. In contrast, trails that evolve organically – those that, in the jargon of the field, are *self-organized* – curve and merge smoothly. Because of this difference, planned paths in open green spaces are often augmented and subverted by new trails worn into the grass where people have followed their own instincts. In one Stuttgart park, Molnár noted the futile attempts of park officials to obliterate such trails and reinstate the original ones that the planners had specified. Yet without any concerted intention to undermine this design, pedestrians soon re-established their preferred routes.

How much more effective it would be if planners could anticipate the walkers' wishes and build accordingly. A model like this one might help them to do just that, producing routes which are natural and comfortable. It is a simple matter, says Helbing, to gear the model to constraints such as a budget limit on the total trail length, so as to find the best compromise between economy and efficiency, or to figure out how best to extend existing walkways. In this way, planning can be fitted to human nature.

THE LANGUAGE OF SPACE

In open park spaces, pedestrians have the freedom to ignore the official paths and collectively define their own trails. In the city or a building that is seldom possible – you can't walk through a wall if the fancy takes you. Can urban and architectural spaces also be planned to conform with human needs and impulses, rather than the other way round?

Geographer Michael Batty and his colleagues at University College in London have simulated the way visitors move around the Tate Gallery, an art museum (now called Tate Britain) in London. They compared the results with observations of people moving through the real gallery on a single day in August 1995. Visitors tended to circulate among the rooms containing the classical and British collections, to the left of the gallery's central axis, in preference to the modern collection on the right. Was this because people prefer classical and homegrown art to modern works? Not at all, said the researchers. Their simulations, in which every room in the gallery exerted an equal 'attractive force' on the visitors, showed the same left–right asymmetry, implying that the apparent preferences were simply being dictated by the way the rooms were laid out: the room structure in the modern section was more convoluted. In other words, what people view at the gallery depends not only on what they like but also on how the gallery is arranged. The ground plan of the Tate was dictating the experience of visitors in ways that were not foreseen or intended by the gallery designers.

Bill Hillier, a specialist in urban planning who also works at University College, believes that there is a kind of logic – a 'syntax' – in the way people relate to and navigate space. Hillier and his co-workers have used computer modelling to look at how people use and navigate a built environment at scales ranging from individual room arrangements to entire city networks. They have applied this kind of modelling to help plan shop interiors and malls, galleries and museums, commercial office buildings, hospitals, schools, airports and railway stations. These models can provide diagnoses of problems that arise in existing street networks and building layouts.

But the bigger challenge is to figure out the *rules* – Hillier's space syntax – that pedestrians are using in these places. Hillier believes that lines of sight play a significant role in this visual language of space. If the language can truly be decoded, it could be programmed into the motivations of Helbing's peoploids to allow them still more realistic ways of moving in space, making possible more powerful predictions about how people would like their environment to be laid out.

Walkways in new city spaces have in the past often been planned through a mixture of guesswork and modernist architectural thinking that imposes an arbitrary aesthetic on places in which people have to live. Hillier and his colleague Julienne Hanson say that, whereas there are distinct patterns of pedestrian movement in traditional urban environments, people in newer developments such as modern estates seem to use the space randomly: they look, and most probably feel, lost in it. Few urban dwellers will have trouble identifying places like this, in which the design feels at odds with the way we instinctively navigate and use open spaces.

At the Elephant and Castle in South London, close to where I live, there is a sprawling estate of concrete tower blocks built in the 1950s. At that time it was considered to be a model project, a shining example of how huge numbers of people could be efficiently and comfortably housed. Shops and community centres were incorporated into the buildings; walkways took people across the roads out of the reach of the traffic below. Now, many of those shops have been closed for years, and the walkways have been demolished – they were convenient escape routes for muggers. The problems of this estate run far deeper, of course, than a failure of design. But one does not need to spend long in this environment to know that it is not a design for living, for making one feel part of a community.

That is an old and familiar lament, and some of the salutary lessons of post-war architectural design have now been learnt. Yet Hillier and Hanson argue that there is more to all of this than poor planning. Urban design, they say, is political: 'the nineteenth-century dreams of a social order, in which the benefits of capitalism are retained through the creation of a quiescent working class, are dreamed in a strongly spatial form.'[9] In other words, urban communities were redesigned in the Victorian age to reproduce and reinforce social hierarchies. By developing mathematical and computer models of the spatial patterns of communities ranging from traditional villages to modern towns, Hillier and Hanson have shown that urbanization tended to increase and diversify people's interactions up until the time when new templates for planning were introduced during the Industrial Revolution. Whether consciously or not, these templates reduced social encounters and fragmented communities, discouraging collective activity and keeping people passive under an imposed authority. High-rise blocks, for example, pack living spaces together densely while reducing the frequency of encounters which generate a sense of social solidarity. 'It is wrong to say that high-rise estates are unsuccessful', Hillier and Hanson argue. 'For their unmanifest purposes of community reduction they are extremely successful.'[10] The 'soft' solution of creating garden cities in which urban spaces are separated into small, relatively isolated zones looks and feels more benign but has the same result.

If Hillier and Hanson are right, the barriers to congenial urban design are created by more than simple ignorance of how people use space. Nevertheless, the more we can understand and predict the ways in which we instinctively want to move around our environment, and the better we are able to model the diverse human movements that these spaces must accommodate, the better we are likely to be able to create places where people feel relaxed, comfortable and considerately housed.

QUICK EXIT

On 28 November 1942 the Coconut Grove nightclub in Boston was heaving. American soldiers and servicemen were making the most of their leave, beer and jazz helping them to briefly forget the war that was spreading across the world. Then the fire started. There was only one exit, and everyone fled to it. But the first people to arrive at the doors discovered that they opened inwards. Before this could be done, the pressure of the crowd at the back pushed the first-comers against the closed exit, and there was then no way out. In the panic and smoke, no message could be relayed to those at the back feverishly trying to move forward. Four hundred and ninety-two people died in the blaze.

On 15 April 1989, football supporters in Sheffield, England, filled the streets around the Hillsborough stadium, desperate to get inside and claim their places for the FA Cup semi-final between Liverpool and Nottingham Forest. The steady trickle of supporters through the turnstiles did little to ease the crush, so ten minutes before the game started the police decided to open an exit gate. The consequent rush onto the terraces pushed fans at the front up against riot-control fences, where ninety-six of them were crushed to death.

At face value, predictability seems to vanish when panic strikes a crowd. Safety measures planned on the assumption that people will use them rationally can rapidly become useless when they must serve a horde of terrified people. What hope is there of anticipating human behaviour in these situations? Yet irrational does not mean unpredictable. Quite the contrary: a panicking crowd has only one objective in mind – to escape as quickly as possible. Fear, though, is not the only cause of stampedes. Injuries and deaths have also resulted from the rush for seats at concerts, for example when eleven people were killed at a concert given by The Who in 1979 in Cincinnati.

In 1999 Helbing went to Budapest to work with Tamás Vicsek. They were intending to use the pedestrian model to study the formation of animal trails, but Vicsek felt there was still more to learn from it about human crowds. The researchers found that if people's motions became too erratic (too 'noisy'), this could lead to a crowd jam in a corridor. Viewed as a problem in physics, this is a counter-intuitive result. Making the movements of peoploids more erratic is like raising the temperature of a group of particles, making them jiggle more frantically. Yet the consequence is that the crowd 'freezes'. In other words, the 'people fluid' could be frozen by 'heating' it, whereas a normal fluid, like water, is frozen by cooling.

Vicsek noted that this jamming of an over-excited crowd looks something like the effects of panic. So he, Helbing, and Vicsek's colleague Illés Farkas of Eötvös University in Budapest began to use the model of pedestrian motion to study what happens when crowds get out of control. They reasoned that one thing in particular makes panic different from normal motion: people lose their inhibitions and come into contact. So much so, in fact, that the pressure can become life-threatening. The force of a surging crowd has been known to bend steel barriers and collapse brick walls. This does not mean that people's natural aversion to physical contact ceases, just that it no longer dominates the motion. When people are touching, their movement is restricted. In a densely packed crowd it can become impossible to slip past others or even to turn round. There is a kind of friction between people that hinders movement.

So the researchers gave their peoploids frictional properties, as if they were billiard balls coated with sandpaper. They included another grim ingredient too. If the pressure on a person became too great, it was assumed that that unfortunate individual would become injured and unable to move. This only makes matters worse, of course, as the injured person then becomes an obstacle to the movement of others. By estimating the level of pressure likely to result in injury, the researchers hoped to derive some indication of when and how casualties arise in dense crowds.

There are some experiments that should be done only on a computer, and lighting a fire in a crowded room is clearly one of them.* Helbing and colleagues put a crowd of peoploids in a room with just one exit and gave them a good reason to head for the door: a fire encroaching steadily from the opposite wall. If the peoploids are able to contain their fear and move at a calm speed – less than a metre and a half per second – then they are able to evacuate the room in an orderly manner. Clustered around the door, they give one another enough space to allow a steady stream to pass through.

---

* Researchers at the University of the Philippines have run real-world tests of models of human escape panic – by watching mice swim to safety through a doorway in a flooded chamber. They found modes of behaviour similar to those seen in their computer models, which were closely related to the model developed by Helbing, Farkas and Vicsek. In particular, clogging at the doorway could make the mice's escape inefficient and sporadic.

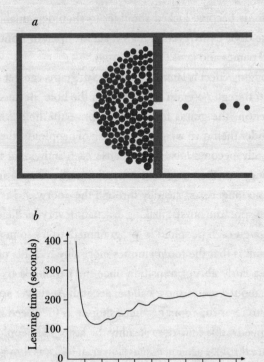

Figure 6.10 If people try to leave a crowded room too quickly, they push up against one another and become jammed around the doorway. This is a panic state (*a*). The time taken for a room to empty of people initially decreases if people move faster. But above a certain threshold speed, corresponding to the appearance of a panic state, the emptying time increases as people try to go faster (*b*). See 'Pedestrian simulations' at http://angel.elte.hu/~panic/

If the peoploids try to move at speeds greater than this, the outcome is chillingly different. Converging on the doorway, the crowd of peoploids press against one another, and friction takes hold.

The peoploids become locked shoulder to shoulder, unable to pass through the door easily even though it stands open in front of them. The crowd panics and jams (Figure 6.10*a*).

This clogging effect is familiar enough: salt grains can get stuck in a shaker even though no grain is bigger than the hole. Because of their mutual friction, the grains form arches above the hole which hold together under their own weight. In a crush of peoploids, these arches can eventually become loosened because each individual is able to keep moving, but their repeated formation and collapse mean that peoploids no longer pass steadily through the doorway. Instead they come out in sporadic bursts, making evacuation very inefficient.

So, although each peoploid is 'programmed' to try to move faster, the end result is that the room empties more slowly. This panic state appears in a fairly abrupt transition once the peoploids try to move faster than about a metre and a half per second. At slower speeds, the time taken to clear the room becomes shorter as the speed increases; for faster speeds, this time rises steadily the faster the peoploids want to go (Figure 6.10*b*). Faster becomes slower. It looks as though there is some form of non-equilibrium phase transition between a relaxed and a panicked state. Once the desired velocity exceeds about five metres per second – a running pace – the pressures in the clogged region become so great that peoploids are injured.* The number of casualties rises steadily with the speed. Faster, then, means not only slower but more dangerous.

Many nightclubs are dark enough to begin with, and once smoke starts to fill a hall it can become impossible to see more than a metre or so ahead. Under these conditions, people do not even know in which direction to run. Helbing and colleagues wondered what might happen then. One's instinct might be to rush around madly and

* This onset of injury depends on the size of the crowd: it happens for slower speeds if the crowd is larger.

randomly looking for a way out. But if you see several people rushing past in a certain direction, you might conclude that they know something you don't. There is a natural tendency to follow the crowd in a herd-like way. The researchers assumed that people would search for obscured escape routes with a mixture of individualistic (random) and herding (collective) behaviour.

A little herding, they found, is a good thing. It reinforces success: when someone finds an exit, others are more likely to follow them to it. This effect is self-reinforcing: the more people move to one place, the more inclined others are to follow them. But too much herding leads to problems. It can persuade nearly everyone to head for a single exit and forget about finding others, even if they know that others exist. Helbing and his colleagues added a herding element to their simulations, in much the same way as Vicsek allowed for flocking in his self-propelled particles: the peoploids had a tendency to match their direction to the average direction of the group immediately around them.

As the herding instinct increases (which might be expected to happen as panic rises), peoploids in a room with several invisible exits at first begin to leave more efficiently, because they are able to capitalize on one individual's success in finding a way out. But beyond a certain herding level the reverse effect kicks in rapidly. A single exit then becomes clogged with individuals who have followed the majority, while other exits remain unfound or barely used. The most efficient use of exits, therefore, happens for a certain optimal level of herding: too much or too little, and exits are wasted.

For safety engineers, these are telling results. Often, says Helbing, safety measures are planned on the assumption that doors are used uniformly. Engineers note the rate at which people file out of a single door, and use that to calculate how many doors are needed in a room with a certain capacity. The crowd is treated rather like water flowing

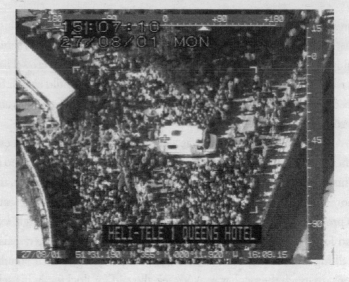

Figure 6.11 An ambulance attempts to manoeuvre through the crowd at the Notting Hill carnival in London, 2001.

through a sieve: all the holes are used equally. But if the doors are used very unevenly by a panicking crowd, this calculation might seriously underestimate how many exits are needed. A simulation of panicking peoploids might be a far better way to deduce how long it will take to clear the building.

Michael Batty has shown how pedestrian modelling might be used pre-emptively to avoid the kind of uncontrolled and dangerous crowd dynamics seen in these simulations of panic. He and his colleagues have simulated the distribution and movements of crowds during London's Notting Hill Carnival, an annual two-day event that now draws up to a million people into a small (3 square kilometres) neighbourhood in north-west London. The carnival poses immense safety problems: in 2001 there were over five hundred accidents, a hundred of which required hospital treatment.

Getting ambulances in and out of the densely packed crowd is a big challenge (Figure 6.11).

Batty's model incorporated a tendency for the pedestrians to flock and swarm, creating mass movements between the various attractions around the carnival route. There was a total of thirty-eight widely used different entry points into the carnival area, of which the five underground stations were the most important. The researchers searched for crowd-control measures, such as barriers, street closures and capacity limits imposed on entry points, that would reduce the risk of overcrowding, which tends to become focused at particular places along the circular route of the carnival procession.* London police, politicians and a body called the Carnival Review Group, run by the Greater London Authority, are making an ongoing assessment of the various options for changing this route to alleviate such dangers. Pedestrian modelling provides a means of testing these alternatives without having to use each annual festival as a huge and hazardous one-off experiment.

Batty and his colleagues point out that this kind of modelling has the potential to reorient the way we think about controlling urban systems. 'In the past', they say, 'there has been an unwitting assumption that cities can be modelled as quasi-natural systems in which control is something that is applied after the fact.'[11] Of course, an event like the Notting Hill Carnival is already highly constrained by the route plan, the entry points, and so forth, before any additional crowd-control measures are proposed. Yet none of these constraints is a 'given': routes can be altered, entry points can be closed. With an ability to predict crowd behaviour, the entire planning process becomes much more malleable: prediction, say the researchers, becomes blended with prescription. Planning, rather than being imposed from the 'top', becomes interwoven with its outcome in an iterative and interactive

* The results of the model simulations can be seen at http://www.casa.ucl.ac.uk/research/urbanstudies/index.html

manner. There is nothing in these models of pedestrian motion that dictates what people *ought* to do. Rather, the aim is to find out what people *will* do, using some simple assumptions about what motivates them and allowing for the constraints they encounter. That is in the true spirit of the modern physics of society.

These are still early days in our attempts to understand human motion, but already we can see that the seeming complexity of human behaviour does not extinguish our ability to understand and predict at least some aspects of it. Equally important is the demonstration that abrupt changes in group behaviour do not necessarily require concerted changes in everyone's intentions. These switches of collective motion can instead emerge spontaneously even as individual predispositions are altered only incrementally.

There are potential pitfalls too. Parrish and Edelstein-Kreshet warn about the hazards of physical modelling of animal group dynamics:

many sets of rules can lead to lifelike group behavior, so that the results, though visually appealing, may be uninformative – it is not always possible to deduce individual behavior from emergent properties.[12]

In other words, just because you get what looks like the right kind of collective motion, that doesn't mean you have the right rules – which in turn means that you won't necessarily make the right prediction under different circumstances. The 'social force' model of pedestrian dynamics does seem versatile enough to cope with situations other than those for which it was originally designed, but it pays to remember that such models of collective motion may not furnish unique explanations for a given phenomenon, nor are they necessarily suited to every situation. They are just a beginning.

CITY LIMITS

Once there were three hills, named Tothill, Penton Hill and the White Mound or Tower Hill. Tracks wound their ways between them, and the tracks became lanes and roads, and settlements grew up around them, and the place was called London. Or so say some legends, which are probably fanciful. But one way or another, dwellings multiply and the byways grow until hamlets become villages and towns and great cities. Not all of them grow as organically as London, of course, with its dense web of streets like a warren practically begging to be populated by Dickensian ruffians. It would be fascinating to know whether pedestrian models have anything to tell us about the evolution of urban geography, but that question has not yet really been explored.

Models of city growth *have*, nonetheless, emerged from physics. In the nineteenth century a group of sociologists that has become known as the 'boosters' argued that cities exert an attractive force on people and on trade which is entirely analogous to gravitational attraction. The boosters envisaged a science of city growth every bit as mechanical and deterministic as Newton's laws describing bodies moving in gravitational fields. But modern physics-inspired theories recognize the complex, 'organic' character of urban expansion, and take their cue from studies of non-equilibrium growth processes of the kind we encountered in the previous chapter, not least those displayed by bacterial colonies.

And why not? The modern city is, after all, alive. It is what makes the city so thrilling, so terrifying, so liable to swallow its inhabitants. London, Tokyo, Delhi and Los Angeles pulsate, they groan and sigh and spread their many tendrils. 'Whether we consider London as a young man refreshed and risen from sleep, therefore,' writes the city's biographer, Peter Ackroyd, 'or whether we lament its condition as a deformed giant, we must regard it as a human shape with its own laws of life and growth.'[13]

To see this shape as human requires a novelist's imagination. Rather, it is an amorphous mass that straddles the Thames. But 'laws of growth' are there for sure; and no one, it seems, can control them. Disturbed by the grotesque overcrowding within and around the city walls, Elizabeth I prohibited 'any new buildings of any house or tenement within three miles from any of the gates of the said city of London'.[14] Did this stem the city's outward sprawl, a process of accretion that had long since burst the bounds of the encircling stone walls? You might as soon try to contain a wildfire.

In 1787 Henry Kett compared the expansion of London to an epidemic, and complained that:

Mansions daily arise upon the marshes of Lambeth, the roads of Kensington, and the hills of Hampstead . . . The chain of buildings so closely unites the country with the town that the distinction is lost between Cheapside and St George's Fields. This idea struck the mind of a child, who lived at Clapham, with so much force, that he observed, 'If they go on building at such a rate, London will soon be next door to us.'[15]

Today, all these districts are in inner London; some people consider them appealingly central.

Every major city has similar tales to tell. And while the consequences of urban sprawl are perhaps not quite as grim as they were in Elizabethan England, still they supply cause for deep concern both in developed and developing nations. Faced with soaring urban air pollution, noise, traffic congestion, lack of open space and the tensions of overcrowding, many US citizens identify urban sprawl as their most significant local anxiety. Containing urban expansion has become a hot topic on the political agenda. Donald Chen, director of Smart

Growth America, an organization that seeks to improve urban conditions, argues that urban sprawl is 'undermining America's environment, economy and social fabric'.[16]

If urban planners are to mitigate sprawl, they must first understand its causes. But as Chen points out, 'The theories explaining why sprawl occurs are as numerous as they are politically controversial.'[17] For example, the accusation that sprawl is encouraged by government spending on infrastructure such as highway construction and subsidies for power and water supplies is somewhat undermined by the fact that the decrease in US public subsidies in recent years has not diminished the growth of American cities. Those amoebae go right on growing.

The organic, formless shape of big cities was evident to Lewis Mumford in the 1930s:

Circle over London, Berlin, New York, or Chicago in an airplane, or view the cities schematically by means of an urban map and block plan. What is the shape of the city and how does it define itself? As the eye stretches towards the hazy periphery one can pick out no definite shape, except that formed by nature: a banked river or a lakefront: one beholds rather a shapeless mass, here bulging or ridged with buildings, there broken by a patch of green or the separate geometric shapes of a gas tank or a series of freight sheds. The growth of a great city is amoeboid: failing to divide its social chromosomes and split up into new cells, the big city continues to grow by breaking through the edges and accepting its sprawl and shapelessness as an inevitable by-product of its physical immensity.[18]

The comparison with the spread of a microbial colony is more than just an emotive metaphor. In the 1990s Michael Batty recognized in the clumpy, irregular outlines of cities something akin to the shapes of clusters of particles formed by diffusion-limited aggregation (DLA clusters; see the previous chapter) – like those Mitsugu Matsushita saw in bacterial growth.

This was a challenging image. City planners strive to see the *regularities* in urban form – naturally enough, since those are the things that they and their predecessors have put in place. The grid structure of American cities – an urban schema first developed by Imperial Rome – is the first thing you notice from city maps or in the view from a Manhattan penthouse. But the city boundaries become anything but neat and formal as they reach out into the map's emptier expanses. Envisaged as DLA clusters, cities shed any pretence of rational planning and begin to look like truly organic things, growing with a life of their own – a 'crystallization of chaos',[19] in Mumford's memorably apt phrase.

Batty and his colleague Paul Longley adapted a theory of DLA-like growth, called the dielectric breakdown model (DBM), to describe the shapes of cities. Whereas DLA clusters grow by the accumulation of particles at their wispy edges, the DBM embodies a process in which the ragged tips advance by pushing their way outwards into some surrounding medium – arguably a more realistic description of urban sprawl. Batty and Longley tuned the model to adjust the density of the clusters – crudely speaking, whether the branches were thick or thin. They found that it could mimic, in an approximate way, the manner in which a city expands while constrained by natural obstacles such as rivers and coastlines (Figure 6.12). In general, as a city grows it becomes increasingly dense, filling up more of the available space: its fractal dimension (see page 137) increases. Batty and Longley calculate that between 1820 and 1962 the fractal dimension of London increased from 1.322 to 1.791.

In the universality of these fractal growth models we can find an echo of Herbert Spencer's words from 1876:

When we say that growth is common to social aggregates and organic aggregates, we do not thus entirely exclude community with inorganic aggregates: some of these, as crystals, grow in a visible manner.[20]

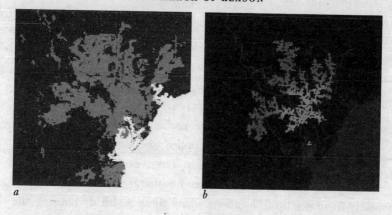

a   b

Figure 6.12 A fractal growth model developed by Michael Batty and Paul Longley can give a reasonable imitation of the shape of the Welsh city of Cardiff, confined by rivers and sea. *a* shows the real city, and *b* is the computer simulation 'grown' from the model.

But the sad truth is that DLA-type clusters don't look much like real cities. For one thing, the former consist of one continuous mass, dense in the centre and getting progressively more diffuse as one moves outwards. But real cities are more clumpy: they tend to grow by linking to and then engulfing outlying settlements, just as London has swallowed Kensington, Clapham and Hampstead. Moreover, the presence of the city promotes the formation of such satellite communities in the first place: the suburbs, whose residents enjoy their proximity to the bright lights without actually having to live there. Businesses in these small communities feed off the trade provided by the city as well as that from locals.

This more complex picture of city growth has been simulated in a model of non-equilibrium growth developed by Hernán Makse, Shlomo Havlin and Gene Stanley at Boston University in 1995. To Makse, the irregular shapes of cities looked like the forms of fluids percolating into porous rock. Some simple models of this phenomenon (which is important for oil mining) treat the progress of the fluid through the tortuous pore network as a random process, so that each advancing tendril is independent of the others. But Makse found that a better description emerged if he assumed that one region of the fluid 'feels' the effect of others over relatively long distances. In the terminology of physics, this means that they are correlated.

Makse and his colleagues applied the model of 'correlated percolation' to the growth of cities. They reasoned that clusters of development are correlated too: new housing and businesses are more likely to spring up in locations where there is already housing and commerce nearby. They regarded city growth as a process of the same type as DLA: it happens by the accretion of new particles ('development units'), generally around the periphery. But whereas in DLA particles stick at random, in correlated percolation they are apt to go where other particles are. And furthermore, the model allowed for new centres of development to appear which are not physically connected to the main cluster (while still feeling its influence through the correlations).

The shapes that result depend on how strong these correlations are. Within a certain range of correlation strength, the growing clusters start to look very much like real cities (Figure 6.13). One way to make a more precise, quantitative comparison is to consider how many small towns there are around the main city. (In the model, towns correspond to outlying clusters of connected particles.) Clearly there are more small towns than large ones, and the data for London and Berlin show that there is a precise mathematical relationship between the size of a town and the total number of towns of that size. This relationship (it is, in fact, the *probability distribution* of town sizes – see page 76) is accurately mimicked by the model.

The striking thing about this relationship is that it is the same for Berlin in 1920 as for Berlin in 1945 – and even for London in 1981, even

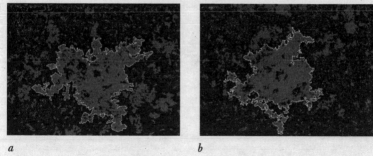

a   *b*

Figure 6.13 The shape of a city like Berlin (*a*), here shown as it was in 1945, is mimicked by the clusters of particles that develop in a model of 'correlated percolation' (*b*). The detailed features are quite different, but the two structures have a similar appearance. In fact they share rather precise mathematical properties in terms of the distributions of 'particles'.

though these two cities have very different sizes and have no doubt been subject to very different planning regulations. There seems to be an inevitable 'shape' to big cities, which the correlated-percolation model can reproduce and which does not fundamentally change as the city grows. Planning seems to have had no effect on this law of growth: looked at in these terms, the 'shape' of London in the 1980s bears no imprint of the Green Belt policy implemented in the 1960s in an attempt to contain its sprawl within bands of protected parkland. 'You could say', argues Gene Stanley, 'that the lawmakers do what they want to do, but people will live where they want to live.'[21] This collective process creates its own physical laws of shape and form.

But planning is not always invisible in the overall structure of a city. Rui Carvalho and Alan Penn at University College in London say that a slightly different way of measuring a city's shape reveals a division into two classes, which show the influence of different planning regimes. They use a mapping scheme introduced by Bill Hillier and Julienne Hanson in which open spaces in the urban environment

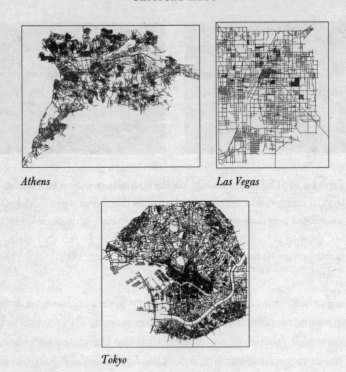

Athens

Las Vegas

Tokyo

Figure 6.14 Another way to depict the underlying spatial structure of cities is to represent them as 'axial maps' in which straight lines, designating open spaces, are extended until they cross another such line. Here we can see the rather different structures of Athens, Las Vegas and Tokyo. The probability distributions of these axial maps for many cities fall into two distinct classes of city structure, corresponding to different planning regimes. A few cities do not fit into either class.

are represented as straight ('axial') lines that are extended until they cross another open space. This creates a so-called axial map (Figure 6.14). Carvalho and Penn looked at the probability distributions of the lengths of the axial lines for thirty-six cities in fourteen different countries, and found that twenty-eight of them fitted onto two different 'master curves': that is, there seem to be two distinct types of city structure. One class corresponds to relatively 'open' structures, with many axial lines that span the whole urban space – this group includes Bangkok, Eindhoven, Seattle and Barcelona. The other group has a denser web dominated by short lines, and includes London, Hong Kong, Athens and Dhaka. The researchers think that the growth of the first group of cities has been governed by 'global' planning of the large-scale structure, while the second group has been guided only by 'local' planning, so that there are fewer 'city-scale' features such as long avenues. A handful of cities, such as New Orleans and The Hague, fall in between these two classes, indicating a mixed influence of local and global planning.

The US economist Herbert Simon points out that an absence of central planning does not necessarily mean that all cities are poorly 'designed'. On the contrary, they are (or at any rate, they once were) often remarkably effective in arranging for goods to be transported, for land to be apportioned between residential, business and manu-facturing districts, and for a lot of activity to be fitted into a small area:

I retain vivid memories of the astonishment and disbelief expressed by the architecture students to whom I taught urban land economics many years ago when I pointed to medieval cities as marvelously patterned systems that had mostly just 'grown' in response to myriads of individual human decisions. To my students a pattern implied a planner in whose mind it had been conceived and by whose hand it had been implemented. The idea that a city could acquire its pattern as naturally as a snowflake was foreign to them. They reacted to it as many Christian fundamentalists responded to Darwin: no design without a Designer![22]

Sadly, the laws that dictate city growth are today swelling many of them to proportions that few would perceive as 'marvelous'. That this process is beyond the reach of planners may seem a pessimistic conclusion for those who worry about urban sprawl.

Maybe so; but maybe not. If cities are going to grow come what may, perhaps it is better to focus attention not on how to fence them in but on how to make them more attractive places in which to live. Good public transport and services, low-emission vehicles, protected green spaces, varied local shops, attractive architecture – all seem perfectly feasible if the will exists, and all seem more appealing than grand plans for imposing futile designs on places which, as one nineteenth-century traveller said of London, already seem to have 'no beginning and no end'.[23]

7

On the Road

The inexorable dynamics of traffic

A social planner can usefully contemplate traffic signals. They remind us that, though planning is often associated with control, the crucial element is often coordination. People need to do the right things at the right time in relation to what others are doing.

Thomas Schelling (1978)

One of the principal objects of theoretical research in any department of knowledge is to find the point of view from which the subject appears in its greatest simplicity.

J. Willard Gibbs

> Down to Gehenna or up to the Throne,
> He travels the fastest who travels alone.
> Rudyard Kipling (1890)

*

How would you feel about spending a whole working week in every year sitting in a box the size of a large cupboard with nothing to do, unable to get up, stretch, sleep, read or watch TV? Well, if you live in Washington DC, Boston or Denver you may be doing that already.

And in Los Angeles you are likely to be experiencing a full week and a half of it.

I'm talking about sitting in traffic jams. As cities sprawl and commuters have farther to travel, as many public transport systems are eroded and cycling becomes a life-threatening activity, sitting in a car in stationary traffic has become a regular part of urban life. The time wasted in this largely solitary and stressful way has more than trebled, on average, in large American cities over the past two decades. In Los Angeles you can anticipate 56 hours of jam-time every year: one and a half times the length of a normal working week. And it is sure to get worse. Londoners never tire of telling one another (usually when stuck in a traffic jam) how the average vehicle speed through the city today is as slow as it was a hundred years ago, when horse power meant exactly that. The city has recently introduced congestion charges in a desperate attempt to ease the flow in the gridlocked centre.

In holiday seasons it is not unusual for jams in Europe to grow to over a hundred kilometres long. The hidden cost of motor transport caused by traffic jams is phenomenal, and would surely give us pause if it were reflected in the price of a vehicle or a tankful of fuel. The economic cost of all the wasted time is estimated at around £60 billion a year in Germany, and that's without considering the environmental impact. The price of the delay and wasted fuel in, say, Houston (where fuel is cheap!) adds up to around $850 (£500) per person per year.

Emissions from road vehicles are one of the biggest sources of pollution. In a densely urbanized country such as Germany, 60 per cent of the noxious carbon monoxide and nitrogen oxides pumped into the atmosphere comes from road traffic. Pollution from road vehicles is the most likely cause of the recent increase in childhood asthma in cities. The greenhouse effect of carbon dioxide emissions from car exhausts is probably helping to break up the polar ice caps. Engines idling in a traffic jam add to these problems while fulfilling no useful function. In Seattle over 350 litres of petrol are burned up per person per year by cars which are motionless or barely moving.

One option is to build more roads. But this simply attracts more traffic, and you are soon back where you started from. 'Building more roads to ease traffic', says Richard Moe, head of the US National Trust for Historic Preservation, 'is kind of like trying to cure obesity by loosening the belt.'[1] The ideal solution is to encourage greater use of public transport, more cycling, more walking. All these things are possible, but our addiction to the internal combustion engine is going to be hard to kick, and it seems to be getting worse in both the developing and the developed world. If reducing the volume of traffic is a daunting task, might we at least in the meantime direct it more efficiently? In short, never has there been a more urgent need to understand how traffic flows – and why sometimes it doesn't.

Despite the use of computers to optimize the design of traffic control measures and street networks, every city has its black spots where traffic planners and road engineers have clearly not got it right. Perhaps they are simply faced with an impossible task, trying to make a decades-old infrastructure cope with twenty-first-century traffic volume. And how well we all know the perversity of traffic flow: the way, for instance, 'phantom' jams on a motorway can form for no apparent reason.

Statistical physics can now help us make sense of how traffic moves and how it clogs. Researchers worldwide are devising conceptual models derived from the physics of liquids and gases which can predict when and where congestion occurs and what form it will take. Traffic has been found to have its own peculiar laws of motion, which are often surprising and occasionally infuriating. The physics of traffic won't solve all our transport problems, but it can guide us to better safety measures, improved highway design and more accurate fore-casting of tailbacks. It will render traffic control less of an empirical activity and more of an exact science. And not a moment too soon, for in the world of transport planning, laissez-faire is no longer an option.

KEEPING TRACK

Navigating around busy road systems is like trying to play the stock market: you are always making choices on the basis of incomplete information. Traffic reports on local radio and in-car computerized route planners give only a limited picture, and not necessarily as quickly as you need it. But the German city of Duisburg shows what we might expect in years to come. On the Internet you can call up a map of the instantaneous traffic flow throughout the city, updated every minute.

Traffic flow can be measured by wiring up roads with monitors which register passing vehicles. A single monitor such as an induction loop, a pressure-sensitive wire pinned across the road, can clock the number of cars passing per minute. The most useful measure of traffic flow, however, is not numbers but density. Ten cars could pass an induction loop in a minute either nose to tail in a slowly moving queue, or more widely spaced at a higher speed. In the former case the traffic density is high; in the latter it is low. To deduce the density, one needs to know not only the traffic flux in cars per minute but their speed. And to measure that requires two induction coils, closely spaced, at each monitoring point.

In principle a city's traffic density can be fully mapped by staking out every road with pairs of induction loops. But this is completely impractical. In Duisburg, the measurements are made only at a small number of key points; most of the streets are not monitored. The gaps are filled by using a computer simulation of traffic flow, a model of moving particles much like those we encountered in the previous chapter. The real data are used to keep the simulations honest: the calculations are continually tweaked to make sure that they give the right values of traffic density at the places where this is known. The assumption is that the bits in between will then be not far from the true picture.

The model was developed in the early 1990s by Michael Schreckenberg, a physicist at the University of Duisburg, in collaboration with Kai Nagel, then at the University of Cologne. Nagel has worked on a similar project in the USA to create real-time traffic maps of Dallas and other American cities.*

WAVES AND PARTICLES

The Nagel–Schreckenberg (NaSch) model was by no means the first attempt to use physics to understand traffic. James Lighthill (1924–98), one of the great twentieth-century experts on the physics of fluid flow, proposed in the 1950s that the behaviour of traffic moving on a road was rather like that of a liquid flowing down a pipe. In collaboration with Gerald Whitham at Manchester University, he fashioned this insight into a rough-and-ready theory of traffic flow.

In the Lighthill–Whitham model, the individuality of drivers is entirely submerged beneath average driving behaviour, just as the theory of fluid motion ignores the vagaries of individual molecules. This is ironic, for Lighthill himself was anything but average in his driving habits. He was a persistent speeding offender, but would explain in court that as Lucasian Professor of Mathematics at Cambridge (the chair once occupied by Newton) he was fully aware both of the laws of mechanics and of his social duty not to waste

* See http://www.traffic.uni-duisburg.de/OLSIM/ and
http://www-transims.tsasa.lanl.gov/

energy. As a result, he told the hapless judges, he felt obliged to desist from braking when going downhill. It seems that this defence was occasionally successful.

Perhaps the monotonous, beetle-like procession of vehicles on a highway might make us less nervous about the apparent psychological naivety of constructing a physics of traffic, as opposed to a physics of pedestrians. But behind every wheel is an individual in command of a machine, and that can do strange things to human behaviour. If pedestrians have their quirks, how much less predictable are those of a car driver, trying to cope with anything from exhaustion to squabbling children, from the influence of narcotics to surging hormones?

Nevertheless, we can still look at the problem in terms of averages and fluctuations around those averages. Most drivers manage to drive sensibly and predictably, without suddenly changing speed or colliding with everything in their path. Most variations from the average are minor: some people like to drive at 10 k.p.h. below the speed limit on a motorway, others at the same amount above the limit. Few make the journey from Birmingham to Southampton entirely in third gear (although I'm told my grandfather used to try).

Various researchers have tried to improve on Lighthill's work by allowing for the responses of real drivers. A group at General Motors in Warren, Michigan, in the 1950s devised one of the first so-called car-following models. This model treated vehicles as discrete objects, rather than regarding traffic as some quasi-continuous fluid, and it assumed that each driver modified his or her speed in response to what the car ahead did. The driver would accelerate or brake (decelerate) depending on two factors: the distance to the car ahead and the relative speed of the two cars. (Drivers tend to brake more sharply if they are both travelling at 100 k.p.h. rather than at 30 k.p.h.) In 1974, Rainer Wiedemann at Karlsruhe University went further, devising a model in which each driver was governed by a complex suite of 'psychological' rules. But the more complex the model, the harder it becomes to know what outcomes are in any sense 'fundamental' aspects of traffic flow and which follow from the choice of rules.

The model Kai Nagel developed in collaboration with Schreckenberg certainly had the virtue of simplicity. It is basically a kind of cellular automaton. The road is divided up into a series of cells, each of which can be either empty or occupied by a vehicle. The vehicles move from cell to cell in each time step, like pieces on a ludo board. So the flow evolves in a succession of freeze-frames. As in Dirk Helbing's pedestrian model, each driver wants to reach a certain preferred speed. Every car will, on a clear road, accelerate until it reaches this speed. The drivers also aim to avoid collisions. This entails maintaining a safe distance from the car in front, which gets larger as the relative velocity of the two vehicles increases. The third ingredient of the model is an element of randomness or 'noise'. No driver estimates accelerations and decelerations perfectly – in particular, he or she might overcompensate when slower vehicles come into view ahead by braking more than is necessary. Fluctuations in speed are also commonly caused by roadside distractions – we have surely all braked at some time when something has caught our eye.

These were the rules of the road. How did the traffic move? Nagel and Schreckenberg found that there were two distinct types of flow, distinguished by how the flow rate varies as the traffic density increases. The flow rate is a measure of how many vehicles pass a certain point each hour (or minute, or whatever time interval you choose). The traffic density is the number of vehicles on each kilometre (or mile, or whatever) of road. As the traffic gets denser while remaining light enough for drivers to drive as they please, the flow rate increases as the density increases: you pack more cars into each kilometre of road without them needing to slow down, so more vehicles pass any given monitoring point each hour. But at a particular

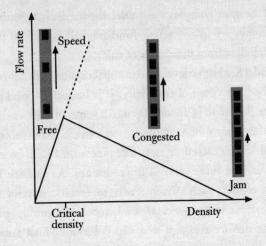

Figure 7.1 The transition from free to congested traffic can be seen on a graph of flow rate plotted against traffic density. Above a certain (critical) density, the flow rate begins to decrease as density increases, because the cars are forced to go slower. But there is an alternative, metastable 'free' state above the critical density (indicated by the dashed line).

'critical' density, this tidy pattern ('free flow') breaks up. The cars start reacting to one another's presence by slowing their speed. Then the increased throughput owing to greater traffic density is compensated by a decreased throughput owing to slower motion. At the critical density, the flow rate begins suddenly to decrease rather than increase as the traffic density rises (Figure 7.1). In the parlance of traffic studies, we have gone from free to congested traffic.

RISK AND CONTINGENCY

In a version of this model explored by Nagel and his colleague Maya Paczuski in 1995, each driver's preferred speed is maintained rigidly whenever possible, mimicking the kind of cruise-control facility available in some vehicles. In this scenario the transition between free and congested flow can be postponed: instead of undergoing a transition to congested flow at the critical density, the traffic continues to flow freely, increasing its flow rate as the density increases (the dashed line in Figure 7.1). It is as if the drivers have collectively decided to take the *risk* of maintaining their speed as the traffic gets heavier.

In this cruise-control model, the critical density represents a bifurcation point at which two options become available. One is the safe bet: all slow down. The other is the gambler's choice: all stay fast. As long as no one loses their nerve or concentration, this dense, fast-moving traffic can survive without collision. But it is a perilous and precarious state. If one person brakes, the car behind must do so too, and the one behind that . . . and all of a sudden the high-flow state collapses, and we are on the congested, slow-moving branch. Of course, these simulated drivers do not *know* they are taking a risk if they keep moving fast, just as molecules do not know that they are in a gas or liquid state. The point is simply that the fast-moving state is a feasible collective state, even above the critical density.

But this state is so fragile that it falls apart at the slightest provocation. Any random fluctuation converts it, in a Grand Ah-Whoom, to a congested state. It is not, in other words, stable. Physicists know all about states like this: they call them *metastable*, which literally means 'next to stable'. Metastable is not the same as unstable – the fast-moving state can persist above the critical traffic density provided no one rocks the boat. Gases, liquids and solids can be metastable too. They can each exist under conditions where another state is actually more stable. A liquid can be cooled below its freezing point without it seizing into rigidity – it may pass by the phase transition as though there were nothing unusual about that point at all. Such a liquid is said to be supercooled.

A metastable liquid survives because freezing has to start somewhere. Water does not freeze everywhere at once: it begins with a few tiny ice crystals, which then grow steadily through the liquid. Typically these seed crystals are germinated by irregularities in the liquid, such as particles of dust or scratches on the walls of the vessel. At such places it is slightly easier for water molecules to come together in an ice-like configuration. If the water is kept scrupulously free from impurities and other sites of ice 'nucleation', the only way ice can begin to form is for the randomly moving water molecules somewhere to come together by chance in an ice-like arrangement. That is to say, some random fluctuation must trigger the transition.

In principle this becomes possible the moment water falls below freezing point. But in practice it may not happen for a very long time – although the chance that it will do so rapidly increases the further the temperature falls below freezing. So far, the record cooling for super-cooled water stands at around minus 39°C. Below this, it is virtually impossible to prevent water from freezing.

We might think of the nucleation phenomenon as rather like the spread of group chants through a football crowd. One sometimes gets the impression that an incoherently yelling crowd is metastable with respect to one that is chanting in unison. But the chant has to start somewhere – the spectators won't all pick it up at the same instant. Many small groups of fans might try to start a chant or song, only to give up when it doesn't catch on with their neighbours. Once the chant spreads to a group of some critical size, however, it seems to take on a life of its own and soon the whole terrace is voicing it as one.

Incidentally, the freezing of the oceans triggered by a shard of ice-nine in Kurt Vonnegut's *Cat's Cradle* (Chapter 4) implies that liquid water is metastable relative to this imaginary form of ice – that, given the chance, water would freeze to 'warm ice'. In this picture the oceans are just waiting to freeze solid, except that a seed of ice-nine is too difficult to form by relying on random fluctuations alone. Only when a pre-existing seed is added does the transformation become possible. In the late 1960s a group of Russian scientists thought they had discovered a new, 'gummy' form of water that was more stable than liquid water under everyday conditions of temperature and pressure. Some feared that this so-called 'polywater' might indeed congeal the oceans if it came into contact with them. Fortunately, polywater was just a figment of a few experimenters' delusions.

Here I must make a slightly technical point about metastability which is important for what follows. I distinguished earlier between first-order and critical phase transitions. Freezing, boiling, and so forth are first-order transitions. The onset of magnetism at the Curie temperature, and the cooling of a fluid below its critical temperature so that it separates into liquid and gas, are critical transitions. It turns out that only first-order transitions can support metastable states, which temporarily 'ignore' the transition. Critical phase transitions, on the other hand, cannot be avoided, for something special happens at a critical point to ensure it is not overlooked.

The switch between free and congested traffic in the NaSch model is thus a kind of first-order phase transition. The existence of the metastable 'free-flow' branch has a further implication. Suppose the density of traffic is gradually increasing as rush hour approaches on a city street. The critical density for congestion may be exceeded without a congested state appearing, so that the traffic persists in a metastable free-flow state. But before long a fluctuation caused by some nervous driver triggers the collapse to congestion. The flow rate plummets to nearly zero and the traffic density surges.

Then the traffic eases as rush hour passes, and the density begins to fall. But the traffic cannot make the transition back to the free-flow branch until the density falls below the critical density, because until this point is reached congested flow is always more stable (Figure 7.2).

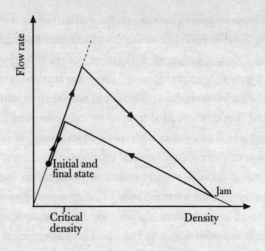

Figure 7.2 If free flow persists into the metastable regime, it can be triggered into switching to congested flow at any moment by a random fluctuation. As the traffic subsequently lightens, free flow cannot be regained until the critical density is reached. So increasing the traffic density has different consequences to decreasing it: there is a 'loop' in the flow history which can be traversed only in one direction.

In other words, metastability is a one-way affair. We can get to a metastable free-flow state by increasing the traffic density from low values, but not by decreasing it from high values. In the same way, one can make supercooled water by cooling the liquid but not by warming up ice close to but below the melting point.

Thus the state of the traffic depends not only on its density but on its *history* – on whether it was previously denser or less dense. As the traffic density rises and then falls, the flow rate follows a loop like that shown in Figure 7.2, which can be traversed only in one direction. Physicists call this one-way behaviour *hysteresis*.

The NaSch model gives an inkling of how traffic jams can form without any apparent cause. Suppose that, while a column of traffic is moving in the metastable free-flow state, one driver decides for some reason to brake suddenly. Perhaps there is a genuine cause: a dog runs out in front of the car, or maybe the driver slows instinctively as he reaches for his ringing mobile phone. This situation can be idealized in a simulation by programming a vehicle to suddenly change its speed from that of the surrounding traffic to some low value, before immediately accelerating again to the same speed as the others. It is the briefest of disturbances – but look what it does to the flow (Figure 7.3).

Here each line of dots rising steeply from lower left to upper right is a plot of how a single car's position changes over time in such a simulation. A straight, sloping line corresponds to a car moving with constant speed. A car brakes at the top left, and then quickly resumes its previous upward 'climb'. But the cars following behind – the lines to the right – respond by slowing their speed to avoid collision. The wave of kinks in the sloping lines corresponds to a wave of cars that abruptly slow their speed. You can see that many cars are affected, including those that enter this stretch of road long after the braking car has vanished from the scene. The thicker the web of black lines, the denser the jam.

And there's more. If the jam stayed at the point where it was initiated, the wave of disturbance would trace out a horizontal line in this diagram. But instead the wave slopes downwards as time increases to the right, indicating that the jam moves upstream – in the opposite direction to the traffic. In other words, a jam initiated at one place can move spontaneously to another place in a stream of traffic. And the initial disturbance splits into several branches, so that a car arriving later (towards the right of the diagram) at this stretch of road will encounter not one jam but a succession of them. These jams show no sign of dying away – if anything, they proliferate as time goes by. So a single, small and brief fluctuation creates several moving waves of nasty snarl-ups.

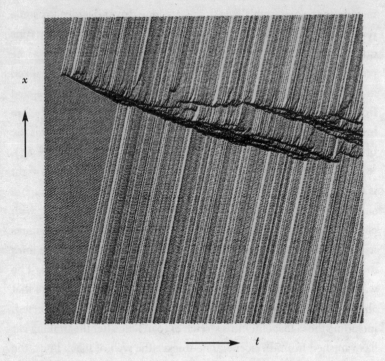

Figure 7.3 Jams triggered by a single spontaneous fluctuation in metastable flow. The figure shows how the distance x along a highway for vehicles varies with time t in a computer simulation. A straight line of dots sloping steeply towards the top and the right denotes a vehicle moving at constant speed. Each vehicle entering the highway (at distance zero, along the bottom edge of the plot) begins a new line. The dark bands sloping down from the upper left correspond to jams. Here a vehicle's smooth progress is disrupted, creating a kink in its 'timeline'. A single disturbance at the top left, caused by sudden braking of one vehicle, triggers a whole set of congestion points as time progresses.

These events probably sound all too plausible, and indeed they do arise in real traffic. But the NaSch model is not ideal for describing them. It turns out to be too simplistic and excessively sensitive to small disturbances. Before looking at more accurate ways of modelling traffic, however, let's see what observations tell us about real-world behaviour on the road.

THREE FORMS OF TRAFFIC

In 1965 a team of researchers at Ohio State University in the USA tracked a number of cars in aerial photographs as they moved along a highway. They saw precisely the kind of 'jam without a cause' that the NaSch model can produce, moving upstream in a persistent wave (Figure 7.4). This type of jam is a consequence of driver overreaction: people braking harder than they really need to.

The general picture described by Nagel and Schreckenberg received more substantial support in 1996, when German researchers Boris Kerner and Hubert Rehborn of the Daimler-Benz research laboratories (now DaimlerChrysler) in Stuttgart made detailed observations of traffic flow on a section of the A5-South autobahn, which connects Giessen in Germany to Basle in Switzerland. The road is particularly busy where it passes close to Frankfurt, and this stretch was fitted with double induction-loop detectors to measure the velocity of every passing vehicle. Kerner and Rehborn saw precisely the behaviour forecast by the NaSch model. The flow rate increased as the traffic density increased, up to a point where the stream could suddenly collapse into a jam. The experimental data staked out a graph very much like the bifurcating plot of the model (Figure 7.5).

What exactly does this graph imply for an individual motorist? In Figure 7.5 a series of data points are sequentially numbered. These

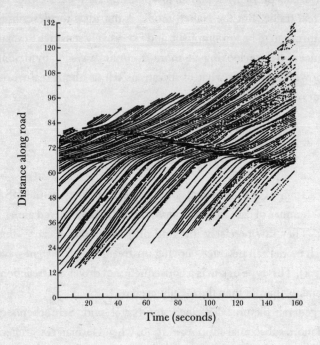

Figure 7.4 A spontaneous jam in real traffic data. This is the same type of graph as Figure 7.3, with each line representing the time line of a single vehicle. The kink sloping to the lower right is a moving jam.

points represent consecutive measurements of the flow every three minutes at one location on the autobahn. At a time corresponding to point 1, the traffic is freely moving but at a density greater than the critical density of around twenty vehicles per kilometre. Subsequently it begins to slow down, carrying the flow rate off the metastable branch (points 2 and 3). Then suddenly a jam coalesces, rendering the vehicles almost immobile. Things stay that way for the next ten minutes or so – points 4, 5 and 6 – before the traffic gradually starts moving again and the flow rejoins the free-flow branch. Notice that free flow is regained at

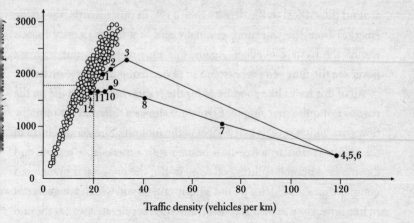

Figure 7.5 Free, congested and metastable free flow show up in observations of real traffic made on a German autobahn in 1996. The numbered data points represent one-minute averages over a twelve-minute flow sequence at one point on the autobahn.

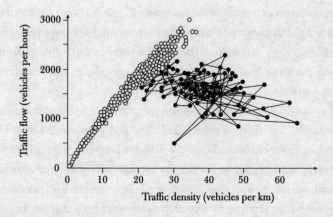

Figure 7.6 Vehicles diverging from the 'free flow' branch can sometimes follow a much more erratic timeline, experiencing a wide and unpredictable range of different traffic states.

around the critical traffic density (point 12). In other words, the traffic emerges from the jam more gradually and at a lower density than it enters: the hysteresis effect. Again, this time history is one-way: we never see the time sequence from 1 to 12 occurring in reverse order.

All of this looks like good news for the NaSch model. But life on the road is not quite that simple. Figure 7.6 shows a different set of traffic flow data which reveals a new facet of the motion. The general form of the results is similar: a free-flow branch with a metastable region, and a congested branch splitting off from it. But the congested branch is a complete mess! Vehicles do not simply enter a jam which congeals out of free flow, slow to near standstill, and then accelerate away at the end of it. Instead, their speeds (and thus the overall flow rates) vary considerably once free flow has turned into congestion.

Kerner and Rehborn proposed that *two* distinct flow states can be identified in congested traffic. In jams movement is minimal, the traffic is very dense, and the flow rate falls close to zero. This corresponds to the right-hand extremity of the 'congested' branch: points 4, 5 and 6 in Figure 7.5. But congested traffic can keep moving at a respectable flow rate even at high density if all the vehicles (in all lanes) move with more or less the same speed. In this case, they said, the traffic has become synchronized.

Thus, said Kerner and Rehborn, there are not two but three basic flow states: free flow, synchronized flow and jams. In the transition from free to synchronized flow the vehicles keep moving and the flow rate stays quite high, but the density increases sharply. In a transition either from free or synchronized flow to a jam, the vehicle speeds drop sharply to near zero, and the density is about as high as it can be, with vehicles nose to tail.

Does this sound familiar? In the phase transition from gas to liquid, the particles remain mobile but the density jumps to a higher value. But when gases or liquids freeze to solids, the particles become very densely packed and immobile. The three states of traffic look remarkably like these three states of matter. Moreover, Kerner and Rehborn say that the transition from free flow to jam rarely happens directly; synchronized flow normally appears as an intermediate state, in the same way that for matter to change from a gas to a solid it normally has to pass through the liquid state.

Thus, say the researchers, what really happens when traffic density exceeds the threshold for stable free flow is that this state becomes metastable relative not to a jam but to synchronized flow. A fluctuation can tip the traffic into the dense, slower-moving synchronized state. They claim that the jump from free flow to synchronized flow is accompanied by a sharp drop in the probability of overtaking in multi-lane traffic. In free flow, drivers can overtake more or less at liberty. In synchronized flow, with all lanes moving at about the same speed, overtaking is almost absent.

DISTURBING THE FLOW

Exactly how free flow gives way to synchronized flow is still debated. Kerner and Rehborn are convinced that synchronized flow is a fundamental traffic state, just as liquids are a fundamental state of matter. Others, such as Dirk Helbing, suspect that its appearance in traffic may be contingent on external influences: perturbations to the free-flow state. That is to say, the synchronized state does not congeal spontaneously out of nothing, but has to be triggered by disturbances such as bends, hills, bottlenecks or entry and exit lanes. These could act like the particles of dust that nucleate a solid from a supercooled liquid – or, more aptly in this case, a liquid from a supercooled gas.

Kerner and Rehborn agree that synchronized flow does seem to be commonly triggered by 'impurities' on the road, such as entry and exit points at junctions where vehicles enter and leave. (These are usually called on-ramps and off-ramps in traffic studies.) They suggest that when synchronized flow appears spontaneously it is a transient state which persists for typically only half an hour or less, whereas it can last for several hours upstream of on-ramps. In some models synchronized flow is itself metastable and apt to congeal into a jam if perturbed. But Michael Schreckenberg and his colleagues at Duisburg maintain that in real traffic synchronized flow is actually quite robust and results from a factor not accounted for in many traffic models: the drivers' desire for smooth and comfortable driving.

Most models assume only two things of drivers: that they aim to reach a certain preferred speed and that they avoid collisions by slowing down. In principle this could lead to a very jerky ride: drivers speed up whenever they have enough room, only to brake when another vehicle comes into view ahead. Schreckenberg and colleagues argue that few people actually drive this way; instead, they prefer to avoid sharp acceleration and deceleration. When this rule is added to the cellular-automaton model, synchronized flow emerges as a fundamental and stable state, which persists even if an upstream-travelling jam moves through it.

This sort of refinement promises to enhance the predictive power of traffic models, but it also shows us something deeper: these collective flow modes are an irreducible aspect of traffic flow. Making the models more psychologically complex may alter the precise conditions under which the flow modes appear, but it will not alter the fact that traffic, like atomic matter, really does seem to have fundamental states. Are there truly just three such states? Well, yes and no. It appears that any one vehicle will indeed find itself at any moment either in light and uncorrelated traffic, dense and synchronized moving traffic, or dense and near-motionless jams. But Dirk Helbing and his collaborators have found that there are many ways in which these flow modes can be mixed and permuted over the course of time.

In collaboration with Martin Treiber at Stuttgart, Helbing devised a model in 1998 which had features in common with both the cellular automaton of Nagel and Schreckenberg and the fluid-flow models originally conceived by Lighthill and Whitham. They dispensed with individual cars and treated the traffic as a smooth fluid. But it was a most peculiar fluid. In the traditional theory of fluid motion, called hydrodynamics, each little 'parcel' of fluid affects those around it through viscous drag: it exerts a frictional force which slows down the movement of the surrounding fluid. In Helbing and Treiber's model the interactions between parcels of 'traffic fluid' are more complex, capturing the same kind of driver responses as the particles in the NaSch model. Again, drivers are assumed to speed up and slow down as they react to what is happening ahead, aiming to reach a particular velocity and to avoid collisions. It is a fluid with a mind of its own; indeed, with a multitude of minds.

At low densities this model generates free flow. As the density increases, free flow becomes metastable: small fluctuations can be absorbed and dispersed, but larger ones will create localized jams which propagate upstream. As the density increases further, these jams can evolve into cascades, like those seen in Figure 7.3 (page 206): congested waves separated by free flow. This stop-and-go motion is familiar to anyone who has sat behind a driving wheel. And at even higher densities the traffic becomes uniformly congested in a slow-moving jam.

Helbing, Treiber and Ansgar Hennecke at Stuttgart looked at how their traffic fluid behaved in the presence of an irregularity such as an on-ramp. They instigated a localized wave of congestion in the traffic ahead of the ramp and watched what happened when this wave, moving upstream, drew level with the ramp. For low traffic flow the congestion simply dispersed. In heavier traffic it could trigger all kinds of behaviour: waves of localized jams separated by free flow, waves of

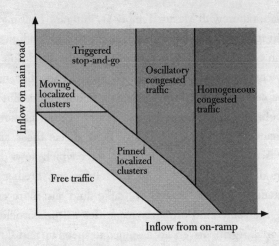

Figure 7.7 Traffic states in the 'fluid' model developed by Helbing and colleagues. Homogeneous congested traffic is a jam that grows steadily in length. In oscillatory congested traffic the jam also lengthens over time, but comes in waves, so that individual vehicles in the jammed region alternately slow to a halt and then move again. Moving localized clusters are knots of dense traffic that move steadily upstream, in the opposite direction to the traffic flow. Triggered stop-and-go is a similar state in which such knots occasionally throw out a smaller dense cluster that moves downstream before triggering a new upstream-travelling knot. Pinned localized clusters are knots that stay fixed at one point on the road, as vehicles move into and out of them.

congestion with almost no let-up in between ('oscillatory congested traffic'), congestion pinned to the position of the on-ramp, uniformly congested traffic behind the ramp, and so forth.

The researchers illustrated this diverse behaviour on a kind of phase diagram that shows the conditions under which each state forms (Figure 7.7). This can be regarded as analogous to the 'morphology diagram' drawn up for bacterial growth patterns (page 139) – after all, traffic flow, like bacterial growth, is a non-equilibrium process. The jumps between different flow states occur quite suddenly as the 'control parameters' – the flow of traffic on the main highway and the influx at the ramp – are altered in the model. Traffic, it seems, is at the mercy of a series of non-equilibrium phase transitions.*

JAMS TOMORROW?

Are these various models truly capturing the behaviour of real traffic, or are they little more than fancy computer games? Only if the model effectively mimics reality can it be trusted to provide accurate traffic forecasts. With that in mind, Helbing, Treiber and Hennecke have compared their predictions with real traffic data taken from various German and Dutch highways fitted with induction loops. They were heartened to find that all the various states they predicted could be identified in these traffic flows. What's more, they found that by feeding into their model the basic features of the real-life traffic (such as the ratio of cars to lorries), they could predict with uncanny accuracy how it would develop as the hours passed – even when the flow pattern was by no means simple or regular (Figure 7.8).

This gave them some confidence that what comes out of this kind of traffic model does not depend too delicately on what goes into it. Provided the drivers are assumed to share certain general characteristics – accelerating to a preferred speed on an open road, braking to maintain a velocity-dependent distance from the vehicle in front, and having imperfect responses that make overreaction likely – it does not matter exactly what one assumes about details such as the reaction time, the shape of the road or the number of lanes. The same general

---

* Helbing's team has shown that basically the same 'phase diagram' of traffic states arises in a particle-based model in which each vehicle is modelled individually. So these states don't seem too dependent on the details of the model.

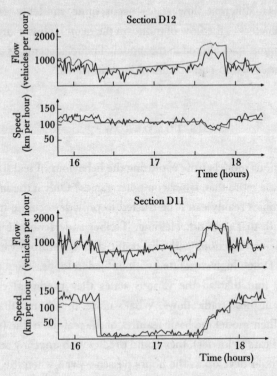

Figure 7.8 The traffic-flow model of Helbing and colleagues can predict how various traffic states will evolve over the course of several hours. The model is fed the traffic density and speed measured at some upstream location, and from this it calculates the likely speed and density at several points farther down the highway. Here I show the traffic flow (speed and density) at two different points on the A5 South autobahn near Frankfurt, over the course of about two and a half hours. Congestion was caused by a lane closure at some point along this section. The black line shows the real data measured by induction loops; the grey line shows the model predictions for these locations. The model misses much of the fine detail, but correctly predicts the basic features of the flow, such as the time and place of jams and their alleviation.

flow modes will emerge in any case. The observations have convinced Helbing and his team that most jams are caused by anomalies: bottlenecks, on-ramps, hills, or perhaps just an aberrant manoeuvre by a single driver. In over a hundred real jams that they studied, all but four (at most) were caused by some kind of stationary bottleneck.

Traffic in cities does not behave like traffic on open roads – in particular, its flow is dominated by intersections. Ofer Biham and colleagues at the Hebrew University of Jerusalem in Israel have studied a cellular-automaton model in which vehicles moved from south to north and from west to east on a square grid, a fair approximation of any American city. The motions of vehicles were timed to mimic the alternation of traffic lights at intersections. Ominously, they found that as the traffic density increased there was an abrupt phase transition from free flow to a stationary jam – to total gridlock, in other words.

Traffic physics is already proving its worth. Nagel's Transportation Analysis and Simulation System, based on his cellular-automaton model and developed at Los Alamos National Laboratory in New Mexico, has been used to plan road networks in Dallas. Knowing the likely effects of intersections, crossings and lane narrowings should help planners place them wisely. Schreckenberg has plans to expand his own system in Duisburg to encompass the whole autobahn network of North Rhine-Westphalia, generating hourly traffic forecasts.

Modelling studies of freeway traffic can also provide useful insights into driving regulations. A model developed by Helbing in collaboration with Bernardo Huberman of Xerox's research laboratories in Palo Alto, California, shows that a mixture of cars and lorries in multi-lane traffic can produce what they call coherent flow, in which all the vehicles move at the same speed without making any lane changes. This is reminiscent of, but not identical to, synchronized flow. Coherent flow is a robust state, quite stable in the face of fluctuations, and even though it prevents all drivers from reaching their preferred velocities it is nevertheless very efficient: the high traffic density and moderate speeds mean that the flow rate is almost as high as it could possibly be. Moreover, coherent flow is a safe state because it reduces two of the main causes of highway accidents – differences of speed and lane changes. So it is worth looking for ways to promote the transition to coherent flow when traffic becomes heavy, for example by imposing appropriate speed and lane-changing restrictions.

Helbing and Huberman say that their model also reveals how American lane rules, which allow drivers to travel at any speed in any lane, can make roads more efficient than European rules, under which the lanes are graded from slow to fast. Because lorries stay largely in the slow lane, car drivers tend to avoid it even when there are no lorries immediately ahead, reducing the capacity of the highway by up to 25 per cent. Helbing and Treiber, meanwhile, have shown how real-life congestion and delays on a German autobahn could have been eliminated by introducing speed limits that alter in response to changes in the traffic density. A limit imposed only during the rush hour could prevent jams and reduce average travel times for all vehicles.

One of the major triggers of jams in dense traffic is fluctuations. Drivers lose concentration, get too close to the vehicle in front and then slow abruptly. If random perturbations like this could be smoothed out, the threat of jams would be much reduced. One way is to design traffic regulations which have the effect of forcing drivers to pay more attention in areas containing potentially jam-forming obstructions such as bottlenecks. Another possibility is to implement traffic-control measures at on-ramps that adapt to changes in traffic density.

A more far-sighted option is to replace the human driver with one that never tires or miscalculates. Nearly all traffic models are collision-free, since the rules are designed explicitly to avoid them. The 'virtual drivers' do not need to be granted any particularly sophisticated responses or capabilities in order to avoid crashes with other vehicles; they need to know little more than how far away the car in front is and how fast it is approaching. With this information, cars could apparently be piloted safely by robotic autopilots – real automata!

That possibility is already being explored by some automobile manufacturers. They are developing 'driver assistance' systems which use radar and other sensors to track what is going on around the car and respond accordingly. One of the main aims is to avoid crashes, or collisions with pedestrians or cyclists, by allowing automated braking or manoeuvring to kick in with a faster response time than any human can achieve. But once such systems are installed in vehicles, they could help with the more mundane driving tasks too, such as lane changing and cruise control.

As we've seen, jam-inducing fluctuations can be caused by drivers overreacting in heavy traffic. This could be eliminated by driver-assistance systems. Traffic simulations by Treiber and Helbing have shown that some jams in heavy traffic can be smoothed away completely if just 20 per cent of the cars are equipped with automated systems which enable them to respond optimally to changes in traffic flow.

Think of linking these autopilots to satellite navigation systems, digital road maps and an up-to-the-minute traffic forecast provided by simulations pegged to real data, and you begin to have a prescription for getting efficiently and safely from A to B without ever having to put your hands on the wheel. Motoring organizations, which are fond of lobbying for greater 'driver freedom', will need to accept that it is sometimes this very freedom that makes the roads hell. Robotic drivers could be not only more considerate, law-abiding and skilful, but also better informed and more foresighted – in short, far better placed to use the roads wisely.

8

Rhythms of the Marketplace

The shaky hidden hand of economics

> There is nothing which requires more to be illustrated by philosophy than does trade.
>
> Samuel Johnson

> If all economists were laid end to end, they would not reach a conclusion.
>
> George Bernard Shaw

> Nobody should be rich but those who understand it.
>
> Wolfgang von Goethe

*

When the Scotsman Adam Smith (1723–90) invented political economy as a field of philosophy, he had no past giants' shoulders on which to stand. No one had previously pondered the question of how a market economy sustains itself, for trade was considered a worldly affair and hardly the stuff of philosophy. But Hobbes's *Leviathan* was essential reading for Smith, since its blueprint for constructing a stable Commonwealth could not afford to neglect matters of how trade was to be managed, or wealth and land were to be distributed. (All the same, Hobbes has been criticized for paying insufficient attention to economics as the real basis of power. In *Oceana* (1656), Hobbes's rival James Harrington argued that the unrest of the Civil War stemmed largely from the changes in land and property ownership that had taken place in the early seventeenth century.)

Yet the world in which Smith wrote *An Inquiry into the Nature and Causes of the Wealth of Nations* (1776) was utterly different from Hobbes's world a century earlier. There was a new social order, and its sovereign was Mammon. For Hobbes, wealth and production meant land and agriculture; for Adam Smith it meant industry. While Hobbes allowed that trade could be an asset to a sovereign, for him the true measure of a nation's wealth was gold, gathered by whatever means necessary – plunder, conquest, royal marriage. Smith, on the other hand, knew that the rise of the mercantile system had made commerce the determinant of a nation's treasury.

Smith had no rosier a view of human nature than did Hobbes. Men, he concluded, are basically acquisitive, out for what they can get. And so he found in Hobbes's political model the foundations for a theory of how the market system works. Just as Hobbes's individuals seek to accumulate power (primarily by buying it from their fellows), Smith's merchants and traders aim to amass wealth.

For Hobbes, the lust for power can be curbed only by surrendering all authority to a supreme sovereign; the alternative is a life nasty, brutish and short. But no one ruled the marketplace: it was a free-for-all. Here was the nub of the conundrum that Adam Smith sought to explain. How, in a society awash with greed and lacking any centralized means of market regulation, do the common people nevertheless generally manage to obtain goods at prices which do not bankrupt them? What is to stop traders from charging whatever they will for their wares, if not an all-powerful authority?

The answer, in a word, is competition. Traders who inflate their prices will lose their customers to others who maintain prices at an affordable level, and so will go out of business. The pressure of competition in a free market economy, argued Smith, should ensure that goods will always be sold at their true value or 'natural price' – an honest measure of the quality and quantity of materials and labour that went into their making. Anyone charging more can always be undercut by their competitors. Smith argued that this self-regulating market will ensure that all the things that people need will be supplied to them. There is no need for a central authority that assigns people to jobs so as to guarantee an adequate number of tailors, cobblers, bakers, dairymen, and so forth. Wherever there is a demand, there is profit to be made – and someone will step in to make it.

Smith's classic treatise was the first real analysis of the workings of the new market economy of the Age of Enlightenment, and it contained an idea which not only pervades all of economic theory today but also resonates profoundly with the physics I have described in the earlier chapters. Not only are there indeed laws of the market, but these laws *emerge* from the push and pull of trade, from the interaction of myriad agents. They are not imposed from outside. Two opposing forces – self-interest and competition – suffice to generate a self-regulating steady state, a kind of equilibrium. This is the 'hidden hand', as Smith put it, that keeps order in the market. As economist Robert Heilbroner says, 'One may appeal the ruling of a planning board . . . but there is no appeal, no dispensation, from the anonymous pressures of the market mechanism.'[1] In Smith's view, 'good' (meaning a fair and comprehensive market) can emerge from selfishness. The message of *The Wealth of Nations* is one of laissez-faire: leave the market alone, and it will take care of itself. It is precisely the opposite of the dictatorial straitjacket that Hobbes proposed to curb humankind's lust for power, and was a welcome message to industrial capitalists seeking to expand their empires without interference from governments.

Adam Smith's world contained the seeds of ours, but it was not the same. His capitalists were all small fry: shop owners, merchants and small-scale industrialists. In the 1770s a 'manufactory' with a dozen workers was a large one. The guild regulations, which had governed business practice in the Middle Ages, were dissolving, and few external rules and laws had been introduced to replace them. There were no labour organizations or trade unions, and no large corporations – let alone multinationals. It was a highly fragmented market, some would say an 'atomistic' one. But even on its own terms, Smith's economic theory was too simplistic to cover the whole story. It was just a beginning, still lacking the kind of certitude that Newtonian mechanics had given physics.

All the same, we can see already that this primitive economic theory betrays an implicit aspiration to be a science of the same stature. Enlightenment intellectuals had long suspected that this was possible; 'Nothing', said Daniel Defoe in 1706, 'obeys the Course of Nature more exactly than Trade, Causes and Consequences follow as directly as Day and Night.'[2] Adam Smith never wrote of market forces as such, but the concept of causative forces analogous to Newton's gravity is evidently there in his great work. His contemporary Josiah Tucker, the Dean of Gloucester, was more explicit: 'The Circulation of Commerce may be conceived to proceed from the Impulse of two distinct Principles of Action in Society, analogous to the Centrifugal and Centripetal Powers in the Planetary System.'[3] By the early nineteenth century the idea that there were fundamental and immutable laws of economics, just as there were in physics, had taken firm root. That belief has, by and large, remained in place ever since.

Smith's hidden hand became a part of the lore of the business world, and to many it was a force that society hindered at its peril.

Here is Ralph Waldo Emerson in 1860:

Wealth brings with it its own checks and balances. The basis of political economy is non-interference. The only safe rule is found in the self-adjusting meter of demand and supply. Do not legislate. Meddle, and you snap the sinews with your sumptuary laws.[4]

Smith's economic principles had begun to take on the aura of an immutable physics of the marketplace, and seemed to be promising the enticing prospect of a calm and stable economy. If the market bucked, someone must have had been tampering with it. Emerson again:

The laws of nature play through trade, as a toy-battery exhibits the effects of electricity. The level of the sea is not more surely kept, than is the equilibrium of value in society by the demand and supply: and artifice or legislation punishes itself, by reactions, gluts, and bankruptcies. The sublime laws play indifferently through atoms and galaxies.[5]

The trouble is, even now no one has discovered what these laws are. In spite of the elegance of Emerson's words, considerations of supply and demand evidently do not suffice to explain the sometimes wild behaviour of the market. Economic models have been augmented, refined, garlanded and decorated with baroque accoutrements. Some of these models now rival those constructed by physicists in their mathematical sophistication. Yet they still lack their 'Newtonian' first principles, basic laws on which everyone agrees.

This is not because economists are incompetent. On the contrary, the finest economic minds have equalled any in natural science. But it does mean that economic models are persistently incompetent, in the sense that they repeatedly fail to make the accurate predictions of which scientific models are routinely capable. It is not only scientists who point this out – some within the economics community admit as much. As John Kay of the London Business School says, 'Economic forecasters . . . all say more or less the same thing at the same time; the degree of agreement is astounding. [But] what they say is almost always wrong.'[6] And it was ever thus. One week before the Great Crash of October 1929 – the largest in history – Irving Fisher of Yale University, perhaps the most distinguished US economist of his time, claimed that the American economy had attained a 'permanently high plateau'.[7] Three years later the national income had fallen by more than 50 per cent. No one, not a single economist, had seen it coming.

Over the past decade, statistical physicists have begun to suggest that economists might want to rethink some of the basic assumptions upon which they construct their models. By importing ideas from physics, say the physicists, economists can start to make sense of the erratic and so far unpredictable behaviour of the world's markets. This endeavour has even coined its own neologism: *econophysics*. How have economists reacted to this encroachment on their territory? Some econophysicists feel they have been met with the deafening silence of indifference, punctuated only by bursts of irritation. Some economists, meanwhile, regard the apparent isolation of econophysicists as self-inflicted. They point out that economics has always been an eclectic, catholic discipline, little concerned with professional affiliations. What matters, they say, is not academic credentials but whether new ideas are *useful*. They wonder how much of the physicists' contribution is genuinely useful to them.

It is true that the physicists have not always made their task easier. As a group they are not renowned for their tact and humility in pointing out the shortcomings, as they perceive them, of other areas of human enquiry. And the things that interest physicists are not always those that interest economists. Moreover, physicists are used to systems whose fundamental laws remain the same in all places at all times. Economic laws are almost certainly not like this, which is one reason why economics is harder than pure physics. (Economist Paul Krugman consoles himself with the thought that 'luckily it is not quite as hard as sociology'.[8]) Yet it seems inevitable that in time these tensions will ease, the misunderstandings will be resolved, and each side will come out of its trench to negotiate pacts and truces. Several leading economists are already listening respectfully, if cautiously, to what the physicists are saying. And most econophysicists are only too eager to engage in discussion, to be put right on their shaky grasp of economic principles and concepts, to find common ground.

The first instinct of physical scientists is one not always shared by economists: to learn from experiment. First they want to know what the fundamental economic phenomena are that one should seek to explain. In economics there is really only one experiment against which to compare theories, and it is vast and ongoing. What happens in the real marketplace?

THE IRON LAW

Smith's model of the market was purely descriptive. But by the nineteenth century confidence in science, and specifically in a mechanistic description of the world, was so buoyant that it seemed possible to bring everything within its embrace. And so economists sought rigour in mathematics. Jeremy Bentham (1748–1832) was one of the first to propose that the ebb and flow of trade and production could be couched in terms of pure mathematics. His utilitarian philosophy, which he called the Felicific Calculus, took a more beneficent view of humanity than did Smith or Hobbes, proposing that people aim to maximize not their power or profit but their 'pleasure' – albeit by a kind of cold-blooded profit-and-loss reckoning.

Building on this image of man as a 'pleasure machine', the Irish economist Francis Edgeworth (1845–1926) brought the mathematical approach to an early pinnacle in 1881. Despite its now exotic-sounding title, his book *Mathematical Psychics* was an attempt to develop a precise theory of political, social and economic behaviour replete with abstruse differential calculus and a good dose of statistical analysis. It was the precursor to a welter of theories which, drifting ever further from any foundation in the real world, tried to subsume human behaviour beneath pages of complex calculations. According to Paul Krugman, this tradition is alive and well in modern economics: 'It is cynical but true to say that in the academic world the theories that are most likely to attract a devoted following are those that best allow a clever but not very original young man to demonstrate his cleverness.'[9]

But the most influential of 'scientific' economic theories in the nineteenth century, at least politically speaking, was that due to an intemperate and almost permanently poverty-stricken German Jew named Karl Marx (1818–83).

In Adam Smith's world-view, labour was simply another commodity on the market. The labourer was a trader, and he traded in his own time and sweat. This commodification of labour was a product of the Industrial Revolution, from which there emerged a class of men and women who, in contrast to the trained craftspeople of the Middle Ages, were just so many pairs of hands. These hands were ready to operate machinery, shovel coal, or do the increasing number of low-skilled but specialized tasks demanded by the division of labour in a factory system. In return, these 'proletariat' workers gained their wages.

Smith proposed that wages reach their 'fair' value just like any other market price: the cheapest labourer captures the market. This meant that the 'true' or 'fundamental' value of labour was a subsistence wage – a worker could not afford to work for less than this, whereas those who asked for more would be undercut. So, as economist David

Ricardo (1772–1823) pointed out in the early nineteenth century, the industrial capitalist could always manage to employ a workforce for no more than the cost of keeping it alive: this was Ricardo's grim Iron Law of Wages. Generally, employers did just that; in fact sometimes not even subsistence was guaranteed to the proletariat.*

Marx considered the future prospects for this capitalist system. His conclusions spelled hope for the oppressed workers and despair for their employers. Capitalism, he said, was doomed to exhaust itself and to be overthrown by proletarian revolution. This was not mere wishful thinking or the desired outcome of any moral imperative; he believed he had proved his forecast with scientific rigour. Marx's approach seemed the epitome of scientific model-building: he idealized, he simplified, he removed irrelevancies. His economic landscape held a two-tiered society consisting only of workers (who sell their labour in return for a wage) and capitalist factory owners (who buy labour and sell goods).

The capitalists seek profit. But as Smith had pointed out, profits are always being eroded: by rising wages, and by the force of competition, which constantly pulls the market price of goods down towards the cost of producing them. The only way a company can maintain a profit is by expanding. And so, said Marx, it must constantly bid for additional labour. This increase in demand for labour means that workers can ask for higher wages, which bites back into the capitalists' profit margins. To escape this cycle, Marx assumed the factory owners

* Adam Smith pointed out that subsistence wages are not equal merely to the cost of keeping a worker alive: 'they must upon most occasions be somewhat more; otherwise it would be impossible for him to bring up a family, and the race of such workmen could not last beyond the first generation.' Smith does not endorse the grinding poverty implicit in his words, but that was the reality of his world – a world where 'one-half the children born . . . die before the age of manhood.'[10] The industrial world was not relentlessly ruthless, however: real wages more or less doubled during the Victorian era.

would do what he saw them doing all about him: introduce labour-saving machinery. Marx's economic model thus embraces technological change. This has come to be regarded as a crucial element in some modern economic theories – and rightly so, for new technology continues to change the market demand for workers at a pace at least equal to that in Marx's time.

But here's the catch. In a working day (which in the mid-nineteenth century could be sixteen hours long), workers can produce goods worth more than a subsistence wage. This 'surplus labour' is the source of the capitalists' profits. Machines, however, do not offer surplus labour – in a competitive market, industrialists will have to buy them at a cost equal to the value of the goods they can generate. So mechanization renders workers unemployed, but does not rescue the capitalists' profits.

The result is an economic recession: wages are low and unemployment is rife. By the mid-nineteenth century it had become clear that the market was prone to recessions, such as that caused by the notorious South Sea Bubble of 1720. But they were dismissed as the consequence of some external factor perturbing 'business as usual'. The belief was that, left to itself, the market attained equilibrium in the manner suggested by Smith. Marx, in contrast, implied that recessions are an inevitable part of the way the market works.

They are, however, transient. As wages fall, the profit margin rises once more. It becomes more viable for capitalists to start employing workers again and to expand their businesses, and the economy is revitalized. Recession provides its own cure. By the same token, an economic boom is self-limiting because it boosts wages and reduces profitability. This is again, then, a self-regulating market. But it is not the steady, equilibrium market of Adam Smith. Instead it is plagued by cycles of boom and bust.

A physicist or engineer will see at once that Marx had identified a mechanism of *negative feedback*: the process in which a change moderates its own cause. In general, negative feedback promotes stability, since it returns the system to a stable state if fluctuations drive it away. So a scientist will rightly object that Marx's scheme does not in fact guarantee boom–bust cycles, as he supposed. This will happen only if the negative feedback overcompensates for the change, swinging the balance back in the opposite direction. There is nothing in Marx's formulation that makes this inevitable.

Yet Marx's economic vision contained the crucial concept of a market that was potentially unsteady and apt to oscillate between boom and bust. And in Marxist theory these fluctuations are an intrinsic, irreducible element of the system, not something imposed from outside. Why, though, should a fluctuating market be destined for collapse, rather than simply for perpetual fluctuation? Here one cannot avoid the impression that Marx's political predispositions overtook his analytical thinking. Each recession, he said, will be worse than the last, for each one bankrupts ever larger firms. In the end, the misery caused by lack of production, unemployment and low wages during a deep recession will prompt a proletarian revolution. 'What the bourgeoisie therefore produces, above all,' said Marx, 'are its own gravediggers.'[11]

Come the revolution, private ownership will cease and the means of production will pass into communal control. The *Communist Manifesto*, which Marx wrote with Friedrich Engels, sought to hasten this end with a call to arms: 'Let the ruling classes tremble at a Communist revolution. The proletarians have nothing to lose but their chains. They have a world to win.'[12] In fact, the proletariat had rather more to lose, as they discovered in France, Austria and Germany, where the attempts at revolution in 1848 were violently suppressed. In Russia, of course, it was to be a different story.

CAN THE MARKET BE STEADY?

If Marx's historical prognosis was flawed, he had nevertheless found the pulse of the capitalist economy. It was a stuttering pulse. The wild fluctuations of the market seem to taunt economists trying to construct models of how trade and commerce work. These chaotic leaps between boom and bust reveal, more than any other measure, the disparity between economic theory and practice. A great deal of the theory developed since Marx can be bluntly characterized as a futile attempt to tame the untameable.

Nothing is more revealing of the desperation this engenders than the insistence on speaking of 'the business cycle'. To a physicist this is almost offensive. Take a look at the changes in the US economy over the past century or so, as measured by the annual percentage change in national output per capita. Using this (rather than the absolute change) as a measure of the 'size' or 'health' of the economy has the advantage that it removes the effects of inflation and business expansion, which create an overall upward trend in output. In a stable economy, the relative change would stay at zero. In reality, it fluctuates by up to 20 per cent or so (Figure 8.1a).

When economists speak of the business cycle, they are referring to the fact that booms – peaks in the graph – are followed by recessions, or valleys.* But a cycle is surely something that happens repeatedly with the same interval – that is, periodically, like the change of seasons or the swing of a pendulum. There are plenty of peaks followed by valleys in this graph, but can you persuade yourself that you are seeing

---

* To some degree, the word 'cycle' is used simply to indicate that whole economic system is affected in the same way: growth or decline applies across different economic sectors, even if there is no direct relation between them. That does not make the choice of this word any less misleading, however.

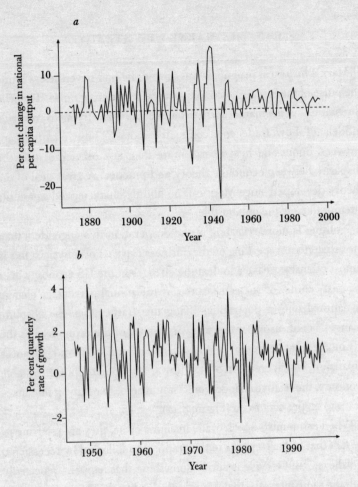

Figure 8.1 (*a*) The national output per capita in the USA fluctuates annually by up to 20 per cent or so, in a manner that looks utterly unpredictable. (Notice the difference between fluctuations before, during and after the Second World War, however, which shows that external events can have a big impact on the economy.) (*b*) The quarterly growth rate of real US gross national product shows how economic measures fluctuate on a finer timescale.

something rhythmic here? The graph looks more like the random static of a detuned radio signal, traced out on an oscilloscope. Like noise, in other words. Actually it is worse than that. The data in this graph are annual – there is only one datum point for each year. Figure 8.1*b* shows the quarterly growth rate of real US gross national product, a similar measure of the size of the US economy. Now the data points come every three months, and we see even finer-scale randomness in the static. Business cycles? Where?

Yet the business cycle is a part of standard economic dogma, and the very term betrays a desire to impose order on the chaos. Some economists have claimed to perceive periodicity – regularity – in the recurrence of boom and bust. By doing so, they hoped to be able to forecast when the next crash was coming, and so avoid the kind of catastrophe that wiped out investments and ruined lives in 1929. The Austrian Joseph Schumpeter claimed in the 1940s that there were at least three business cycles: one short-term, one of 7–11 years, and one with a period of about 50 years. The apparent randomness of the fluctuations was, he said, a result of the out-of-step interactions between these three cycles. The Great Depression was caused by the coincidence of troughs in all three cycles.*

The ingenuity of economists in conjuring cyclical behaviour out of the market's paroxysms has to be applauded, if not lauded. In the 1930s, for instance, a retired US accountant named Ralph Elliot claimed that markets wax and wane in an eight-step pattern of five 'advancing' waves and three 'declining' waves.† This pattern, he said,

---

* Paul Krugman calls Schumpeter's two-volume tome *Business Cycles* 'turgid, almost meaningless'.[13] Indeed, he comments that attempts to account for the business cycle have led many economists to produce their worst work. Clearly, they are as perplexed as they are troubled by it.

† Elliott had good cause to want to understand market fluctuations, having previously lost his job and a good part of his savings in the Wall Street Crash.

could be discerned over several time frames, ranging from the day-to-day fluctuations to 'supercycles' lasting over 200 years. Elliot waves thus divide each cycle into stages whose structure echoes the famous Fibonacci sequence, in which each number is the sum of the previous two: one cycle can be divided into two parts (declining or 'corrective', and advancing or 'impulsive' – the 'bull' and 'bear' phases of the economy), composed respectively of three and five waves, giving the sequence 1, 1, 2, 3, 5, 8, ... Smaller-scale subdivisions of these waves yield higher numbers in the Fibonacci sequence. This piece of numerology, ostensibly constructed from the theory of bull and bear markets devised by Wall Street analyst Charles Dow, leaves market speculators even today juggling with Fibonacci numbers and the 'golden ratio' that the series encodes in a quasi-mystical belief in hidden patterns.

Schumpter listed a whole slew of purportedly periodic economic cycles: Kitchin cycles, Juglar cycles, Kuznets and Kondratieff cycles. The last of these are particularly notorious. The Russian Nikolai Kondratieff asserted in 1926 that capitalist economies go through boom–bust cycles lasting 50–60 years. Whether or not this identified (as hoped) a fatal flaw in the capitalist system, it did not protect Kondratieff from the Stalinist regime when he proposed the 'reactionary' idea that land expropriated by the Party should be returned to the peasants. He was banished to a gulag where, driven to madness, he died in 1938. Some economists insist that Kondratieff waves appear in the long-term history of Western markets: they claim to identify one with the Industrial Revolution from 1787 to around 1842, another with the 'bourgeois' period from 1843 to 1897, a third with the growth in electrical power and car manufacture in the first half of the twentieth century, and a fourth covering the post-war period until the present. But there is simply not a long enough historical record to permit the identification of such a long-period cycle with any statistical conviction. On such timescales, one must suspect that a cyclical model becomes more a way of systematizing history than a way of understanding the economy.

The truth is that dips and peaks in the economy resolutely refuse to recur in any predictable manner, making attempts to construct cyclical theories of economics look increasingly like Ptolemy's elaborate schemes for predicting the motions of the planets while retaining an Earth-centred universe (except that at least there *is* an underlying rhythmicity to the planets). Irving Fisher admitted in 1925 that the business cycle might be inherently unpredictable – an idea that he seems to have forgotten four years later.

Having failed to tame these untidy irregularities with notions of periodicity, economists then tried to banish them from conventional economic theory. Although Marx's pioneering analysis of boom and bust placed the causative factors within the workings of the economic system (it was, in the jargon, an *endogenous* theory), the aim soon became to seek external forces that drive a potentially stable market into these paroxysms. This is still largely true of conventional economic theory today, for apparently the idea of an inherently unstable market remains too unsettling for many to contemplate.

Schumpeter, for example, proposed that the 50-year Kondratieff cycles were associated with episodes of major technological advance, which transformed methods of production and thus altered the economic system that they fed. This idea of technology as an external perturbation on the economic system resurfaces in one of today's most orthodox economic theories, discussed in Chapter 9. The British economist Stanley Jevons, meanwhile, attempted in the 1870s to explain the business cycle as a consequence of the sunspot cycle – not quite as exotic an idea as it sounds, although it is wrong. Sunspots become more numerous every eleven years or so, a symptom of periodic changes in the output of solar radiation. Jevons suggested that these changes affect harvest yields and thus the price of corn – which drive corresponding variations in the rest of the market.

Economists sometimes attribute to market forces a supremely stabilizing influence. They speak in terms of market equilibrium: a stable state in which all indicators are at an ideal and balanced level. Supply adjusts itself to match demand precisely, so that the markets are always 'cleared' – there is no waste and the goods are distributed to the greatest benefit of society.* This state, called the Pareto optimum after the Italian sociologist and economist Vilfredo Pareto (1848– 1923), would allow the capitalist system to run smoothly and efficiently if it were not for the disruptive influence of exogenous factors and the meddling of governments. Even in the absence of such perturbations, however, achieving the Pareto optimum in conventional economic models demands assumptions that bear little relation to the real world.

In a mathematical tour de force, the French economist Léon Walras (1834–1910) put the concept of market equilibrium on a formal footing, and his work gave rise to the 'general equilibrium theory', which dominated economic thinking for most of the twentieth century. The trouble is that this theory relies on suppositions about how trading is conducted and how traders act that are patently absurd; and the theory collapses if these constraints are relaxed. But faith in the Holy Grail of equilibrium is still endemic in economic practice, and it has engendered a strong belief, particularly among recent US economic-policy advisors, that deregulation – freeing the market of all legal

* Adam Smith's laws of supply and demand, which provide the 'hidden hand' that is supposed to engineer this equilibrium, were challenged in 1958 by arguably the first econophysicist, M. F. M. Osborne. He questioned the fundamental assumption that prices depend on consumer demand in such a straightforward manner. This issue is still debated today, although empirically there is evidence that prices do fall as demand grows, just as Smith believed.

hindrances and burdens – is the best way to keep an economy healthy and to enable growth and wealth creation.

One prominent economist who echoed Marx in arguing that fluctuations are intrinsic to the economic system was John Maynard Keynes (1883–1946). In the 1930s he attempted to explain the recurrence of boom and bust by analysing the flow of income. Keynes pointed out that the vigour of an economy was characterized not by how much wealth it represented but by how much money was changing hands. Even in a depressed economy there may be a small proportion of fabulously wealthy people coexisting with a mass of workers on the breadline and a high rate of unemployment. But in that situation, the rich tend to hold on to their money rather than invest it. As long as money keeps flowing, an economy may remain buoyant: businesses expand when there is investment capital to draw on, and employment levels and wages stay high. That in turn encourages more saving, investment and growth.

Keynes suggested that if people 'freeze' part of their income by hoarding it rather than invest it, money is siphoned out of this cycle until eventually the economy is on the point of spiralling into depression. In Keynes's picture, businesses must constantly borrow and expand to keep money flowing and maintain a healthy economy – which means that production and consumption must also increase. In this view, an economy's stability depends on its dynamism: like the Red Queen in *Alice's Adventures in Wonderland*, it must constantly keep moving in order to stay in the same place.*

* The hydraulic metaphors of flow in Keynesian macroeconomic theory were not lost on some of his contemporaries, who even attempted to construct mechanical devices that, by appropriate manipulation of water flows, could generate predictions about the economy. They were the first 'simulations' of economics – cumbersome, quixotic, supremely Newtonian, and later rendered obsolete by the computer.

RANDOM WALKS

Irving Fisher was not the first to suspect the presence of chaos in the 'business cycle'. In 1900 a Frenchman named Louis Bachelier proposed that the fluctuations in the prices of stocks and shares, and thus the underlying structure of the market economy, are effectively random. Bachelier's name does not tend to feature in economics textbooks, for he was not an economist. He was a physicist, studying for his doctorate at the École Normale Supérieure under the eminent mathematical physicist Henri Poincaré, whose work supplied the foundations of modern chaos theory. Bachelier's thesis was unusual: it was entitled 'Théorie de la spéculation', and presented a model of economics based on ideas from physics. For his contemporaries this was simply too strange, and Bachelier made no subsequent impact either on science or on economics.

Yet what Bachelier achieved in his thesis was remarkable. To construct a mathematical description of random fluctuations, he had to devise what amounted to a theory of the random-walk problem – five years before Einstein began to secure fame with his own treatment of the matter in his celebrated study of Brownian motion (page 52). The direction of motion of a particle undergoing a random walk fluctuates unpredictably, and Bachelier assumed that stock prices do the same. The fluctuations are a kind of noise. We saw in Chapter 2 how this background of random 'static' due to erratic particle motions pervades everything, and that its amplitude is a measure of temperature. The hotter a gas is, the more pronounced are the fluctuations of its constituent particles. In other words, there is a characteristic *scale* to the fluctuations of particles undergoing random walks – a typical size of the deviations.

A random walk has a very well-defined mathematical signature which emerges from the statistics of the process. It is impossible to say, at any instant, how big the next random change in direction will be. But if we keep a tally of the size of these fluctuations over a sufficient time span, a pattern becomes clear. Let's say we draw a graph of the size of the fluctuations against the number of times such fluctuations appear. What we find is our familiar bell-shaped curve: De Moivre's error curve, championed by Adolphe Quetelet as the hidden regularity of social statistics, and now generally known as a gaussian curve. As the statisticians of the nineteenth century discovered, any set of quantities whose values are randomly determined will fit onto a curve like this.

Bachelier thus assumed that stock prices fluctuate in accordance with gaussian statistics. It seemed a fair guess, especially in view of the prevalence of gaussian behaviour elsewhere in the social sciences. He went on to construct an economic model that incorporated this random element in prices. Bachelier made no attempt to explain where the fluctuations came from, but simply accepted them as a given feature of the data.

In Bachelier's day the concepts of noise and fluctuations were new ones in physics. Most scientists were interested in what happened *above* the level of the background noise. They needed to know only the pressure that a gas exerted on a confining wall, not the tiny fluctuations in pressure from microsecond to microsecond caused by the very small differences in the number of molecules striking the wall at successive instants. Indeed, these fluctuations were generally too small even to detect using the techniques of that time, as Maxwell implicitly acknowledged in his kinetic theory.

Today it is recognized that noise and fluctuations are among the most subtle and significant aspects of statistical physics. One of the most important considerations to emerge from these studies is that not all noise is gaussian. Just because a series of data points looks wildly erratic and unpredictable, it is not necessarily governed by gaussian

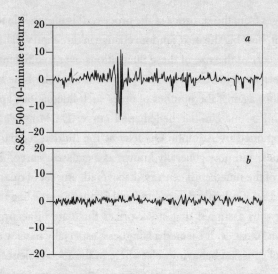

Figure 8.2 (*a*) Fluctuations in 'returns' for the Standard & Poor's 500 market index, a common measure of the state of the US economy (see page 245). A return is the difference between two values of the index a certain time interval apart; here the interval is ten minutes. So a return of zero at a particular time indicates that the S&P 500 index has not changed over the past ten minutes. (*b*) Gaussian fluctuations, which would be expected for purely random changes.

statistics. Had Bachelier carefully measured the stock market fluctuations, he would have discovered that his assumption was wrong. We can see this in the raw data, once we know what to look for. Figure 8.2 shows a typical record of stock price fluctuations over time, along with a graph of gaussian noise. Clearly the market variations are not gaussian – they are punctuated by occasional large peaks, whereas all the gaussian fluctuations tend to stay within a limited range: they have a typical scale, equal to the width of the band of spikes. For the real data, there is no such well-defined scale.

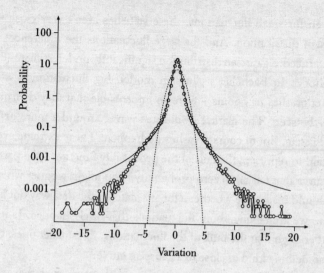

Figure 8.3 The probability distribution function for fluctuations in the S&P 500 market index. Here the time interval for the returns is one minute, rather than 10 minutes as in Figure 8.2a. But the general shape of the curve is the same for time intervals between 1 minute and at least a day. For comparison, the probability distribution for gaussian fluctuations (a random walk) is shown as a dotted line. The solid line shows the probability distribution function for a Lévy flight – see page 243.

Another way of showing the distinction, which allows us to be rather more precise in our comparison, is to draw a graph of how the probability of a fluctuation of a particular size depends on its magnitude – the probability distribution function (Figure 8.3). The error curve is the probability distribution function of gaussian fluctuations. For the non-gaussian market variations of Figure 8.2a, we can see at a glance that small fluctuations are more likely than large ones. The probability distribution function tells us exactly how much more likely they are.

Even for small fluctuations, these statistics aren't very close to a gaussian distribution. And for large fluctuations the differences are pronounced: a gaussian distribution significantly underestimates their likelihood. In Bachelier's gaussian model, big fluctuations – stock market crashes or booms – were so improbable that they'd virtually never be seen. The market would just waver around a more or less steady state. But of course crashes do happen. These extreme events are said to represent the tails of the probability function – the parts of the curve that tail off to zero probability. You might wonder whether we should really care whether a theory matches the data way out here, where we are talking about fluctuations that rarely happen. Is it not good enough that the bulk of the fluctuations, the smaller ones, trace out something kind of close to a gaussian curve?

No it isn't. For the extreme events are the very ones that worry economic forecasters most: the market crashes. Trying to understand economics with a model that cannot describe crashes is like planning river management for the small, normal variations in water height and not bothering about the occasional big changes that cause catastrophic flooding. And the deviations from gaussian behaviour don't just apply to big crashes like those in 1987 and 1997. The statistics of even moderately sized booms and slumps deviate markedly from a gaussian distribution. The simple fact is that market behaviour is not random.

FAT TAILS

If the fluctuations in the market are not gaussian, what *are* they like? In the 1960s, the mathematician Benoit Mandelbrot, popularly known as the 'father of fractals', looked at the fluctuations in cotton prices and recognized that Bachelier's random walks would not suffice to describe them. He proposed that the probability distribution of these fluctuations was 'fat-tailed': close to gaussian for small fluctuations, but with pumped-up tails encompassing the big fluctuations. This demanded a big shift in the way that the dynamics of the markets were analysed and modelled. In 1964 Paul Cootner of the MIT School of Management said:

Mandelbrot . . . has forced us to face up in a substantive way to those uncomfortable empirical observations that there is little doubt most of us have had to sweep under the carpet up till now. He has marshalled evidence of a more complicated and much more disturbing view of the economic world than economists have hitherto endorsed.[14]

Mandelbrot proposed that instead of following a random walk through the range of possible values, price variations executed a *Lévy flight*, named after the French mathematician Paul Lévy (1886–1971), who introduced this notion in 1926.

A Lévy flight is rather like a random walk punctuated by occasional big leaps.* Some animals forage in this way. They explore a small area by wandering at random, but if they find no food they will move rapidly to a new area and do the same there. This can be more effective than merely following a random walk over the entire terrain, since it gets one out of unpromising areas more efficiently. Systems showing this kind of dynamical behaviour are said to be Lévy-stable processes. Mandelbrot suggested that the fluctuations of economic markets are governed by such a process, and that the occasional big leaps account for the fat tails of the probability distribution functions of the fluctuations. The important thing to appreciate about this claim is that it is

* Technically speaking, a Lévy flight is a random walk in which the step size has a power-law probability distribution function (see page 285): the steps can be of any size, but their probability decreases as the steps get bigger. In a normal random walk all steps are the same size.

descriptive, just like Bachelier's formulation of random walks. It is a way of describing the fluctuations, without proposing to explain how they come to be that way.

Purely descriptive models were alien to mainstream economics at that time. The approach of most academic economists was to work not from data but from first principles. They would construct models based on assumptions about how the market operates, and see what they predicted. The model predictions were rarely compared against hard data. Economist Paul Ormerod argues that the scientific tradition of gathering data against which theories can be tested did not permeate to economics until the past few decades. So handling real data was a notion foreign to many economists; *starting* with it would have seemed bizarre.*

THE SHAPE OF CHANGE

Since the mid-1960s, Mandelbrot's Lévy-stable description of market fluctuations has been gradually accepted by many academic economists. Practitioners, on the other hand – traders and their advisers, who seek to use economic models to make real forecasts – have tended to persist with the idea that fluctuations are gaussian. This is partly a matter of practicality: purely random, gaussian noise is relatively easy to understand intuitively and to handle mathematically in economic models, whereas Lévy flights are not. These hands-on economists trust that the precise nature of market fluctuations will not matter too much to their calculations.

But the fact is that even Lévy flights provide an imperfect

* By no means all economists have shared this attitude. Indeed, the economist Eugene Fama proposed independently two years after Mandelbrot that price fluctuations follow a Lévy-stable probability distribution.

description of real market fluctuations. In 1995, physicists Rosario Mantegna and Gene Stanley at Boston University performed an analysis of over a million records of a standard market index – five years' worth of economic data – to uncover its statistical behaviour. They looked at the statistics of the 'returns': the differences in the value of the index at times separated by some specified interval: from one day to the next, say, or from one hour or one week to the next. The returns are a measure of the market fluctuations: when the index remains steady from one time step to the next, the return is zero. Mantegna and Stanley used Standard and Poor's (S&P) 500 index – the sum of the market capitalization of 500 top US companies, selected for the size of their market share, their state of liquidity and the diversity of industries they represent. The index is supposed to offer a single-figure summary of how the US economy is faring.

In Figure 8.3 I showed what Mantegna and Stanley found. It was certainly no random walk. The smallest fluctuations fit the probability distribution of a Lévy-stable process rather well. But the larger fluctuations deviate from this curve, falling instead somewhere between a gaussian and a Lévy process. In other words, whereas Bachelier's description of the fluctuations underestimates the frequency of large events, Mandelbrot's description of the fat tails overestimates them. There seems to be a crossover from a Lévy-stable process for small fluctuations to some other kind of behaviour for large ones.

This statistical distribution of market fluctuations seems to be the same over a wide range of different time intervals between successive data points. Mantegna and Stanley found that the statistics of the returns in the S&P 500 index from minute to minute are just the same as those from hour to hour and from day to day. In other words, the behaviour of the market looks the same at different levels of 'magnification' in time (within limits, as I explain below). This is a way of saying that there is no characteristic size of the fluctuations: they are *scale-free*, unlike gaussian fluctuations – which, as we saw earlier, have a certain typical size. If we were to magnify a part of a graph like Figure 8.2*a* (page 240), we'd see a wiggly line that looks just as jagged as the original one. We can magnify just one day of a week's fluctuations, and then just one hour of that day, and then just one minute of that hour – and in each case we see the same kind of jumpy graph.*

This similarity of form or shape at different levels of magnification is a characteristic of the fractal structures mentioned in Chapter 5. It is a characteristic of many natural shapes, such as mountain ranges and coastlines. Mandelbrot gave fractals their name, and he was the first to deduce that the ups and downs of an economic market over time have fractal properties. But with his Lévy-stable distribution he over-simplified the precise mathematical form of those properties.

Just what *is* this form, then? The harder one looks, the subtler the question seems to be. No single kind of curve appears to fit the entire statistical distribution of market fluctuations – it depends, for example, on how large a variation (in percentage terms) you are looking at, and on what timescale. Although the distribution shown in Figure 8.3 seems to hold over a wide range of timescales, for very long time steps (several months, say) the distribution begins to look more like a gaussian curve. (That's one reason why we should be sceptical about any single theory that claims to explain how the market works.) Yet this statistical behaviour, however slippery, is consistent across the markets: the S&P 500 index shows much the same behaviour as the Japanese Nikkei index and Hong Kong's Hang Seng index. There

* More precisely, the *magnitude* of the variations does depend on the duration of the time intervals: a large return is more likely to accumulate over a day than to happen all at once. To allow for this, we need to rescale plots with different time steps. This is just like rescaling two probability distribution functions of children's heights – for a single school and the whole country, say – so that they both lie on the same gaussian curve (see page 116). Such rescaling doesn't affect the mathematical form of the curves – whether they are gaussian, Lévy stable or whatever.

does seems to be a universal statistical signature of the way capitalist markets work.

What remains contentious is whether major crashes are typical or atypical of market fluctuations. Clearly they are atypical in the sense that they don't happen very often. But that's exactly what the probability distribution of fluctuations implies: very big events are rare. The question is, do crashes fit onto the same curve as smaller fluctuations? Because crashes are uncommon, it is not easy to give a definitive answer. Some say that these extreme events fit the distribution perfectly, which means that crashes, while rare, are not anomalous – they are an inherent feature of the way markets behave, arising from the same fundamental processes that cause smaller fluctuations. Others are not so sure. For example, Mantegna and his colleague Fabrizio Lillo have shown that the short-term statistical distributions of returns on days when a market crashes, or when it rallies after a crash, differ from those on normal days – that is, there seems to be something unusual about the way the market is functioning on those days.

But whatever the precise statistics of market fluctuations, it is indisputable that they are not fully random (gaussian), and are more prone to attaining large values. What does that mean for economic theory?

THE RIGHT WIGGLES

One of the dreams of entrepreneur economists is to develop a theory which can predict the ups and downs of the market before they happen. This would allow traders to make profits indefinitely, buying and selling exactly when their calculations tell them the time is right. Real traders – at least, those with a serious understanding of market dynamics – have long since given up on this idea. They know that it cannot be achieved. Indeed, the impossibility of making completely accurate forecasts is enshrined in one of the central tenets of economics: the 'efficient market hypothesis'. One way of stating this hypothesis is that it is not possible to predict future stock prices from previous values. Note that it is couched as a hypothesis – a belief, if you will – for there is no rigorous proof that it is true.

Analysing the statistics of market fluctuations has shown that, empirically, the efficient market hypothesis holds water. The question of predictability hinges on the idea of correlations. To make an accurate forecast of the price of a stock based on its previous price, there would need to be some mathematical relationship between the two prices: technically speaking, they would have to be correlated. Imagine, for example, that a declining price over the past day invariably meant that the price would continue to fall at the same rate over the next day. Sometimes, of course, this relationship does hold – a stock can devalue steadily over the course of many days. But equally obviously, it is not a general, invariant relationship – if it were, then a stock which began to fall in price would inevitably decline until it was worthless. There is no telling whether a stock which is falling on one day will recover or continue to fall on the next.

Physicists have a mathematical tool for measuring correlations: it is called a correlation function. They use it, for example, to deduce to what extent the motion of a particle in a fluid is predictable on the basis of the motion of other particles. If the correlation function between two particles has a high value, the trajectories of the two will be closely related. For example, a mother and child holding hands and moving through a crowd are highly correlated in their motion. We can predict where the child goes by tracking where its mother goes.

The *auto*correlation function of a particle characterizes how its motion at any point in time is related to its *own* motion in the past. The autocorrelation function of a particle in a liquid has a high value over very short times, because the particle continues to move along its previous trajectory. But this value drops rapidly to zero at longer times, because collisions with other particles quickly randomize the motion, destroying all trace of the earlier trajectory. It is possible to calculate the autocorrelation function of an economic index or stock price, which tells us whether its future value depends at all on its past. Typically this analysis shows that the correlation falls to zero within the space of about five to fifteen minutes. For times longer than this, the efficient market hypothesis is therefore valid: the price has completely 'forgotten' its past.

One might imagine that by buying and selling in the few minutes during which correlations exist, one could use the past to forecast the immediate future and operate at a sure profit. Practically, however, this is not feasible – partly because of the finite time needed to make a forecast and then complete a transaction, but also because even the small transaction costs that are imposed will quickly erode any potential profits from exploiting very short-term correlations. Thus there is no magic formula that will let one play the market without risk. This fits with experience and with common sense. Yet it is a telling comment on the state of conventional economic analysis that at least one celebrated theory suggests the contrary. Robert Merton and Myron Scholes were awarded the 1997 Nobel Prize for Economics for the theory of options pricing that they developed with Fischer Black, who would surely have shared the award had he not died several years earlier. The Black–Scholes model, extended by Merton, more or less defines the thinking of dealers in options.

Options are basically a kind of insurance contract that protects dealers against the vicissitudes of the market or of world events. The buyer of an option pays a small sum for the right to buy or sell something at a specified price some time in the future. If that price falls short of the actual value at the time of the transaction, the 'writer' of the option makes up the difference. Options are a way of diffusing risk – or of passing it on to someone else. At least, that's the theory. Options can be used to insure against almost anything: variations in the weather (which might ruin a crop or an outdoor event), changes in exchange rates, or whatever might bring the buyer future losses through ill luck or acts of God. They are one form of the class of assets known as derivatives, which depend on the value of something else. The derivatives known as futures are agreements to buy or sell something at a given time for a price agreed now. They are essentially a form of gambling.

Derivatives have developed notoriety in recent years because of their high-risk behaviour, which has led to massive losses incurred by organizations or institutions such as Barings Bank and Proctor & Gamble. In 1997 alone, derivatives are estimated to have cost traders $2.65 billion worldwide. Options, however, are supposed to be relatively tame derivatives – thanks to the Black–Scholes model, which has been described as 'the most successful theory not only in finance but in all of economics'.[15] Black and Scholes considered the question of strategy: what is the best price for the buyer, and how can both the buyer and the writer minimize the risks? It was assumed that the buyer would be given a 'risk discount' which reflects the uncertainty in the stock price covered by the option he or she takes out. Scholes and Black proposed that these premiums are already inherent in the stock price, since riskier stock sells for relatively less than its expected future value than does safer stock. Based on this idea, the two went on to devise a formula for calculating the 'fair price' of an option. The theory was a gift to the trader, who had only to plug in appropriate numbers and get out the figure he or she should pay.

But there was just one element of the model that could not be readily specified: the market volatility, or how the market fluctuates. To calculate this, Black and Scholes assumed that the fluctuations were gaussian. Not only do we know that this is not true, but it means that the Black–Scholes formula can produce nonsensical results: it suggests that option-writing can be conducted in a risk-free manner. This is a potentially disastrous message, imbuing a false sense of confidence which can lead to huge losses. The shortcoming arises from the erroneous assumption about market variability, showing that it matters very much in practical terms exactly how the fluctuations should be described.

The drawbacks of the Black–Scholes theory are known to economists, but they have failed to ameliorate them. Many extensions and modifications of the model have been proposed, yet none guarantees to eliminate all the problems. It has been estimated that the deficiencies of such models account for up to 40 per cent of the 1997 losses in derivatives trading, and it appears that in some cases traders' rules of thumb do better than mathematically sophisticated models. Econophysicists such as Jean-Philippe Bouchard in France have proposed new models of options pricing that go beyond the Black–Scholes theory by taking into account the non-gaussian nature of fluctuations. These models banish the illusory notion of risk-free pricing. Bouchard's confidence in the role that physics can play in economics is reflected by his setting up a consultancy company, Science & Finance, which exposes his theories to the exacting demands of the marketplace.

Options pricing is just one area in which a better understanding of the statistics of the marketplace could transform economic theory. Another active area is the improvement of schemes for optimizing portfolios. Physicists use their methods to search for correlations between different stock prices and thus to find the least risky way of spreading investments over a market. If one invests in strongly correlated stocks, for instance, a downturn in one of them could sink the lot. For some of these purposes it is enough to have a more accurate description of market fluctuations. But in other cases, particularly for devising models of economic behaviour, one also needs to know where the fluctuations come from in the first place. It is to this question that we now turn.

9

Agents of Fortune

*Why interaction matters
to the economy*

If economics could become a true branch of science, it would
enormously increase our capacity to predict the course of events, as
well as the outcome of attempts to change that course . . .
[E]conomic science . . . would increase our ability to foresee the
consequences of changing the workings of the economic system,
and thereby to choose the most favourable course of action.

Robert Heilbroner (1999)

Napoleon was fond of telling the story of the Marseilles banker, who
said to his visitor, surprised at the contrast between the splendour
of the banker's chateau and hospitality, and the meanness of the
counting-room in which he had seen him, – 'Young man, you are
too young to understand how masses are formed, – the true and
only power, – whether composed of money, water, or men, it is all
alike, – a mass is an immense centre of motion, but it must be
begun, it must be kept up'; and he might have added, that the way
in which it must be begun and kept up is by obedience to the laws
of particles.

Ralph Waldo Emerson (1860)

Once, long ago, the economist Kenneth Boulding asked me, 'What would you like to do in economics?' Being young and brash, I said very immodestly, 'I want to bring economics into the twentieth century.' He looked at me and said, 'Don't you think you should bring it into the eighteenth century first?'

W. Brian Arthur

*

If Maxwell's statistical physics borrowed from social sciences, economists of the late nineteenth century had no qualms about returning the favour. But could economics really aspire to the accuracy and certainty of physics? Surely the precision possible in the world of atoms seldom extends to the sphere of fickle human behaviour? Was the phrase 'economic science' oxymoronic? When the historian Thomas Carlyle called economics 'the dismal science',[1] he did not mean that it was bad science but that often it led to unpalatable conclusions. Economists, however, soon enough found plenty that was dismal in the discrepancies between their models and the real world of trade and industry.

Francis Edgeworth was undeterred, proclaiming cheerfully that economists should not set their sights too narrowly. The goal, he said, was 'not so much to hit a particular bird, but so to shoot among the most clustered covey as to bring down the most game'.[2] In his *Mathematical Psychics* (1881) he confessed that belief in a 'social mechanics' – a mathematical economics – was partly an act of faith, which drew moral strength from the evident successes of physics:

'Mécanique sociale', in comparison with her elder sister [*mécanique céleste*: celestial mechanics], is less attractive to the vulgar worshipper in that she is discernible by the eye of faith alone. The statuesque beauty of the one is manifest, but the fairy-like features of the other and her fluent form are veiled.

254

But mathematics has long walked by the evidence of things not seen in the world of atoms (the methods whereof, it may incidentally be remarked, statistical and rough, may illustrate the possibility of social mathematics). The invisible energy of electricity is grasped by the marvellous methods of Lagrange; the invisible energy of pleasure may admit of a similar handling.[3]

'Pleasure', according to Edgeworth, was the force that propelled his hedonistic 'charioteers', the individual agents of society who (in his characteristically colourful imagery) interact like so many atoms in the void. He began to glimpse an economics that treated people like 'the multeity of atoms which constitute the foundations of the uniformity of Physics'.[4]

In this respect Edgeworth resembled his contemporary Alfred Marshall, John Maynard Keynes's tutor at Cambridge, in seeking an understanding of the economy from first principles. Marshall did for economics what Maxwell and Boltzmann did for thermodynamics, holding up to it a magnifying glass and attempting to reveal what individual particles were doing on the microscopic scale. This was the start of microeconomics, which underpins most economic theory today. 'Economics, after all,' says Robert Heilbroner,

does concern the actions of aggregates of people, and human aggregates, like aggregates of atoms, do tend to display statistical regularities and laws of probability. Thus, as the professoriat turned its eyes to the exploration of the idea of *equilibrium* – the state toward which the market would tend as a result of the random collisions of individuals all seeking to maximize their utilities – it did in fact elucidate some tendencies of the social universe.[5]

In other respects, Edgeworth and Marshall were very different. Marshall was a practical man, whereas in Keynes's evaluation Edgeworth was concerned not so much with using theory to formulate policies or to determine how we should conduct our affairs, as with discovering 'theorems of intellectual and aesthetic interest'.[6] Some would say that this tendency is alive and well today.

Economists in the early twentieth century were perfectly prepared, even eager, to accommodate analogies from the statistical mechanics devised by Maxwell, Boltzmann and Gibbs. The trouble was that they were the wrong analogies. As we saw earlier, physicists focused on the idea of equilibrium states. Likewise, economists wanted to believe in a steady market, gently ruffled by random noise. But the economic system is very clearly not in equilibrium. Its theorists are still coming to terms with this distinction. Nevertheless, by invoking images of particles in motion, economists were implicitly suggesting that they might emulate physics and formulate a description of their object from the bottom up. That is the subject of this chapter.

THE RATIONAL TRADER

Throughout much of the twentieth century, the scientific community was sometimes said to be afflicted by a disease called 'physics envy'. Scientists in disciplines outside physics wished their own subjects could boast the intellectual profundity, the mathematical agility and the foundational rigour that was evident in physics. Economists, failing perhaps to appreciate that theirs was the more difficult task, were not immune. Paul Krugman tells of an Indian economist who explained to his students his unorthodox theory of reincarnation: 'If you are a good economist, a virtuous economist, you are reborn as a physicist. But if you are an evil, wicked economist, you are reborn as a sociologist.'[7]

Edgeworth's dazzling mathematics may be seen as a compensatory ploy, as if aiming to show that even though economics could not attain the precision of physics, it could dress in the same clothes. Marshall's approach to economics, although rather different, also sounds a lot like physical science: theories are developed from fundamental postulates. But he cautioned against taking this analogy too far. 'Economics', he said, 'cannot be compared with the exact physical sciences, for it deals with the ever-changing and subtle forces of human nature.' Heilbroner expands on this:

[T]here is an unbridgeable gap between the 'behaviour' of [subatomic particles] and those of the human beings who constitute the objects of study of social science . . . aside from pure physical reflexes, human behaviour cannot be understood without the concept of volition – the unpredictable capacity to change our minds up to the very last moment. By way of contrast, the elements of nature 'behave' as they do for reasons of which we know only one thing: the particles of physics do not 'choose' to behave as they do.[8]

There are several problems with these assertions. Heilbroner is of course right to say that humans make choices whereas particles do not (although quantum physicists sometimes speak as though particles do). Yet the resulting element of indeterminacy need not in itself prevent social science from devising models of mass behaviour, as we have seen already. The early statisticians realized that regularities can arise in large populations even when their individual motives are unknown. And Heilbroner forgets that in many situations people are faced with only a very limited number of choices, and can therefore generate relatively predictable long-term patterns of behaviour even when each individual makes those choices quite freely. If the pedestrians in Dirk Helbing's 'human trail' simulations (page 168) were to arrive at random at any point on the perimeter of an open space, and cross to any other random point, no clear trail pattern would emerge. It is because the entry and exit points are constrained that order appears. The same is true in economics: traders on the market can exercise free will, but their choices are restricted to buying or selling at any point in time.

Heilbroner's caution about whether economics can be an exact science is quite understandable, and to some extent perfectly justified – for we have seen how haphazardly the markets fluctuate. Yet ironically, conventional economics has gone to the opposite extreme. Rather than throwing up their hands in despair at ever capturing the capriciousness of human behaviour in an economic model, theorists have assumed that humans act like perfectly predictable, rational automata. A part of the agenda of physicists is to inject a *more* realistic and less idealized description of human behaviour into the field.

How can it be that economists came to regard humans as automata? If I say that this is the only way they could make their models work, I do not mean that to sound cynical. It is simply the truth: there was no obvious means, until relatively recently, of dealing with the indeterminacy of human conduct.

But mathematics also played a part in squeezing the human element out of economics. The early economists, from Adam Smith to John Stuart Mill to Karl Marx, regarded themselves as moral and political philosophers. They mined their studies of the capitalist market system for implications for social and political behaviour. Many of the most eminent twentieth-century figures, such as Keynes and Schumpeter, did the same. But mainstream academic economists, following Edgeworth's lead, began to construct ever more elegant and abstract mathematical models in which there was no room for the messiness and clamour of the real world. Some of them asserted that to include noise in their models was to insult the intelligence of the market.

By what legerdemain, then, was free will spirited out of conventional economic theory? This theory has long assumed that individual market agents – the traders who buy and sell stocks and shares – are Hobbesian to a man or woman. They are out for all they can get, and seek only to maximize their own gains, whether this be the profit they make from their transactions or some more generalized well-being that economists call *utility*, which includes factors such as risk and security.

This seems reasonable enough. Why trade on the stock exchange if not for gain? The problems become apparent once we consider how these agents go about maximizing their utility. The traditional view makes them pre-programmed, omniscient computers. It assumes that each agent has a fixed objective, and pursues it in a wholly rational way. He or she assimilates, at every instant, all the information available about the state of the market, and uses that to calculate the next best move, like IBM's Deep Blue computer playing Garry Kasparov at chess. The agents are, in other words, perfectly informed 'rational maximizers'.

The attraction of this assumption is that it makes it easy to see what agents will do next. Their behaviour is predetermined by a set of rules. Traditional models then go on to assume that the factors influencing prices are unpredictable and external to the market itself, so that the fluctuations are random (that is, generally gaussian) and beyond the control of the traders. The traders simply respond to these imposed ups and downs in whichever way allows them to maximize their utility at that particular moment.

Although prices fluctuate unpredictably, this does not mean (within the conventional theory) that the underlying factors influencing prices are wholly mysterious. Rather, changes in asset prices are assumed to reflect the (unpredictable) changes in the 'fundamentals' – the causative forces that determine value. The fundamental value of a company, for example, is the value of all future dividends it will pay. The price of a company share is assumed to reflect this value.

Does that seem a little strange? How on earth can one know how the company's fortunes, and thus its dividends, will fare in the future? Of course, one cannot know. This is one of the leaps of faith that traditional microeconomic theory is forced to take to keep the maths tractable. A second such leap can be identified in the assumption that information about fundamentals is simultaneously available to all agents in the market, and is incorporated instantly into asset prices.

These beliefs underpin the efficient market hypothesis (page 248), a precondition of which is that asset prices change only when fundamentals change – that is, when new information becomes available. And because this information is available to all, and because all agents know exactly how best to maximize their utilities, no one can exploit it to the detriment of others. So there is no sure-fire way to secure a market advantage. Two other aspects of this picture are worth mentioning, since we shall revisit them later. First, all agents are identical: all are rational maximizers playing by the same rules. Second, all have fixed preferences: they never change the beliefs that govern their trading decisions.

No one will deny that these assumptions are all simplistic. The question is whether their simplicity matters. Finding ways to simplify complex problems is the hallmark of good science. Often scientists make approximations that would seem absurd to outsiders, yet find that the resulting theories work surprisingly well. Einstein once said that scientific theories should be made as simple as possible, but not simpler. That, in many respects, is the test of a good theory. So we should not be too quick to disparage traditional microeconomics for presenting this cartoon view of the market. Cartoons often carry all the information one needs in order to follow a story. But the critical test is whether the theory measures up against reality. The answer to that depends on what you are measuring.

The concept of rational individuals maximizing utility and profits and operating with perfect foresight is one that dominates conventional microeconomic theory, which is called neoclassical micro-economics. This picture is enshrined in the so-called real business cycle (RBC) theory, which seeks to explain the ups and downs of the market. It is a rather bland, not to say dissembling, picture that RBC theory offers. It proposes that the causes of the business cycle are exogenous: they are imposed from outside the market system in the form of a series of random shocks delivered by technological progress. The market simply responds to these shocks.

The most unsatisfying aspect of the theory is that it seems designed to give us back only what we put in. The statistical properties of price fluctuations depend heavily on what one assumes about the random-ness of the shocks. Inevitably, perhaps, the usual assumption is that they obey gaussian statistics. This then generates economic fluctu-ations with particular characteristics, which can be compared with market data. Superficially, RBC theory gives plausibly erratic price curves, but on closer inspection the statistics come out wrong. For example, the autocorrelation function (see page 248) does not match that of the real data.

A deeper criticism follows from the demonstration by economist Robert Shiller in the 1980s that asset prices are not necessarily tied to their 'fundamental' value (the supposed 'true' value that they would attain in an equilibrium market), as RBC theory assumes. The prices change more erratically than the fundamentals: in market terminology, they are more volatile. If asset prices vary while the fundamentals do not, what drives this change? Why are traders paying different prices for goods even though their underlying value has not altered? Clearly this is at odds with the idea that agents always act rationally.

In fact, we hardly need a detailed analysis of the statistics to tell us that agents sometimes act irrationally. If prices rise above the level that the fundamentals should dictate, they are 'overvalued', and common sense should tell traders not to buy. But buy they do. This prevents the market from regaining its instant equilibrium through the conventional laws of supply and demand. Irrationality, it seems, is rife. Again, this comes as no surprise to economists. We should not imagine that they have somehow been deluded into a deep misconception of human behaviour. They know that markets are not complete, that people are not rational, that they make myopic decisions and lack full information, and that they are different from one another. The question is, what can one do about it?

IGNORANCE AND BELIEF

There is an unfortunate tendency for outsiders to deride economists' neglect of irrationality, as though to imply that they are psychologically unsophisticated. One recent critic (not a physicist) claims that:

Mainstream economic theory, as set out in countless textbooks, stipulates that a perfectly knowledgeable man or woman will always make a rational and fully informed decision on what to buy and what to sell and at what price . . . this starting proposition is dotty.[9]

But to this accusation, economist John Kay rejoins:

Indeed it is. This is why mainstream economic theory today does not stipulate anything of the sort. The economics of imperfect information has been an important research theme – perhaps *the* most important research theme – of mainstream economics in the last 30 years.[10]

This is a valid defence. Many economic models are now commonly couched in terms of 'bounded rationality': rational decisions made in the face of constraints, such as incomplete information, that simply try to make the best of circumstances.

All the same, we must be careful to clarify exactly what we mean by 'imperfection' or 'uncertainty'. You can still make a rational best decision on the basis of incomplete information, by weighing up probabilities about the nature of what it is you don't know. The calculus of trade and risk in the face of an uncertain future or incomplete information is central to the work of many leading economists, such as Nobel laureates Harry Markowitz, Joseph Stiglitz, George Akerlof and Michael Spence. Yet much of the effort in this area has nonetheless gone into finding optimal (and thus rational) strategies in the face of the market's unpredictable fluctuations.

Some economists accept that the uncertainties faced by traders and businesses are so great that they give up the futile quest for an optimal solution, and instead accept one that is merely *good enough*, as judged against some set of criteria. They are said to 'satisfice' rather than to maximize. Again this assumes that decisions are basically rational, albeit laced with a strong dose of reality. Some believe that the difference between maximizing and satisficing is in any case rather small.

Keynes, however, doubted that humans are readily capable of optimal or even particularly rational choices:

a large proportion of our positive activities depend on spontaneous optimism rather than on a mathematical expectation . . . a spontaneous urge to action rather than inaction, and not as the outcome of a weighted average of quantitative benefits multiplied by quantitative probabilities.[11]

In other words, you can devise all the complicated formulas you like, but in the end most of us – including traders – are guided by instinct and impulse, by what Keynes called 'animal spirits'. Top-level management decisions, for example about a company's policies on recruitment, investment, diversification or specialization, rely a great deal on such subjective judgements, based less on the mathematics of economic rationality and more on their chief executive officers' years of experience.

One of the prophets of this perspective on economics was the American Thorstein Veblen, whose thinking was shaped by the harsh, pugilistic climate of the New World economy. He envisaged the business world as irrational to the point of barbarity – a world where behaviour was governed as much by custom and foolishness as by planning and logic. Entrepreneurs, he believed, positively welcomed the unpredictabilities of the market, since these offered opportunities for profit that a staid economy did not.

It is indeed rare to find the wolfishness demanded of most market traders unaccompanied by a degree of impulsiveness that pays no heed to cool calculation. But perhaps the most productive way of accounting for the irrational element in trading behaviour is simply to concede that different individuals, armed with exactly the same information, will tend to act in different ways. These ways are by no means unconsidered. Rather, irrationality tends to manifest itself in terms of different beliefs about how to respond to circumstances. Thus there is heterogeneity in the market: not all agents are alike. There is no consensus as to what the best course is. Several economists have recently begun to build this aspect into their models. In 'discrete choice theory', for example, each agent chooses one of several alternative actions, each with a certain probability.

By and large, however, views about the role of rationality in economics diverge along the fault line between academic economists and market traders. The academics have traditionally preferred the idealized abstractions championed by Edgeworth. Their models describe a world quite alien to that inhabited by the traders themselves, for whom Veblen's hawkish and opportunistic picture rings true. Neoclassical economic theorists cleave firmly to the notion of an equilibrium market where prices always reach their proper level and there are virtually no opportunities to exploit market trends. For their omniscient traders, a strategy which works better than others would be adopted instantaneously by all, squeezing out any advantage it offers. Market traders, on the other hand, live by their wits and cunning, each believing they can discover an unforeseen avenue for profit. They speak as if the market has a personality of its own – sometimes nervous and jittery, at other times confident and brash. Some traders dismiss academic economics as fantasy. George Soros, an undoubted expert on market-floor 'practical' economics, says that 'it may seem strange that a patently false theory [conventional neoclassical economics] should gain such widespread acceptance'.[12]

One of the first economic theorists to introduce belief and choice into modelling was Alan Kirman of the University of Aix-Marseilles in France, who in the 1990s proposed that the inclinations of traders could be encapsulated within two general categories. The *fundamentalists* are the traditionalists: those who adhere to the rationalist dogma that prices reflect fundamentals, and who will buy and sell accordingly. The *chartists*, on the other hand, take a more empirical approach. They think that future prices can be predicted from their past behaviour. Some might use rules of thumb, or intuition; others will rely on complex formulas from probability theory. But all are essentially optimists, believing that they need not be at the mercy of a market fluctuating purely at random. Chartists are also sometimes called 'noise traders', reflecting their conviction that there is valuable information in the market fluctuations (noise) which can be teased out if you know how.

In order to model the irrational tendencies of agents, Kirman did not consider it sufficient simply to introduce heterogeneity into microeconomics. He needed something else as well. Once several options are open to us, we cannot ignore the fact that our own choice is inevitably influenced by what others do. We have to take on board the factor that represents probably the most egregious omission of conventional neoclassical theory, and which brings us most firmly into the realm of statistical physics. That factor is *interaction*.

FOLLOW YOUR NEIGHBOUR

The economic system is a supremely interactive one. Traders influence one another directly: a rush to buy or sell a particular asset can prompt others to do the same. It seems intuitively clear, simply from looking at pictures of faces on the trading floor, how major crashes are stampede phenomena in which individuals respond to the mood of the market in a herd-like and sometimes panic-stricken manner. Yet microeconomic models which ignore interaction insist on a different interpretation: one in which crashes are driven by some exogenous fluctuation that is beyond control, or in which agents all decide independently and simultaneously on the same course of action.

Moreover, agents influence one another indirectly. The choices they make have a direct effect on prices – which in turn affects the choices of others. As an engineer would say, there is strong feedback at work. Whereas traditional models assume that agents adopt their (rational) strategies in response to an externally imposed evolution of prices, in reality the agents help to make those prices as well as responding to them.

Here again there is the danger of caricaturing the way that economists think. As John Kay points out, 'The idea that the behaviour of participants in markets is influenced by what happens in markets' has been explored in 'literally thousands of books and articles written by economists.'[13] What physics has to offer microeconomic modelling is not new insights into the factors that control the markets, but new tools with which to accommodate them. Physicists have been dealing with systems of many interacting particles for over a century. It would be foolish to assume that these tools can necessarily be translated directly into economic terms. But equally, it would be surprising if some of the phenomena already well understood in physics should not turn up in some form in economics.

The man who first introduced interactions into microeconomics was a mathematician familiar with both physics and economic theory. In 1974, Hans Föllmer of the University of Bonn in Germany concocted an 'interacting-agents' model of the economy which was based on the principles of the Ising model of magnets (page 111). As we have seen, in this simplified description of a magnetic material the magnetic atoms all sit on a regular grid, and they 'make choices': they align their spins either in one direction or in the opposite direction. These choices are interdependent: the alignment of each atom depends on those of its neighbours, since their magnetic fields exert forces on one another. In Föllmer's model, each atom represented an agent faced with choices about how to trade. The same idea is now widely applied by economists and econophysicists who, like Alan Kirman, are seeking to extend traditional microeconomics by using interacting-agents models. The predictions of these models depend on the rules that govern the interactions. Föllmer found that his model generated more than one stable state – more than one macroeconomic landscape – just as the Ising model offers two magnetically aligned states. This was already food for thought for economists weaned on the idea that the market has a particular, unique equilibrium.

Föllmer's approach was extended in the 1990s by US economists William Brock and Steven Durlauf, both literate in modern physics. They formulated their model of interacting agents making binary choices in a manner which paralleled the way Pierre Weiss had modelled the magnetic phenomena studied by Pierre Curie (page 108). The Curie–Weiss model makes a so-called mean-field approximation: it assumes that each atom is influenced by the average effect of all the others, not just by its near neighbours. In physics this is something of an oversimplification, the consequences of which become particularly apparent near a critical point. But in economics, 'global' interactions are more feasible. Traders on the London Stock Exchange are influenced by others on the same floor; but they can also, thanks to telecommunications, be aware of the trading patterns on the New York or Tokyo exchanges. So for economic modelling, a mean-field theory might be quite a realistic approximation.

The idea of a 'mean field' that enables each individual to know and respond to what all the others do is also apparent in attempts by some economists to handle interactions using game theory – a mathematical description of how people make choices in competitive games, which we shall investigate in Chapter 17. In such models, traders try to second-guess what other traders might do. Game theory has yielded some useful insights into the non-rational aspects of market trading, but it has not furnished an explanation of why the market fluctuates the way it does.

Yet even in economics the mean-field assumption cannot be entirely sound. No trader can possess an instantaneous global over-view of the market. There is bound to be some degree of localization of interactions between agents. In particular, the choices each agent makes are more likely to be influenced by others trading in similar assets, no matter where they are geographically, than by what a distantly related market is doing.

With such issues in mind, Alan Kirman has focused attention on the question of exactly how information propagates on the trading floor. The key issue, he says, is what the structure of the information network is like: 'It seems to me that the question of how economic networks evolve is one of the most important if we are to begin to understand how markets come to be organized.'[14]

Several microeconomic models of interacting agents have assumed that the links between agents are random. That is to say, each agent has an equal probability of being 'connected' to any other random member of the trading community. Individuals are influenced only by those others to whom they are connected. Kirman proposed this kind of communication network in 1983. He found that network effects in interacting-agents models can lead to clustering: groups of traders develop who buy and sell primarily among themselves, having very few interactions with others outside the group. We shall look at network structures in Chapter 15, where we shall see that this assumption of randomness is not the only, and not necessarily the best, way to describe many of the social and business networks that people form.

In addition to clustering, interactions between agents can lead to an effect known as *herding*: the tendency for traders to mimic one another. Episodes of apparently irrational economic behaviour prompted by a collective mania that sweeps through the market are notorious in economic history, not least because they defy the basic assumptions of theorists. In seventeenth-century Holland, for example, tulip prices ballooned absurdly, while conventional economic theory says that the laws of supply and demand should have stabilized the price.

John Maynard Keynes again got here first. In the 1930s he likened economic markets to the kind of beauty contests that were then appearing frequently in the popular press. Readers would be presented with a selection of 'beauties' and asked to guess which of them would gather the most votes from other readers. This is, of course, subtler than asking readers simply to nominate the 'most beautiful' – it requires some intuition about how others will cast their votes. Keynes supposed that this sort of competition would create herding behaviour, but he could not find a way to describe it theoretically.

In the 1980s Robert Shiller considered how herding behaviour might influence market dynamics in a quantitative way. He was interested in what controls the moment-to-moment variations in the amount of trading. Economic data commonly show a property called volatility clustering, in which big fluctuations come in bursts separated by relatively quiescent periods. During the bursts, the market is very active. It seems possible that these bursts are the result of herding behaviour, which prompts increasing numbers of traders into frantic buying or selling. But there remains the underlying question: where do the fluctuations come from?

NOTHING FUNDAMENTAL

We have seen how modern theories, as represented by the real business cycle theory, remain resolute in their determination to banish fluctuations from the model and place them outside as a 'given', the result of, say, changes in technology. Interacting-agents models, on the other hand, can supply an endogenous explanation for the fluctuations, showing that the 'spiky' behaviour evident in share prices or economic indices (see Figure 8.2a on page 240) can arise even in a system disturbed by nothing more than gaussian noise. As Alan Kirman put it:

Models which take account of the direct interaction between agents allow us to provide an account of macro phenomena which are caused by this interaction at the micro level but are no longer a blown-up version of that activity.[15]

In other words, the fluctuations in an economic index due to the activities of thousands of traders need not be simply a scaled-up version of the random forces to which each individual trader is subject.

This has been demonstrated by the economist Thomas Lux, formerly of the University of Bonn and now at Kiel in Germany, and physicist Michele Marchesi of the University of Cagliari in Italy. In 1998 they used Alan Kirman's model of fundamentalists and chartists to probe the origins of variations in asset prices. The chartists were divided into two groups: optimists, who will buy additional units of an asset in hopeful anticipation of rising prices, and pessimists, who sell more units than their charts dictate in gloomy expectation of sinking prices. Both groups of chartists take the behaviour of the other traders into account when they make their calculations. Chartists can switch between an optimistic and a pessimistic outlook, displaying herd-like tendencies which are dictated by the majority opinion: when most chartists are optimists, for example, the pessimists are more likely to convert. Moreover, chartists of either persuasion can become fundamentalists and vice versa. The agents make these choices by considering which strategy happens to be the most profitable at the time: if another camp is doing better, there is a certain probability that they will join it.

The changes in asset prices in this model are determined by what the traders do, according to the normal laws of supply and demand. The driving force for change, meanwhile, is the variation in funda-mental values. Lux and Marchesi assumed that these fluctuations are gaussian. Whether that is so in reality is immaterial for these purposes; what the researchers wanted to test was the idea, central to conven-tional microeconomic theory, that fluctuations in prices reflect those in fundamentals – the efficient market hypothesis. If this idea holds, the prices that emerge from the model should also show gaussian variations.

It was entirely possible that the model would merely generate economic nonsense. But in macroeconomic terms it turned out to be

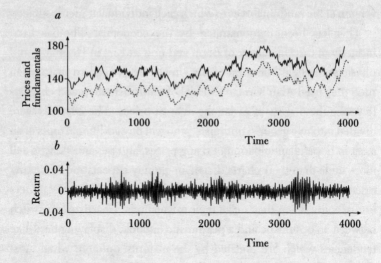

Figure 9.1 The economic model developed by Lux and Marchesi shows how random variability of the 'fundamentals' that drive price changes gets converted by trader interactions into non-random price fluctuations. (*a*) The *long-term* variation of prices (solid line) shadows that of fundamentals (dotted line), showing that the model is 'properly behaved'. But . . . (*b*) The variations in *short-term* returns (for a definition of returns see Figure 8.2) has a non-gaussian statistical distribution – that is, the fluctuations are non-random, with more large spikes.

perfectly well behaved. Over the long term the market was 'efficient' – prices and fundamentals both varied considerably but stayed more or less in step (Figure 9.1*a*). In the short term, however, the story was quite different. The returns, which reflect the price fluctuations, varied in a decidedly non-gaussian way (Figure 9.1*b*). In other words, interactions between the traders were converting a gaussian 'input' (the fundamentals) into an 'output' (prices or returns) with quite different statistical properties, in which extreme fluctuations were much more common than the input alone would seem to require.

Moreover, the non-gaussian probability distribution for short time steps changed into a gaussian one for long times – just as is found in real market data.

Lux and Marchesi found just the kind of volatility clustering seen in reality. Periods of high volatility (large fluctuations) coincided with periods when most traders were chartists. In other words, a predominantly chartist strategy destabilizes the market, causing bursts of trading. But the model contains a compensating mechanism which restores stability. When fluctuations are large, prices can differ substantially from those dictated by fundamentals. A chartist will simply trust to whatever trend happens to be prevalent at the time, whereas a fundamentalist will exploit this divergence as an opportunity to profit. Thus the fundamentalist strategy will fare better and will come to be seen as preferable, prompting chartists to change camps. This sets a limit on how wild the market can get.

Thus Lux and Marchesi showed how the characteristic fluctuations of the market can arise endogenously – from inside the system. Paul Ormerod has constructed a similar interacting-agents model in an attempt to account for variations in the industrial productivity of nations, typically defined by their gross national product (GNP), the total output of all the nation's businesses. In this model the autonomous agents are not individual traders but businesses. Just as market traders keep an eye on what one another are doing, so do businesses. They adjust their output for each economic period (generally a quarter year) according to short-term forecasts of demand which take into account the perceived 'mood' of the markets.

Fluctuations in productivity are essentially a surrogate for the business cycle, the erratic series of booms and slumps reflected in swings in financial indices such as the S&P 500. We saw earlier that the conventional 'real business cycle' predicts the wrong statistical properties for these fluctuations. Ormerod's model, in which variability is driven by nothing more than gaussian noise, produces a much more realistic result: the fluctuating total productivity displays statistics close to those seen in reality. If Ormerod modified his assumptions about how firms are distributed in size, the fluctuations retain the same statistical character but may change in amplitude: the 'mountains' get flatter or more rugged. The amplitude gets larger if the distribution is skewed in favour of a few big firms, while the variations are smoothed by an industrial profile composed of many small firms. This is a sober message in the light of the tendency for small companies to get swallowed up by a handful of big ones: in such an economy, we must expect deeper, more severe recessions. In this sense at least, a 'healthy' market is a diverse one.

There is no unique way to model the shifting and sometimes irrational beliefs of market traders or businesses. The psychology behind these beliefs, no doubt a complex mixture of blind faith, wishful thinking, careful data analysis, past experience, and much else, is probably impenetrable, or at least impossible to express in mathematical terms. But that need not matter. As economist Brian Arthur and his co-organizers of a recent workshop on 'complexity' in economics have said,

How individual agents decide what to do may not matter very much. What happens as a result of their actions may depend much more on the interaction structure through which they act – who interacts with whom, according to what rules.[16]

What most of these interacting-agents models reveal is that, once rational maximizers are jettisoned, the myth of an equilibrium economy vanishes too and is replaced by something that looks much more like the real world: a market that fluctuates wildly and is prone to crashes. Arthur and his colleagues, for example, have developed a model in which heterogeneous agents trade on the basis of a whole range of strategies and expectations, which the agents revise continually. Successful strategies are retained; those that perform badly are discarded. If the agents rarely switch between alternatives, the economy resembles that predicted by rational neoclassical theories. If the agents revise their behaviour with a frequency which the researchers considered to be more realistic, the market descends into the messy, ever-changing and unpredictable state that it seems to inhabit in real life. In other words, the economy does not appear to be guided very much by rationality.

LET IT BE?

If interactions between traders or between firms can provide an explanation for the furious ups and downs of the market, it is sheer folly to imagine that the market can be engineered to iron out blips on timescales of days or weeks, or perhaps even longer. Yet many governments continue to believe that this kind of manipulation is both possible and desirable. Most probably it is wasteful of resources, if not positively harmful to the economy.

It goes without saying that market fluctuations, particularly recessions, have serious social as well as economic consequences. Economic growth is linked intimately with employment: in a slump unemployment is high, and in a boom it is low. Keynes advocated that during a major, extended slump like the one that followed the Wall Street Crash, governments should inject money into the economy to stimulate it back into a growth phase. Keynes feared that, left to itself, the economic system might spiral into a decline which would ultimately leave it frozen, unable to bounce back through the normal business cycle.

There are good reasons for believing that Keynes was correct:

government intervention may be the antidote to a recession. Certainly, it seems that by pumping money into the economy after the 1987 crash, the US Federal Reserve may have helped to avoid a slump. But the questions about precisely how Keynesian intervention might (or might not) work are complex, hingeing on such issues as the lags between increases in monetary supply and their effects on the market, the question of how to define 'monetary supply' in the first place, and the expectations generated by interventionist policies. These complexities are beautifully explained in Paul Krugman's *Peddling Prosperity* (1994).

Recessions, however, are extreme events that call for extreme measures. The microeconomic models we have discussed suggest that governments that worry about short-term, moderate fluctuations in unemployment might be like meteorologists fretting about a wet week in July. Such variations are simply part of the system, and the most effective response is not to try to micro-engineer the economy but to alleviate the temporary miseries of unemployment. If there were greater public understanding of the inevitability of such fluctuations, governments would no longer have to fear the opportunism of opposition parties seeking to make currency from transient economic downturns (while ignoring the occurrence of the same phenomena the last time they were in power). This is not defeatism, but a proposal for effective focusing of resources. By the same token, we would do well to be more sceptical when governments claim spurious credit for short-term economic improvements which, in all likelihood, have nothing at all to do with their policies.

On the other hand, there is perhaps a greater danger today of falling prey to the belief that the market should not have to endure any interference whatsoever. As I indicated in the previous chapter, free-market fundamentalists argue that total non-interventionism is the best way to let the economy reach equilibrium. In that blissful state all prices are supposed to reach their 'true' value as determined by Adam Smith's laws of supply and demand (the 'market-clearing' price), and all goods are assumed to be allocated efficiently to their optimal social uses.

The appeal of laissez-faire economic policies is long-standing. Decades before *The Wealth of Nations*, Charles Davenant concluded that 'Trade is in its Nature free, finds its own Channel, and best directeth its own course'[17] – a hydrodynamic metaphor which carries with it the notion of prices finding their own level. While Smith himself felt that trade should not lose sight of elementary justices, he nevertheless asserted that 'we may often fulfil all the rules of justice by sitting still and doing nothing'.[18] If that is not laissez-faire, what is! To the conservative Edmund Burke, all trade regulation was 'senseless, barbarous and, in fact, wicked'.[19] That is a view by no means more extreme than those of contemporary right-wing economists.

We can now see that the idea of an equilibrium market is absurd. No process or system which fluctuates in such an undisciplined manner can be considered to approach anything like equilibrium. What is more, it seems no longer possible to blame these disturbances on external influences which perturb an otherwise well-behaved system. The interacting-agents models of microeconomics suggest that the economy is intrinsically unstable.

All the same, some might say, the market responds so flexibly and so rapidly that it is instantaneously optimal even as it changes. In other words, it still distributes goods with maximum efficiency from moment to moment. But even this pervasive idea, the last refuge of Smith's alluring and enduring notion of an 'efficient market', has now been challenged. A highly complex and sophisticated interacting-agents model called Sugarscape, devised by Joshua Epstein and Robert Axtell of the Brookings Institution in Washington, DC, and described in Chapter 14, reveals that, once trade is confined to realistic agents lacking infinite knowledge and access to all other traders (that is, if they are not omnipotent, omnipresent and omniscient), the distribution of goods is less efficient than it could be. In other words, some agents are unable to procure goods which they both desire and can afford.

Moreover, the Sugarscape model suggests that, while trade increases a land's 'carrying capacity' – the free exchange of goods makes a region able to sustain a larger population – it also skews the distribution of wealth towards greater inequality. In other words, trade seems inevitably to place most wealth in the hands of a few – a situation discussed in the next chapter.

This may be the price we must pay for capitalism, for experience tells us that a rigidly controlled economy (like that of the former Soviet Union) works against the interests of both general welfare and efficiency. This is borne out by econophysics. Sorin Solomon and colleagues at the Hebrew University of Jerusalem in Israel have shown that attempts to enforce a worldwide equality of wealth are apt to make that wealth dwindle uniformly to zero. And an interacting-agents model of the economy devised by Zdzislaw Burda at the Jagellonian University in Cracow, Poland, and his colleagues suggests that 'socialist' economies which severely restrict trade are more apt to collapse into a state in which a large proportion of the wealth ends up in the hands of one person. In other words, they are more prone to corruption. Corruption is generally regarded simply as a human failing, but it seems that some economic systems are more likely than others to create the preconditions for it.

Yet extreme capitalism is perilous too. After all, cautionary tales of mega-companies such as Enron and WorldCom hardly encourage the belief that corruption is unique to socialist economies. And Solomon and his team have shown that once markets become global, there is an increased risk of all the wealth getting concentrated in one place. This is not only morally intolerable, but dangerous for the market, which is then much more liable to collapse catastrophically than if such 'wealth condensation' can be spread over many different places. 'Extreme capitalism and extreme socialism', they say, 'are equally counter-productive and possibly disastrous.'[20]

Findings like these are based on models which can justifiably be called schematic, even crude. They should not be seen as anything more than suggestive. Should we believe them at all? Let me pose the question a different way. Can the political right provide any compar-able justification, based on theory rather than ideology, for their advocacy of wholly unfettered markets? Doubtless such markets work out very nicely for some players, but they do not appear to be the utilitarian ideal the (libertarian) right might have us believe. 'Certain economists', say Epstein and Axtell, 'ascribe nearly miraculous powers to markets', insisting that all government intervention simply prevents the market from realizing its self-regulating, beneficial qualities. This 'is also, unfortunately, a position frequently promul-gated in policy circles, especially when there is no econometric or other evidence upon which to base decision making.'[21] It is time to recognize such claims for what they are: expressions of faith, unhindered by facts and based largely on predetermined views about the role of governments, taxation and legislation.

So deeply entrenched is the free-market philosophy in US eco-nomic theory today (I am talking here about the pundits who exert a real influence – the TV analysts, the *Wall Street Journal* op-ed columnists, the think-tankists, and all too often the White House advisers – but not the academic economists*) that the supporters of

---

* All these groups are typically labelled economists, but they have little in common. Paul Krugman prefers to call the non-academics 'policy entrepreneurs', and the academics 'professors'. The professors, he says, include plenty of right-wingers and market fundamentalists, but their ideas remain rooted in theory rather than political dogma.

this creed are hoping even to ride out the catastrophic stock market collapse that is proceeding at full throttle at the time of writing. They place the blame on a few corrupt CEOs, on government policies, on small and fickle investors, on labour unions, on left-wing critics who spread doubt and negative thinking – anywhere but on the market itself. If only all these people would *behave*, say the free-marketeers, stocks would keep rising for ever.

This transparent nonsense will probably always find a voice when times are good: when the economy is buoyant, extreme libertarians will consider themselves vindicated. Slumps will be touted as a thing of the past – until the next bubble bursts. The truth is that economic ups and downs are an intrinsic feature of the capitalist game, and cannot be conjured away by political ideology. They are a part of the natural rules of the economy – for such rules surely exist, even if we have not yet fully uncovered or explained them. If we wish to do so, it seems likely that physics-based models can assist, and so help to prevent us from Canute-like posturing before the waves instead of rescuing those in danger of drowning.

10

Uncommon Proportions

Critical states and
the power of the straight line

... the reader will appreciate the orderliness of the lines ... and he
will see how this orderliness points to the existence of a fundamental
governing principle.

George Kingsley Zipf (1949)

The aesthetics of natural science and mathematics is at one with the
aesthetics of music and painting – both inhere in the discovery of a
partially concealed pattern.

Herbert Simon (1996)

Much of the real world is controlled as much by the 'tails' of
distributions as by means or averages; by the exceptional, not the
mean; by the catastrophe, not the steady drip; by the very rich, not
the 'middle class'. We need to free ourselves from 'average' thinking.

Philip Anderson (1997)

*

One might imagine that the best way to make a fortune on the stock
market is to devote one's life to studying it. For economists John
Maynard Keynes and David Ricardo, knowledge paid: both men made

fortunes as speculators. (It is not clear, however, whether their theoretical ideas played a stronger part in their success than did their shrewd instincts.) By contrast, Ricardo's friend Thomas Malthus, although a professor of political economy, never developed an aptitude for playing the market. By letting excessive caution guide his investments, he missed his chance to profit fabulously from Wellington's defeat of Napoleon.

It would certainly be a splendid way to gauge economists' faith in their own ideas if we were to demand that they test them out of their own pockets. A cynic might expect to see a substantial thinning of the literature on market forecasting if this were a criterion of publication. In 1995 the French scientist Jean-Pierre Aguilar had the rare courage to stake his money on the proposal that there is physics in economics. He was persuaded by a physics-based model of market crashes to buy options on the account of a fund management company which traded on the basis of such models. The model predicted a crash in Japanese government bonds in May of that year. It never happened, and Aguilar needed to engage in some delicate counter-trading to avoid losing his stake. Not surprisingly, Aguilar was among the sceptics when the claim was made in 1998 that a similar econophysics technique had retrospectively 'predicted' the crash of October 1997. This predictive method suggested an intriguing idea: that crashes are closely akin to critical points in statistical physics.

The notion that market dynamics are somehow redolent of phenomena at critical points has a broad constituency. In order to see the connection, we need to look again at this strange location in the landscape of many-particle systems. Many of the most powerful and startling insights in statistical physics stem from investigations of this unique place. The critical point motivated van der Waals' exploration of 'continuity' between liquids and gases. Many physical systems, such as magnets and superconductors, must pass through this portal as they undergo global changes of state. Critical points are like the black holes of statistical physics, for everyone in the field gets drawn to them sooner or later. But usually they emerge on the other side with a far deeper and richer appreciation of how this branch of physics unifies the physical world.

Indeed, it is now fashionable to see critical points everywhere: in earthquakes, evolution, forest fires, even in hospital waiting lists and world wars. The persuasiveness of such claims varies enormously, but it is fair to say that many of the characteristics of critical-point phenomena – such as extreme sensitivity to fluctuations, 'scale-free' events and, in particular, a special form of probability distribution – are found widely both in nature and in the world of human affairs. Critical points are at least a good metaphor, and sometimes rather more than that, for the strange combination of the unpredictable and the rule-bound that governs much of our lives.

PHYSICS ON A KNIFE-EDGE

Van der Waals' theory explained the existence of a critical point at which the liquid and gaseous states of a substance become indistinguishable. But the theory could not account for some of the strange phenomena found there. For one thing, fluids close to the critical point become cloudy (this is known as critical opalescence), and no one knew why.

An experimental peculiarity which the theory did embrace was the extraordinary sensitivity of the critical point. A system near its critical state becomes extremely responsive to disturbances. If you squeeze a substance, it shrinks in volume. The resistance it offers to this compression is a measure of its so-called compressibility. A rubber ball is more compressible than a steel ball, and a gas is typically much more compressible than a liquid – one can squeeze it more easily. At the critical point of a liquid and gas, the fluid becomes absurdly compressible – in fact, more or less infinitely so. In principle, the gentlest squeeze is sufficient to collapse a critical fluid into invisibility. This sounds absurd, and experimentally one can never observe such extreme behaviour because maintaining a substance exactly at its critical point is too difficult – the critical state is too unstable. But one can see the compressibility start to increase very rapidly as the critical point is approached.

The same is true of a substance's response to heat. To raise the temperature of a system, you have to put energy in. The amount of heat energy needed to raise the temperature by one degree is called the *heat capacity*. Water has an unusually high heat capacity, which is why it takes a long time – a lot of heat input – to boil a kettle. At a critical point, a substance's heat capacity becomes infinite. This means that it becomes like a kind of infinite heat sink: you can heat it as much as you like but the temperature doesn't change at all. A critical point separates ultra-cold liquid helium from the weird, inviscid state known as a superfluid (page 118), and the sudden increase in the liquid's heat capacity at around 2°C above absolute zero is the tell-tale signature that it is approaching its phase transition to a superfluid.

These bizarre behaviours are called *divergences*: some property of the substance diverges off to infinity at the critical point. Van der Waals' theory predicts divergences in the compressibility and heat capacity of fluids at the critical point. It explains why things get out of hand there. There is a convenient quantity called the *critical exponent* that specifies these rates of divergence. By measuring how rapidly a quantity such as heat capacity increases as the temperature approaches the critical temperature, scientists can easily calculate the critical exponent. The astonishing thing is that this exponent is the same for all fluids. The critical exponent that characterizes the divergence of compressibility is different from that for the heat capacity, but again it has the same value for all fluids. Liquid–gas critical exponents seem to be 'universal'.

To understand the meaning of critical exponents requires some mathematics, but only a little. The exponents define a mathematical relationship between two quantities that is called a power law. There is no hidden Hobbesian significance here in the word 'power' – it is just a mathematical term. If the value of some quantity y depends on the value of another quantity x according to a power-law relationship, this means that each time x is doubled, y increases by some constant factor. The exponent of the power law is a number that tells us how big this factor is. The bigger the exponent, the more rapidly y increases for each doubling of x. If the exponent has the value 2, for example, then y increases fourfold each time x is doubled. If the exponent is 3, then y increases eightfold.

This is perhaps a simpler concept than you might think. There is, for example, a power-law relationship between the length of a cube's sides and its volume – and the exponent in this case is 3.* If you double the length of its side, the cube's volume increases by a factor of 8. You can fit eight cubes of 5 cm width into one cube of 10 cm width.

Each property of a fluid which diverges at a critical point does so at a pace set by a critical exponent – which is the same for all fluids. Some quantities don't increase to infinity at a critical point, but decrease to zero. The density difference between liquid and gas does this, for example, as does the magnetization of a magnet at its Curie point (see page 110). Here again, there is a characteristic rate at which these things fall to zero. They too can be assigned critical exponents, but the exponents have negative values.

Van der Waals' theory of the liquid–gas critical point predicts these

---

* In other words, the volume depends on the length of the side raised to the power 3: the length cubed, or $(\text{length})^3$. The exponent is simply the superscript in this relationship. A general power law has the mathematical form $y = x^n$, where n is the exponent.

power-law divergences, but it miscalculates the values of the critical exponents. In effect the theory tells us that the divergences loom like mountains at the critical point, but fails to say how steep they are. This became apparent in the 1890s, when Jules Verschaffelt in van der Waals' old laboratory at Leiden analysed very accurate measurements of the critical behaviour of a liquid hydrocarbon called isopentane. He found that the critical exponent for the vanishing of the density difference seemed to be about -0.343, whereas van der Waals' theory predicted a value of exactly -0.5. That is a small shortcoming, one might think – and others at the time thought so too. But if the critical exponents are universal for all fluids, they must be encoding something pretty fundamental about the nature of matter. So it is well worth asking what is missing from van der Waals' theory that prevents it from telling the full truth about the critical point.

THE SHAKY BALANCE

What van der Waals' theory could not tell him about critical points was that the key to their oddness lies in their fluctuations. A critical point represents a forked path: a place where choices are made. This is how, you may recall, a critical phase transition differs fundamentally from the first-order transitions of, say, freezing and melting. When a liquid teeters on the brink of its freezing point, every part of it stares the same fate in the face: to become a solid. But if we cool a fluid through its critical temperature, suddenly it can exist in either of two states, both equally appealing: a liquid or a gas.* The same is true of a magnet. To recall Ising's model (page 111), below the critical temperature (the

* To physicists, a fluid is simply a substance that flows: it could be gas-like or liquid-like. Above the critical point, we can't really speak of a 'gas' or a 'liquid' at all, so the term 'fluid' then becomes indispensable.

Curie temperature) the magnet can align all its atoms' magnetic needles (spins) in either of two opposed directions – and there is nothing to choose between them. So a critical point is like a ball poised on the top of a hill between two identical valleys. It must roll down one or the other – but which? The choice is determined by fluctuations.

A supercritical fluid has, in theory, the same density throughout. But chance fluctuations – the random motions of the atoms – will make it momentarily denser in some places and less dense in others. As the fluid is cooled through the critical phase transition, the denser regions will be inclined to become liquid, since they are already on the way there. The less dense regions tend to opt for the gaseous state. Either of these random choices is self-perpetuating. This is easier to see in the case of a magnet, where a domain in which all the spins are aligned will encourage other spins at the periphery to align the same way too, via interactions between one spin and its near neighbours.

As a consequence the system becomes exquisitely sensitive to random fluctuations. A tiny chance preference will tip the balance. This instability means that the critical state is highly precarious, constantly on the verge of rolling down into one valley or the other. Holding a substance at its critical point is like trying to balance a needle on its tip by blowing on it from all directions at once.

A fundamental peculiarity emerges from this extreme sensitivity to perturbations: events in one part of the system can have a virtually instantaneous knock-on effect in any other part. The orientation of one spin can affect that of another spin a long way away, even though the two are too far apart to feel each other's influence directly. In the parlance of statistical physics, there are long-range correlations in the system. The range of the correlations – the typical distance over which one particle can affect another particle – is another of the quantities that diverges to infinity at a critical point.

This hypersensitivity is a collective effect. The range of the force of

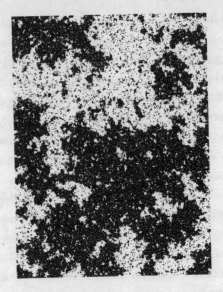

Figure 10.1 At a critical point, fluctuations of all sizes can occur. In this computer simulation, black regions could represent a liquid and white regions a gas. Or the two could denote regions in a magnet where the spins point in opposite directions. The size of these regions spans all scales – from the size of a single particle to the size of the entire system. There is no typical scale to this unevenness: it is *scale-free*.

interaction between two particles or spins in the system does not suddenly get longer at the critical point; in the Ising model, for instance, it extends only to the nearest neighbours. But these interactions can be transmitted from particle to particle over long distances without being overwhelmed by the randomizing effects of thermal motion. Their Chinese whispers are as clear at the end as they were at the beginning. At the critical point, the particles can all act together.

The trouble is that each particle 'wants' all the others to follow its own choice. As a result, the critical state fragments into regions that make one choice or the other, each triggered by localized imbalances that arise by chance. These regions can be of any size, from a single particle to a group big enough to span the entire system (Figure 10.1). They are constantly forming and dissolving, growing and shrinking. The critical state thus conjures up its own special kind of fluctuations from the thermal noise, and they are *scale-free*: of all conceivable sizes.*

This behaviour is the cause of critical opalescence: the milkiness of fluids near their critical point. Under these conditions a fluid separates into regions of liquid and gas of many different sizes. Some of these regions are about the same size as the wavelengths of visible light (a few hundred millionths of a millimetre), and this coincidence means that they will scatter light strongly, like the microscopic fat globules in milk. So the fluid takes on a pearly, opaque appearance.

Van der Waals' theory gets the critical exponents wrong because it doesn't acknowledge this microscopic picture of critical-state fluctuations. Rather, it regards the critical state as, on average, the same everywhere – as though we were to squint at Figure 10.1 from a distance and see only grey, not separate regions of black and white. According to that picture, each particle would feel not the 'whiteness' or 'blackness' of its discrete neighbours, but only some average 'grey' influence of all the other particles. This is known as a mean-field approximation (see page 268). The mean-field theory of van der Waals does a surprisingly good job of describing the critical state, falling down only in the details. As we have seen, Pierre Weiss's analysis of the magnetic Curie point (page 108) is a mean-field theory too, and it predicts the same, slightly incorrect critical exponents for the magnetic transition as van der Waals' theory did for the liquid–gas

---

* Thermal noise arises from the fluctuations in particle energies caused by collisions. It manifests itself as tiny variations in temperature at any location in the system. These fluctuations are random, or gaussian: like those shown in Figure 8.2*b* (page 240), they have a characteristic size.

critical transition. Lars Onsager's analysis of the two-dimensional Ising model (page 113) went beyond the mean-field approximation, allowing him to calculate the exact values of the critical exponents. But to predict these values theoretically for a real fluid, we need to reproduce Onsager's success for a 3D model. This, as we saw earlier, has so far proved impossible.

There is a way around the difficulty: a trick which allows theoretical physicists to sneak up on the 'true' values of critical exponents in models like the 3D Ising model. It is called *renormalization*, and it takes advantage of the scale-free nature of the critical point. The technique was developed by Kenneth Wilson of Cornell University in the 1960s, for which he was awarded the 1982 Nobel Prize for Physics. In effect, renormalization is a mathematical way of squinting at the critical state and *selectively* eliminating fine details: the small blobs blur to grey while the big ones remain. By conducting this renormalization process over successively larger scales, one can calculate the true values of the critical exponents. Applied to the 3D Ising model, this method supplies values very close to those measured experimentally for real fluids.

The Ising model of fluids, with its filled or empty boxes on a grid, is a pretty crude approximation to a real fluid. But the model's critical exponents in three dimensions are, as far as we can tell, exactly the same as those measured experimentally for liquids and gases. This is again a manifestation of the notion of universality: as far as critical behaviour is concerned, the details don't matter. Only the broadest-brush properties of a system affect its critical behaviour – not whether it is nitrogen or isopentane, or a magnetic metal, or a crude model of either, but whether it is two- or three-dimensional, whether its particles interact through long-range or short-range forces. Differences of this kind are enough to put two systems in different *universality classes*, all members of which share the same critical exponents. Unless there are such differences, seemingly different systems look the same at the critical point.

CRITICAL CRASHES

A paper co-authored by Jean-Pierre Aguilar in 1999 opens thus:

It is rather tempting to see financial crashes as the analogue of critical points in statistical mechanics, where the response to a small external perturbation becomes infinite, because all the subparts of the system respond cooperatively.[1]

This is the temptation to which Aguilar succumbed in 1995. It has been proposed that financial crashes correspond to a special kind of critical transition: one that shows so-called log-periodic behaviour. Such critical points arise within certain models in statistical physics, and they have a distinctive signature. This kind of system is prone to oscillatory, periodic fluctuations – in an economic context they would be analogous to periodic business cycles. But log-periodic variations are not like the regular oscillations of a light wave or a tuning fork. Instead, the peaks and troughs of the waves get steadily closer together. At the critical point itself they pile up on top of one another. The approach to such a critical transition is therefore signalled in advance by peaks and dips that follow one after the other at ever shorter intervals – a series of accelerating wobbles that herald catastrophe.

This precursor to the critical point in such systems leads some physicists to believe that, if financial crashes really are log-periodic critical points, they can be predicted by identifying the tell-tale oscillations in an economic index and extrapolating them to the point where they coincide. In other words, it should be possible to calculate

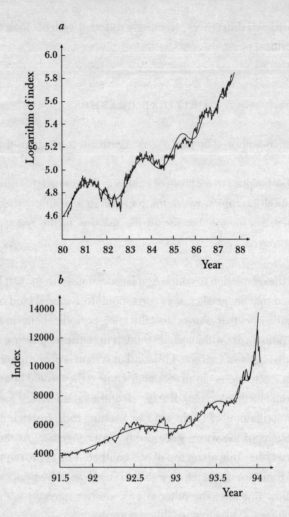

Figure 10.2 In the log-periodic model of market crashes, oscillations in economic indices pile up at a critical point, whereupon a crash occurs. Some researchers claim that they can fit such a log-periodic curve to several crashes. Here are Didier Sornette's fits to the S&P 500 index preceding the crash of 19 October 1987 (*a*) and the Hang Seng index before the Hong Kong stock market crash of March 1994 (*b*).

when a big crash will come. In 1998 Marcel Ausloos and his colleagues in Belgium used this approach to study the October 1997 crash. From their analysis of the fluctuations just before the crash, they claimed they would have been able to tell when it was going to happen.

Their work made headline news. If it was true, it was revolutionary. No longer need investors be ruined by unpredictable collapses of the market. Simply by keeping a careful eye on its ups and downs, they could anticipate the future. It was not a foolproof way to play the market, since the oscillations showed up only near a crash; at other times, prices 'lose all memory' of the past within minutes and prediction is impossible. But it would be a fantastic tool for investment companies.

French mathematical physicist Didier Sornette, based at the University of California at Los Angeles, is convinced that market crashes are log-periodic, and he claims to show how several of them in the recent past fit the model of accelerating wobbles (Figure 10.2). In this picture, an impending crash is foreshadowed in the way a market behaves not just weeks but months or even years in advance. But some other econophysicists remain unconvinced. Aguilar and his colleagues in France have argued that the way Ausloos's group matched the curve predicted by the log-periodic model to the real 1997 data involved a selective interpretation of what constituted a dip. They say that the technique doesn't seem to work when applied to other crashes. And they consider it highly improbable that a crash can really bear any relation to what the market was doing several years beforehand. Fits between the theory of log-periodic critical points and real data (as in the October 1997 crash) which appear to be good are merely fortuitous, they conclude.

Of course, 'hindcasts' are always questionable in any event. Since what did happen is already known, is it possible to remain truly objective in 'predicting' it? Sornette admits with some chagrin that

trying to make a genuine prediction of a crash is a thankless task.*
There are, he says, at least three possible outcomes:

- No one believes the prediction, but the market crashes anyway. Then critics will say it is just a lone, lucky correlation with no statistical significance. Besides, what good is a warning if it fails to avert a crash?
- Many investors believe the prediction, get triggered into panic buying and selling, and thereby cause a crash – that is, the prediction becomes self-fulfilling.
- Many investors believe the prediction and take careful compensatory action so that a crash is averted – that is, the prediction becomes self-defeating.

That's the problem with the dream of predictability in economics: future market behaviour depends on what traders and investors believe that behaviour will be, so the act of predicting the future (if it is taken at all seriously) is likely to change it.

THE SELF-ORGANIZED MARKET

Despite scepticism about the log-periodic model of crashes, the idea that market dynamics are governed by critical-state-like behaviour has found wider acceptance. It is not hard to understand the attraction of this idea. We saw in Chapter 8 that the statistics of economic fluctuations are not gaussian. Instead, the fluctuations seem (at least

* Perhaps we shall see. At the time of writing, Sornette and his colleague Wei-Xing Zhou have bravely predicted a crash (or at least a pronounced change) in the UK housing market, arriving around the end of 2003. Sornette boldly advertises his economic predictions on his web site at http://www.ess.ucla.edu/faculty/sornette/, which is a risk few economists would be likely to take.

for short to moderate timescales) to be scale-free: variations of all sizes are seen. Statistically, the economic data in the 'fat tails' of the probability distribution follow a power law – a characteristic feature of critical-point behaviour.

The power law tells us about the probability of fluctuations of a given size. Suppose we peruse the time sequence of an economic index like the one shown in Figure 8.2 (page 240), and keep note of how often we get a return of a certain size. The erratic graph hovers either side of zero – the most probable return is zero, and we can see that large deviations are comparatively rare. If we measure the relative number of fluctuations of successively increasing size, we find that the probability decreases according to a power law. Just as a system at its critical point is apt to respond to a perturbation by undergoing a fluctuation (of any size), so, we might suppose, this distribution of fluctuation probabilities in the economic market is a sign that it too is poised, in some sense, in a critical state, and is sent hither and thither in unpredictable lurches by the random factors that affect it.

But a critical state is highly precarious and liable to collapse one way or the other at the slightest provocation. How, if the market is critical, can it have stayed that way for so long? In 1987 a group of physicists working at Brookhaven National Laboratory on Long Island in the USA stumbled by chance upon a system which had the strange, seemingly miraculous property of constantly reorganizing itself into a critical state. They called the phenomenon *self-organized criticality*. These researchers – Per Bak, Chao Tang and Kurt Wiesenfeld – hadn't set out to investigate critical points. What they were trying to do was make sense of a recondite problem in solid-state physics, about how electrons sometimes move through crystalline solids in a series of waves called charge-density waves. This turns out to be a kind of correlated motion. The electrons don't move independently, as they do in a normal current-carrying metal wire.

Instead, the movement of one electron strongly affects the others: they are coupled.

Bak, Tang and Wiesenfeld began to think more broadly about coupled systems of many interacting particles. As a crude model of charge-density waves, they imagined many swinging pendulums connected together by springs. The two systems don't sound very similar at all, but mathematically there is an analogy between them, just as a single pendulum serves as a good model for any process which involves simple periodic motion. When the researchers studied the behaviour of this model by writing down its Newtonian equations of motion and solving them on a computer, they found a curious thing. If one pendulum in the coupled array was set swinging, it could set others moving as energy was transferred along the springs. But this influence could extend over any range. Sometimes a single pendulum would do no more than make its near neighbours oscillate. On other occasions an 'avalanche' would sweep through the system, setting thousands of pendulums into motion. There was no telling how big the avalanche would be. When the researchers drew the probability distribution of these avalanches – the dependence of the probability of an avalanche on its size – they found a power law. The signature of power-law graphs is that, when they are plotted on a logarithmic scale (see the caption to Figure 10.3, opposite), the probability distribution curve becomes a straight line.

The Brookhaven team then devised a new and more intuitively appealing model of this coupled behaviour. They replaced their pendulums and springs with a pile of sand. Imagine dropping grains of sand one by one onto a table top. The pile builds up slowly into a little sand mountain. Once the slopes reach a certain steepness, dropping new grains on top can trigger an avalanche. Before this point, the grains are held in place on the slopes by friction, which prevents them from sliding. At some well-defined angle, friction is no

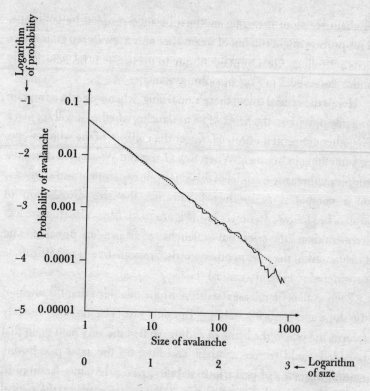

Figure 10.3 The power-law probability distribution of avalanches in a mathematical model of a sand pile. The best way to show the power law is to plot the logarithm of the avalanche size against the logarithm of its probability: the graph then becomes a straight line. The slope of the line is equal to the exponent of the power law. Here the slope is close to –1, which is characteristic of self-organized criticality. Large events, to the right of the graph, are less frequent, so the statistics are less reliable here: the line becomes more jagged. The straight line that fits the data most closely is shown as a dotted line.

longer able to restrain the motion, and the avalanches commence. Once one grain starts to move, it collides with others, triggering a kind of chain reaction: the grain motions become coupled by collisions. This process might run out of steam after only a few dozen grains have been set rolling. Or it might continue to propagate until virtually the entire slope slides in a catastrophic avalanche.

Here's the crucial thing: there's no telling, when we add a new grain to a pile poised on the brink of an avalanche, whether it will set just a few other grains in motion, or trigger the collapse of the whole slope, or something in between. When Bak, Tang and Wiesenfeld devised a simple mathematical model of this sand-pile experiment and studied it on a computer, again they found that the size distribution of avalanches followed a power law (Figure 10.3). Big avalanches are less frequent than little ones, but avalanches of all sizes are possible. The fluctuations of the pile, in other words, are scale-free. The pile can be considered to exist in a critical state.

Each avalanche releases 'tension' in the pile, lowering the angle of the slope and restoring stability. But only just. The avalanche merely returns the pile to the brink of sliding, so that the very next grain that is added could trigger another landslide. So the sand pile hovers around this state of precarious stability, never deviating far from it. Rather than the pile being liable to collapse irretrievably, the critical state is constantly resurrected after each avalanche. This is why the critical state is said to be self-organized. In contrast, the critical state of a liquid or gas can be regarded as self-destructive, ready to transform itself into a wholly different, stable state at the slightest disturbance.

The sand-pile model describes a non-equilibrium situation. The critical state here is a 'stationary state', in the sense that the system never strays far from it. But it is not an equilibrium state, because it is not unchanging: new grains are constantly being added. This dropping of grains onto the heap is the driving force that prevents equilibrium from being attained. Self-organized criticality is a property of non-equilibrium systems.

Self-organized criticality is one of the few genuinely new discoveries to have been made in statistical physics over the past two decades, and it has proved an astonishingly fertile idea. Per Bak has identified a wide range of natural phenomena that seem to undergo power-law fluctuations suggestive of self-organized criticality, or something like it. Earthquakes have long been known to obey a power-law probability distribution: that is, the probability of an earthquake diminishes as its size increases according to a power law. This was discovered in the 1940s by seismologists Beno Gutenberg and Charles Richter from the California Institute of Technology. They consulted a world catalogue of earthquakes and made a graph of the number of events of each size. At the time, they had no way of interpreting their power law. But Bak and others have suggested that it reflects self-organized criticality in geological fault systems. The motions of the Earth's crust build up stresses in rock formations until slippage finally occurs. This releases the stress – but only just enough to restore stability. Then the pressure begins to build up again. Usually the stress is released in small doses, causing minor tremors. But every so often there is a catastrophic release, and disaster strikes a city such as Los Angeles or Kobe.

Bak also saw self-organized criticality in forest fires. Many large forested areas are prone to wildfires, but most fires are of only limited extent. Sometimes, however, a fire can go on spreading from tree to tree until it engulfs almost an entire forest. Volcanic activity, solar flares, the 'starquakes' thought to occur on exotic celestial objects called neutron stars, even the pattern of species extinctions in the fossil record – all, said Bak, show the power-law fingerprint of self-organized criticality.

Ironically, it appears that the sand pile itself might not be strictly in a self-organized critical state at all. The experiment sounds simple in principle, but it is rather hard to measure sand-pile landslides unambiguously in practice, and to gather data over a big enough size range of avalanches that one can distinguish a genuine power law from something looking very much like it. So various experiments have produced conflicting results. But it seems that the original mathematical model constructed by Bak, Tang and Wiesenfeld neglects some important features of real sand piles, such as an accurate description of how the energy of a tumbling grain is dissipated in collisions. True self-organized criticality (SOC) appears to be a rather elusive thing in piles of grains – it crops up, perhaps, in some cases but not in others. For instance, SOC has been claimed for piles of *rice* but not of sand: the shape of the grain matters.

This rather undermines Bak's assertion that SOC is the key to 'how nature works': it appears to be an insufficiently general and robust phenomenon to aspire to such a degree of universality. Nevertheless, there is little doubt that the basic characteristics underlying SOC – a power-law probability distribution of fluctuations, and catastrophic events that 'release tension' while returning the system to the verge of instability – seem to provide a powerful framework for understanding a wide range of phenomena. We should hardly be surprised, then, if these phenomena include aspects of human social interactions.

HARD-LINE ECONOMICS

In 1988 Per Bak was working at the Santa Fe Institute in New Mexico, the centre of the intellectual universe for any researcher interested in complex interacting systems, whether they be in physics, biology, geophysics, social science, or anything else. He was approached by two economists from the University of Chicago, Michael Woodford and Jose Scheinkman, who had heard about SOC and wanted to investigate whether the economy behaved in this way. It was an astute intuition, especially as the concept of SOC itself was barely a year old.

Some errant economists had already become interested in applying ideas from chaos theory to economics – Scheinkman was one of them – but that notion was decidedly out of the mainstream. Woodford and Scheinkman saw more promise in Bak's theory, since it seemed to allow for precisely the kind of extreme events – big 'avalanches' – that traditional economic models neglected.

As a physicist, Bak's introduction to economic modelling was revealing. Physicists generally deal with inanimate objects whose behaviour has nothing of the irrational in it, and even when they make all manner of approximations and simplifications (as they invariably do), the maths becomes too complicated to solve without a computer. Bak anticipated that economical modelling would be even more complex, so he was astonished to find that his economist colleagues preferred to look for 'models that can be solved analytically with pen and paper mathematics'. It was the beginning of what Bak called 'a very productive, though rather painful, collaboration'.[2]

The result was an agent-based model of the economy, along the lines of those discussed in the previous chapter, which was published in 1993 by Bak and his fellow physicist Kan Chen, along with Woodford and Scheinkman. 'Our conclusion', said Bak,

is that the large fluctuations observed in economics indicate an economy operating at the self-organized critical state, in which minor shocks can lead to avalanches [crashes] of all sizes, just like earthquakes. The fluctuations are unavoidable. There is no way that one can stabilize the economy and get rid of the fluctuations through regulations of interest rates or other measures.[3]

The idea of a self-organized critical economy is an appealing one. But sadly, like the sand-pile model, it seems to be right in spirit but wrong in detail. The very essence of SOC is scale-free behaviour described by a power law. But although economic data, such as the rise and fall of the S&P 500 index, can appear to behave in this way within certain limits, these features do not persist in the big picture. The fluctuations observed for time steps of a few minutes look more or less like those on hourly or daily time steps if everything is 'rescaled' appropriately (page 246) – but in fact the statistics are not precisely the same. The larger the time step, the more the probability distribution of fluctuations approaches the gaussian form. Over periods of years, meanwhile, prices grow more or less linearly (that is, proportional to the time elapsed), punctuated by sharp drops. So any model which assumes or predicts a single mathematical form for the statistics of price changes on all timescales cannot be right. By the same token, the probability distribution functions of fluctuations in prices obey a power law only over a limited range in the 'fat tails'; outside this range, a different kind of relationship is observed. So although it provides a nice illustration of how extreme events can represent a natural, if rare, aspect of how a system fluctuates, SOC alone is not the key to how the economy works.

THE SPIRIT OF THE LAW

Power-law probability distributions, which do not discriminate (as gaussian fluctuations do) against extreme events, may turn out to be a pervasive feature of the way people organize their affairs. Physicist Sidney Redner at Boston University showed in 1998 that the statistics of citations in the scientific literature obeys a power law. Scientific papers include a list of references to earlier papers whose findings are mentioned or used in the research reported. Some papers – the SOC paper by Bak, Tang and Wiesenfeld in 1987 is one of them – contain ideas which are very influential and stimulate scores of follow-up studies. These are widely cited. Other publications address highly specialized topics and few subsequent researchers have any need to refer to them, so they garner only a handful of citations. Redner looked at nearly 800,000 papers published in 1981 and found that their citation statistics obeyed a power law: most papers were cited only a few times, but a few appeared in a large number of reference lists. (In fact, about half of the papers were never cited at all, a telling reminder of the modest nature of most research projects. Tennyson was right to say that 'science creep[s] on from point to point'.[4])

One might argue that these power-law relationships are statements of the obvious. Of course big earthquakes are rarer than smaller ones. Of course some research papers are highly influential, while others make an impact on no one. That's true, but a power law says more than that. It describes a particular *kind* of way in which the probability of an event declines as the event gets bigger or more extreme. There is no a priori reason to suspect that, each time you double the size of the event, the probability will diminish by a constant factor – that the probability of four citations will be an eighth of that of two citations, say, and the probability of eight citations will also be an eighth of that of four citations. The general message of a power law is indeed intuitively obvious; the precise mathematical relationship, however, is not at all inevitable.

In his book *Ubiquity*, physicist and science writer Mark Buchanan has made the intriguing suggestion that history operates in a self-organized critical state. His idea is that conflicts and wars are the result of tensions that hold international relations poised at the brink of catastrophe. As a consequence, wars of all magnitudes, from minor skirmishes to global conflagrations, are inevitable. Buchanan offers what he admits is tentative evidence for this proposal in the form of a graph that displays the relation between the number of conflicts and their size, measured according to the number of fatalities that resulted.

Wars which left only a few hundred dead and world wars which have killed millions: all follow a single power law.

This behaviour was first noticed by the British physicist Lewis Fry Richardson. One of the early pioneers of a modern 'physics of political policy', Richardson used concepts taken from meteorology to develop mathematical models of the arms race between antagonistic nations. As a Quaker who served as an ambulance driver in the First World War, Richardson hoped to promote international peace by elucidating the causes of war.* Between the 1920s and the 1950s he amassed data on the statistics of 'deadly quarrels', in which he provocatively included war alongside other kinds of killing, such as murder in peacetime. He found that there was a kind of Gutenberg–Richter law of conflicts in which all events from individual murders ('magnitude 0 conflicts') to the two world wars ('magnitude 7') obey a power-law probability distribution.

The sobering corollary, Buchanan argues, is that there can be no telling how big a conflict might be sparked by the smallest disturbance. The First World War, after all, followed from the assassination of the Austro-Hungarian Archduke Franz Ferdinand in Sarajevo in 1914, an event which hinged on an unhappy and unpredictable set of circumstances on that fateful day. One can immediately find holes to pick in this idea. It seems highly likely, for example, that the Second World War was made inevitable by (among other things) the First World War and the Treaty of Versailles, rather than being triggered by some freak occurrence in 1939. Indeed, the First World War was not really fought over the Archduke's murder at all. But that is not the

---

* Richardson's interest in whether wars between neighbouring nations depended on the lengths of their common boundary led him to realize that the measured length of boundaries and coastlines depends on the step size used to track these tortuous, zigzag lines. This discovery prompted the later development of fractal geometry by Benoit Mandelbrot.

point; Buchanan is saying that if there is 'tension' in a complex system like this, small events can have disproportionate consequences. In any event, Richardson's data imply that Immanuel Kant's notion of natural laws governing the course of history still has some mileage. The question historians will ask is whether this can tell us anything about history *as such*: whether it helps us understand why things turned out the way they did. That is, in a rather different context, a question to which we shall return later.

THE LEAST ONE CAN DO

The American social scientist George Kingsley Zipf (1902–50) helped to revive the Enlightenment belief in natural laws of society and the possibility of a truly scientific sociology with his book *Human Behavior and the Principle of Least Effort*, published in 1949. It is a curious document, at the same time phenomenally prescient and a product of its times.

Zipf's central notion was that people act in ways that require them to make minimal effort. This seems commonsensical enough, and Zipf regarded it as the sociological equivalent of the physical principle proposed in the nineteenth century by the Irish mathematician William Hamilton. The 'law of least action', said Hamilton, determines how any entity moves under the influence of forces. He demonstrated that underlying the laws of mechanics adduced by Newton was the tendency of an object to follow a trajectory that generates the smallest possible 'action' – a mechanical quantity that depends on the path taken. There are many trajectories that take a ball at rest on a table to a position at rest on the floor, but only one of them minimizes the action of the motion, and that's the path the ball follows when it rolls over the edge.

It is straightforward to calculate the action accrued by an object moving along a particular trajectory. But determining the 'effort' involved in human activities is far from trivial. This is not simply the energy expended: as Zipf pointed out, different people might choose different ways to complete a given task, depending on how much they *think* it will cost them – in energy, time, discomfort, money, or whatever. One engineer will link two towns by boring a tunnel through a mountain; another will build a road over the peak. According to Zipf, both act according to their estimates of how much 'effort' will be involved – it is simply that their estimates differ.

This is clearly a shaky principle, given the subjectivity of any assessment of 'effort'. The idea sounds good, but its central parameter is all but unquantifiable. Nonetheless, Zipf attempted to explain a bewildering variety of human traits and behavioural patterns in this way, including the properties and development of language, the structure of music, human demography, the distribution of industries, travel statistics, marriage data, international and civil conflicts, and income distributions.

Zipf's main contribution to the physics of society was, however, empirical. He collected data in all these areas and showed that they all featured probability distributions characterized by power laws, with their distinctive straight-line graphs. At that time, social scientists seldom looked much beyond the statistics of gaussian (random) distributions, and the significance of Zipf's data has come to be appreciated only through the recent emergence of power-law behaviour at the heart of statistical physics. Whereas Per Bak (who died in 2002) believed that SOC is in some sense 'how nature works', Zipf considered that his power-law graphs showed 'how society works'. He believed that the social sciences differed from the natural sciences in that they are dominated by power-law rather than gaussian statistics. We now know that power laws are common in the natural world too.

Be that as it may, Zipf uncovered something of fundamental significance. In 1983 Mandelbrot wrote that 'The failure of applied statisticians and social scientists to heed Zipf helps account for the striking backwardness of their fields.'[5] Zipf asserted that one power law in particular – one in which the straight-line plots have a slope of –1 (see the caption to Figure 10.3), now seen as diagnostic of SOC – was characteristic of phenomena in which people acted in groups rather than as individuals: that is, in which people interacted. This was more or less the full extent of his appreciation of the role of interactions in social phenomena, which we can now see to be crucial. As physicist Philip Anderson has pointed out, power-law distributions in social phenomena destroy the idea that what matters is the 'average' behaviour, or Quetelet's 'average man'. In effect, a power law raises the odds of extreme events, which a gaussian probability distribution relegates to the status of negligible aberrations.

The sociologist Vilfredo Pareto was perhaps the first to introduce power laws into social science, and he did so before anything of the sort had been discovered in physics. He claimed in 1897 that incomes towards the wealthier end of the social spectrum are distributed according to a power law (Figure 10.4). This implies that much of a nation's wealth is held by a few individuals. For the USA, some estimates attribute 40 per cent of the wealth to 1 per cent of the population, and over half of the wealth to 5 per cent of the population.[6] This inequality has been growing since the 1970s – a trend which has been identified in other nations too.

Pareto expressed this imbalance in terms of the so-called 80 : 20 rule: 80 per cent of the wealth is possessed by 20 per cent of the people. He observed that this income distribution held true for many countries, regardless of their political system or taxation regime. The 80 : 20 principle has come to be regarded as a rule of thumb for management decisions: 80 per cent of your benefit will come from

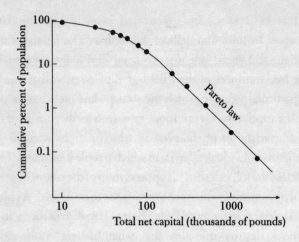

Figure 10.4 National income distributions typically follow a power law, as first observed in the nineteenth century by Vilfredo Pareto. He claimed that the slope of the straight line is always –1, but later studies have shown that it is usually larger. The larger the slope, the more the economy is impoverished. Here I show the wealth distribution for the population of the UK in 1996, according to figures collected by the Inland Revenue. This is a cumulative distribution: each data point shows the percentage of the population with a net capital wealth greater than the corresponding amount on the horizontal axis.

20 per cent of your outlay, 80 per cent of your results will be due to 20 per cent of your workforce, and so forth. But focusing on these particular figures rather obscures the main point, which is not simply that there is inequality between, say, effort and reward, but that the *distribution* follows a power law. The exact figures – that's to say, the slope of the straight-line power-law graph – may vary: it's not always an 80 : 20 ratio. But the straight line remains. This kind of wealth distribution has been inferred even in ancient Egypt in the 14th century BC, from an analysis of the distribution of house sizes in the ruined city of Akhetaten.

Pareto's income distribution later came to be regarded as a fundamental aspect of society, possessing much the same mystique as the gaussian distribution had in the early nineteenth century. In 1940, the economist Carl Snyder said that 'Pareto's curve is destined to take its place as one of the great generalizations of human knowledge.'[7] In truth, however, it is still not entirely clear how rigidly the 'law' applies. Income distributions are surprisingly hard to determine accurately (one way is to look at tax returns; another is to infer levels of wealth from payments of death duties). In 1935 the economist George Findlay Shirras questioned the validity of the Pareto law, saying that there was no place for it in studies of income distribution. But there seems good reason to believe that a power-law distribution does hold at least for the highest-earning bracket of any society, as Figure 10.4 shows. The 'steepness' of such a power-law slope reflects the degree of disparity between incomes: the steeper the slope, the more unequal the distribution of wealth and the more poverty there is. In extremely unequal economies, such as those of Haiti or Zaire (or, indeed, ancient Egypt), a very few have got rich at the expense of a vast majority of very poor people. This is one good reason never to trust the average statistics cited by politicians unless you know the underlying distribution on which they are based.

Econophysicists Sorin Solomon in Israel, Jean-Philippe Bouchard in France, and their colleagues have proposed models to explain how the Pareto law might arise, drawing on ideas developed to explain how the chain-like molecules of polymers move. They compare these motions with the movements of money in investment markets. Solomon, working with and Zhi-Feng Huang of Cologne University, has shown how there is a tendency for trade to increase the steepness of the Pareto slope in unregulated markets, creating an ever-increasing disparity between rich and poor. One consequence of this is that market fluctuations also increase: the market becomes less stable. So, the researchers argue, a social policy which aims to raise the wealth of the poorest members of society 'is not just a humane duty but also a vital interest of the capital markets'[8] – a form of enlightened self-interest.

UNIVERSAL ORDER?

George Zipf hoped that his observations would be used as a guide to policy-making. His dream sounds, in these post-Marxist times, a little unnerving: 'a systematic social science will make possible an objective social engineering'.[9] But he was right, if by social engineering we mean no more than informed planning. His ambitions went even further. In the true spirit of the Enlightenment, Zipf felt that the discovery of order (by which he meant the ubiquitous power law) at the centre of human affairs revealed a kind of natural design operating beyond human volition. In an age of declining religious conviction, Zipf imagined that this might encourage a rational social science to take its place:

For we are finding in the everyday phenomena of life a unity and orderliness and balance that can only give faith in the ultimate reasonableness of the whole whose totality lies beyond our powers of comprehension.[10]

From there, it is perhaps but a short step to Pythagorean mysticism. But that need not distract us from responding with a certain wonder at the universality that organizes many aspects of society in the same way as it directs the properties of atoms. We need not turn this into a 'religion of science', any more than we need regard the whorls of a flowing stream as evidence of divine planning. We can simply celebrate the fact that there are indeed 'laws of large numbers', and that they let us divine order and regularity in an otherwise terrifying diversity.

11

The Work of Many Hands

The growth of firms

I met a vineyard owner in California. He was, he told me, trying to squeeze every cent of margin out of his product 'in order to develop my winery as fast as possible.' 'Do you plan to get very much bigger?' I asked. 'Oh, I don't want to grow bigger,' he replied, 'I want to grow better.' To do that, he needed better equipment, better vines, better people – not more.

    Charles Handy (1976)

Large human organisations always lead to oppression unless there is a democratic machinery to prevent it.

    J. B. S. Haldane (1949)

What scale is appropriate? It depends on what we are trying to do. The question of scale is extremely crucial today, in political, social and economic affairs just as in almost everything else.

    E. F. Schumacher (1973)

*

Three years ago, Brecon was a Welsh market town on the verge of becoming – like many other communities today – a supermarket town.

Safeway, the giant British food retailer, was coming. When environmental campaigner George Monbiot went to the town, he found locals expecting the worst.

'Safeways say they're bringing choice to Brecon,' said butcher Brian Keylock,

but there are nine butchers in town already. You've got nine choices. If we all disappear, you'll have no choice. The fishmonger has already closed down. Clothes shops have closed down . . . Go to Leominster. Since that thing arrived there, the town is dead. They'll do the same to us.[1]

Frankly, Brecon was lucky to have survived for so long. British towns have been succumbing to the death of the high street and the demise of the local shop since the early 1980s. Now chains, banks and estate agents colonize town centres, and it's hard to buy a pint of milk there. To do that, you need to go to the outskirts of town, where supermarkets offer milk in ten different varieties, along with all the other items that once took people from shop to shop.

The problem – if you are inclined to see the disappearance of small shops as a problem – is not just that supermarkets and chain stores exist, but that we all use them. Increasingly we have little choice: the out-of-town Sainsbury's supermarket in the Devon town of Axminster was the only place I could get a cup of tea on Sunday afternoon as I cycled through town. The small tea shops could not compete. And that's the trouble: big companies can afford to do things that small ones cannot. How many local bakers, for instance, could sell a loaf of bread for eight pence, as Safeway did in 1999? This is way below cost price, but supermarkets gamble that by undercutting competitors on staple products, they can create an impression of fantastic cheapness that will draw in customers who will then buy more expensive items too. The Lidl supermarkets in Britain were recently charging *negative* prices (yes, deducting money from your bill) for tins of baked beans.

Capitalism's evangelists tell us this is all for the best. A 1978 British Government Green Paper (a policy document) claimed that:

Unrestrained interaction of competitive forces will usually result in the best allocation of our economic resources, the lowest prices, the highest quality and the greatest material progress, while at the same time providing an environment conducive to the preservation of our democratic, political and social institutions.[2]

Brecon's butchers might beg to differ.

Capitalism's opponents, meanwhile, often portray the free market as a system in which big fish inevitably swallow up small ones, leading to a homogeneous world of commerce dominated by a few major players. It can certainly look that way. Between 1990 and 1996, British shops that sold less than £100,000 worth of goods a year declined by 36 per cent, while the number of supermarket branches almost trebled from 1986 to 1997. In the broader picture, big companies have the economic muscle of entire nations: 52 of the 100 largest economies in the world are those of corporations. There is ample reason to worry about this. Monbiot argues that the growth of mega-corporations threatens not just the cohesion of local communities but the concept of democracy itself. That seems reason enough to wonder about the factors that drive the expansion and contraction of companies.

In this chapter we shall explore whether there are fundamental laws that govern the way firms grow. Many economic theorists have suspected that there are; but if this is so, no one is yet agreed on the form these laws take or how they operate. Nevertheless, signs are beginning to emerge of a kind of 'universality' principle in business which implies that, whether you are looking at steelmaking, publishing or baking bread, the growth of firms follows an 'iron law' which tells us what kind of distribution of small, medium and large we must expect. And that's the strange thing. For all the signs that the business world is becoming top-heavy, the undoubted inequities of globalization are not the whole picture. While a few big corporations exert an inordinate influence on product choice and advertising, there is a far greater number of small firms than big ones. Most of these tiny acorns never get much bigger; few indeed are destined to become mighty oaks. Yet the acorns are not going away – not, at least, if we are to judge from today's figures.

Small businesses know only too well how strongly success depends on chance. But from the fluctuating fortunes of many individual companies, robust laws seem to arise that govern the way firms grow. Whether we can tailor the world of commerce to the shape we prefer (and whose preference?) is an open question, but surely we cannot hope to do so until we can characterize and understand these laws. If we wish, for instance, to limit the size of big corporations, or to help small firms to survive without their feeling compelled to expand, we need models of company growth that allow us to predict the consequences of the legislation or trade rules we might apply. If we were to discover (and this seems likely) that any market permitting a modicum of free trade will inevitably spawn a few very large companies, then at least we know what our choices are.

KEEPING COMPANY

World trade does not *need* firms and companies. In Roman Britain one could buy goods made from Chinese silk. Conversely, Rome exported to the Orient: Roman coins have been found in Vietnam. The Tojaidi temple at Nara in Japan, built in AD 752, contains Byzantine glass. The awesome trade network of the ancient world was sustained by individual craftspeople selling their wares to merchants who passed them on, at a profit, to those who could afford exotic imports from far-flung lands.

So what are firms for? Artisans in the Middle Ages discovered some of the benefits of collective organization. The trade guilds protected their members from exploitation and were able to enforce wage stand-ards for the contracts the guildsmen undertook. But the Industrial Revolution revealed the greatest advantage of collective labour: economy of scale, made possible by mechanization. The advantage was, of course, primarily to the few, not the many: industrialization benefited the capitalist bosses much more than the workers. Even so, a worker on subsistence wages with a big firm was better off, by and large, than the traditional craftsperson, who could not hope to com-pete against industrialized rivals. The former was on the breadline; but unless the latter signed up for the same, he starved.

Economies of scale are achieved in several ways. Big firms are able to purchase expensive machinery which is beyond the means of individuals. They gain efficiency by bringing their workforce together under the same roof and by apportioning tasks among specialized workers. 'The division of labour', said Adam Smith, 'occasions, in every art, a proportionable increase of the productive powers of labour.'[3] And big firms reduce transaction costs by striking a few big deals instead of countless smaller ones.

The coercive employment market of the Industrial Revolution eventually softened in the latter half of the twentieth century into what promised to be a reassuring environment for the worker. The pater-nalistic big company offered job security, employee benefits such as health insurance, transport or accommodation, as well as camaraderie and, in time, the promise of a pension. This, perhaps more than anything else, undermined Marx's 'inevitable revolution', based as it was on a nineteenth-century model of exploitation and Smithian market principles which compelled employers (if they needed compelling) to pay workers only the bare minimum necessary to keep them alive. In Japan in particular the transaction between worker and boss became no longer just that of work for money, but of life-long loyalty for life-long protection. Whether or not one still believes in this rosy picture of capitalism in action, there seems little doubt that most people still prefer to be part of a collective workforce than to be on their own.

The traditional theory of the firm has had rather little to say about the complex relationship between managers and their workforces. Developed by economists, this theory has kept things simple by assuming that agents exhibit the calculated rationality to which the discipline has become accustomed. Firms, it says, exist to maximize profits. They aim to establish a revenue that is as far as possible above costs. Primed with this clear goal, companies then use a well-developed calculus of supply and demand which enables them to chart their optimal choices. They will seek to expand to the point where further increases in revenue are exactly balanced by rising costs, so that no more profit accrues from increased production.*

This is all very neat in the short term. Unfortunately, the conven-

---

* The ideal level of production is that for which the marginal revenue (the change in revenue per unit increase in production) equals the marginal cost (the change in total cost per unit increase in production). Because any given production plant has an optimal level of usage, trying to expand its output beyond this capacity eventually produces diminishing returns of scale (without capital input to make a bigger plant). When that happens, the marginal cost rises steadily as output increases, and this is what eventually curbs the profitability of increased production.

Economists look for this compromise between marginal revenue and marginal cost by drawing theoretical curves that show how they relate to output. It turns out that these idealized curves rarely look much like those derived from real data, which adds to the (often overlooked) problems of the conventional theory of the firm.

tional theory of the firm also predicts that, as a result, in a completely free market no one can make any profit at all in the long run. Capitalism, the theory says, is bad for capitalists. The reason is simple, and Adam Smith foresaw it. In a market governed by perfect competition, where all firms are equally capable and can charge whatever they like for a product, there will always be advantage to be gained over other firms by cutting back on profit margins and selling more cheaply. Thus no firm can increase prices above the break-even point at which revenue is exactly balanced by costs. If it does, someone else will sell more cheaply and no one will buy the more expensive product.

The problem of where profits come from was therefore a major concern for nineteenth-century economists, as we saw in Chapter 8. Marx argued that they result from the 'surplus value' that a worker gives to his employer by working longer hours than his wage warrants: the employer gets some labour 'for free', and this is where he makes his profit. This suited Marx's image of the exploited workers; and in our present climate of shrinking workforces working ever longer hours, one has to suspect that Marx was on to something. But it is not hard to think of many other reasons why companies make a profit despite the operation of free-market forces. Advertising clearly distorts people's buying preferences: they will not necessarily look for the cheapest product, but will be drawn towards those they believe to be the 'best' or the most prestigious. People will pay a high premium for fashionable brand names.

That is not the whole story. The assumption of perfect competition is in fact a way of side-stepping the awkward fact that companies are interdependent: the behaviour of one firm can affect the pricing of goods and the profitability of the others. Economists know this is so, but are not sure how best to deal with it.

The conventional, so-called neoclassical theory of the firm is comfortable only with two extreme scenarios. Under conditions of perfect competition, each firm can be considered to operate within a market that has fixed, externally imposed features. The relationships between manufacturing cost, selling price and quantity of a product are assumed to be immutable: they arise from the average behaviour of all the firms in the market. All a company has to do is juggle with these imposed conditions to seek profitability. The same is true of the other extreme: monopoly, in which the market contains only one firm. By definition, interaction can then be neglected because there is no one else to interact with. But in practice monopolies are clearly special cases, for which specific legislation usually exists to regulate the way the market operates. (Otherwise the firm with the monopoly can charge whatever it likes for its products.)

The truth is that a free market rarely, if ever, operates under conditions of perfect competition between a large number of more or less equivalent firms. Instead, there are usually many companies of varying size, each with overlapping but not necessarily equivalent product profiles. That is where the conventional theory breaks down. But it does have a decent shot at describing a more restricted situation: the oligopoly, in which just a few major players dominate the field. This is not uncommon: in the UK, for example, five supermarket chains account for three-quarters of all grocery sales, and there are only half a dozen or so major national newspapers.

Interaction between companies is critical to pricing and profitability in an oligopoly. Undercutting operates in much the same way as it does in a perfectly competitive market, but with fewer players the feedback between firms is more extreme. This can lead to the kind of instabilities and sudden changes that, in physics, are often the fingerprint of strong interactions. For example, prices can stay steady for long periods of time and then go abruptly into free fall as a price war escalates. Such hostilities have flared in the recent past between manufacturers of personal computers, airlines and national newspapers.

These battles can work to the apparent advantage of consumers, who can find themselves paying way below cost price. But in the long term the struggle can eliminate players, reducing choice and letting the market slide towards a monopoly, as is being threatened by Rupert Murdoch's News International group in the UK.

Economists have tried to build interaction into models of oligopolies in various ways. A crude solution is to recognize that rivalry may change the shape of the demand curve, which relates price to quantity of goods. For example, a firm's rivals might be ready to match any price cuts it makes, but not price increases. But this simply modifies one of the imposed characteristics of the model: the relationship between price and quantity. It is really nothing more than an attempt to reduce the interactive market to one in which each firm manoeuvres within externally fixed circumstances.

Oligopolies do not have to be competitive. It is clearly in the interests of companies to collude in fixing high prices, which consumers are then forced to pay through lack of choice. This creates cartels which can operate as de facto monopolies. Some countries outlaw cartels, but it is nevertheless very hard to prevent collusion at the expense of the consumer. 'There can surely be little doubt', says George Monbiot,

that some of the big chains operate what are, in effect, local or regional monopolies. Tesco and Sainsbury, for example, control 57 per cent of the grocery market in London and the south of England between them.[4]

Cartels rely on their members playing the game and not defecting by lowering their prices and thereby grabbing a bigger market share. This situation of cooperative advantage juxtaposed with the temptation to profit by defection lends itself perfectly to the analytical methods of game theory, described in Chapter 17. Such methods are commonly used in economics to model oligopolies that are open to collusive practices.

None of these traditional approaches, however, takes much account of how the market is really structured – that is, how firms are really distributed in size. Perfect competition, monopolies and oligopolies are all exceptions. In the real marketplace there are generally firms of many sizes. Small firms may modify the behaviour of oligopolies, while big firms constrain the competitiveness of small ones.

Moreover, the motivations of firms are not well described by the conventional theory. Most firms are almost certainly *not* pure maximizers of profit. Each pursues its own agenda, and this can be a brew of many conflicting ingredients. Even the most hardened cynic cannot claim that all firms put profits before all else – remember, after all, that we are talking here not only about multinationals but about local bakeries with three employees. A company for which I once worked provides free lunches for its workers. If it did not, its profits would increase and I doubt that anyone would have quit as a result, although they might have grumbled. Of course, this is partly enlightened self-interest: it is worth a firm's while to keep its workforce happy.

Other companies act as though to maximize not profits but total revenue, figuring that market share is more important in the long run than profitability. Some make growth of the workforce their objective. Some seek whatever course will keep their shareholders happy. Others aim, in the tortured terminology of economics, to satisfice: to achieve satisfactory (not necessarily optimal) performance according to a wide range of criteria (page 263).

Given these diverse strategies, one might expect the growth of some firms to have little or nothing in common with the growth of others. But we must remember the importance of distinguishing individual from statistical outcomes. Just because people have many different motives as they move through their environment, this does not mean that their collective movement is devoid of pattern. Do firms too display statistical regularities beneath their individual idiosyncrasies?

BUSINESS LAW

One might think that a book entitled *Inégalités économiques* ('Economic Inequalities') would be a Marxist denouncement of capitalism. But this was not the intention of its author, Robert Gibrat, at all. In his book, published in 1931, this French economist developed the first universal theory of how firms get to be of unequal sizes.

Gibrat understood that any such theory must be statistical: it must address the mathematical form of the probability distribution of firm sizes. It was clear to him that this distribution is highly skewed – that is to say, there are many more small firms than large ones. This basic fact remains as true today as it was in the 1930s; the economic data from any industrialized country confirm it. Economist Robert Axtell of the Brookings Institution in Washington, DC, says that, 'The stability of this distribution over time makes it . . . perhaps the most robust statistical regularity in all the social sciences.'[5]

Gibrat suspected that the factors underlying the way that firms grow include a considerable element of randomness, a conclusion he drew from the work of the Dutch astronomer Jacobus Kapteyn. Around 1916, Kapteyn showed that 'skewed distributions' in population ecology can arise from gaussian (that is to say, random) processes. Gibrat proposed that a firm grows at a random rate, and that the rate at any given moment is amplified by the firm's size at that moment.

This defines Gibrat's now-celebrated law of proportionate growth, which he envisaged as a kind of Newtonian law of the business world. It works as follows. To predict how much a firm changes in size between now and some future time, choose a number at random from between −1 and 1 (it could be, say, 0.5, or −0.3528, or zero) and multiply it by the current size of the firm. Thus bigger firms tend to change in size more markedly than small ones; but not inevitably so. This element of chance is included because the various factors that influence growth are hard to predict. To put the growth rule another way, the bigger a firm is, the more able it is to capitalize on whatever opportunities come its way.

Gibrat's 'law' leads to a size distribution with a mathematical form called 'log normal'.* He collected data on the plant sizes of French manufacturing firms for the years 1920 and 1921, and showed that their size distribution fitted his prediction rather well. And he didn't stop there. He presented data suggesting that his law held for different times (applying just as faithfully in 1896 as in 1921), for different economic sectors (industry and agriculture) and for different industries within a sector. By the 1940s Gibrat's model was highly regarded, and even today it is often used as a benchmark for theoretical and empirical studies of firm growth.

Nevertheless, the model is basically wrong – and everyone knows it. Whatever Gibrat's model says about the distribution of firm sizes, it is clear, simply from observing how real firms grow, that they do not expand and shrink in steps of random size. And this picture is certainly not consistent with the neoclassical idea of firms rationally maximizing their profits (which implies that firms should respond similarly, rather than independently at random, to changes in market conditions).

Thus, while Gibrat's law serves as a useful idealization, no one expects it to come very close to reality. Indeed, in the 1950s the whole idea of a universal law of growth fell out of favour. Instead, economists became convinced that market structure and dynamics varied from one industry to another – perhaps because different manufacturing

* This means that the probability distribution of the *logarithm* of firm sizes is gaussian.

methods offered different economies of scale, or because of the varying roles of advertising or of research and development. And more careful studies of size distributions began to suggest that Gibrat's 'universal' curves were probably just a fluke produced by inadequate data. Economist John Sutton of the London School of Economics summarizes the prevailing modern view:

there is no obvious rationale for positing any general relationship between a firm's size and its expected growth rate, nor is there any reason to expect the size distribution of firms to take any particular form for the general run of industries . . . [E]mpirical investigations from the 1960s onwards have thrown doubt on whether any single form of size distribution can be regarded as 'usual' or 'typical'.[6]

A rejection of Gibrat's model is vindicated not only by its apparent lack of conformity with how firms really grow but by its failure to provide more than an arbitrary prescription for that process: his law of proportionate growth is not justified by microeconomic principles. Yet the standard economic theory of the firm has nothing better to offer in its place. Robert Axtell of the Brookings Institution points to 'the inability of the neoclassical theory of the firm – with its U-shaped cost functions and perfectly informed and rational managers – to render a plausible explanation of the empirical size distribution'.[7]

This appears to force economists into the frustrating business of having to devise industry-specific models of firm growth which are geared to the particular quirks of that trade. So much for Gibrat's universality. In Sutton's view, 'The evolution of market structure is a complex phenomenon and the quest for any single model that encompasses all the statistical regularities observed is probably not an appropriate goal.'[8]

FIRM PRINCIPLES

Others are unsure about that. In 1996 physicist Gene Stanley, economist Michael Salinger and colleagues at Boston University looked at the growth rates of all publicly traded US manufacturing companies between 1975 and 1991. Encompassing around 8,000 firms, this was a larger data set than had been tackled in previous studies. The researchers found that the growth rates did not fit Gibrat's log-normal distribution, but instead followed a power-law relationship like that seen in critical phenomena. This means that plotting the logarithm of the growth rate against the logarithm of the probability of that rate produces a straight line. Or to be more accurate, *two* straight lines: one for positive growth rates and one for negative, both with the same slope. This generates a tent-shaped graph (Figure 11.1). And the power law holds true for two different measures of firm 'size': sales (revenue) and number of employees.

So there might after all be a general law of the growth of firms, albeit one different from Gibrat's. But there is more to it than that. Gibrat's log-normal distribution followed from the assumption that firms grow independently of one another – that the process is random, but 'weighted' by the size of the firm. Physicists' long experience with power laws, on the other hand, leads them to believe that such laws are the universal signature of interdependence. A power law generally emerges from collective behaviour between entities through which local interactions can develop into a long-range influence of one entity on another.

This idea is supported by a microeconomic model of the growth of firms developed by Robert Axtell in which firms arise by the aggregation of many interacting workers ('agents') each following their own agenda. The aggregation is driven by rules of the model that can

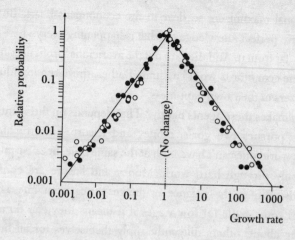

Figure 11.1 The probability distribution of growth rates for all publicly traded US firms between 1975 and 1991. The growth rates are measured by increases (or decreases) in both sales (black circles) and number of employees (white circles). The growth rates are generally bigger for bigger firms, but the figures all fall onto a single graph if one considers the *relative* growth rates – essentially, the percentage change in the company's size. The probability of a certain growth rate decreases as the rate increases according to a power law, so that the figures fall onto a straight line when plotted as logarithms. Decreases in size follow the same power law as increases, making the graph tent-shaped.

make collective labour advantageous. Axtell's model is in the spirit of microeconomics in that it attempts to deduce the overall behaviour of a system from the motivations of the individual agents that constitute it. But unlike many such 'theories of the firm', it starts with no preconceptions about what firms are for or how they behave; indeed, the agents are not forced to aggregate into firms at all. That they do so is an indication that it is in their best interests. These agents do everything with their best interests in mind. In that sense, they are the rational maximizers so dear to the economist's heart. But they don't have perfect knowledge of what is happening everywhere in the system – far from it. And they are not the avaricious 'profit maximizers' that some economists would assume. Rather, they are individualistic maximizers of their own happiness.

What makes these agents happy? That depends on the agent. Each of them pursues two goals: money and leisure. Unfortunately, in Axtell's world one can't have both at the same time, because money is made only through hard work. Money and leisure are conflicting demands. Where the balance is drawn varies from one agent to another: some settle for low wages if it means they can do a lot of lounging about, others diligently apply themselves for all hours in order to swell their coffers. The relative preferences for money and leisure vary across the agent population. So what each agent tries to do is to find a job that will provide the desired compromise between these two things. If it likes to work, it will join a firm of hard grafters. If it likes leisure, it might seek a firm that enables it to free-ride, putting in relatively little effort without this being noticed. The wages might be low, but hey, who cares?

In effect each agent is seeking to maximize its *utility* (page 259) – a measure of 'happiness', here denoting the preferred balance of labour and leisure. Jeremy Bentham invoked a similar concept in his philosophy of Utilitarianism, in which he argued that society should seek out the state that maximizes the collective utility of the population: the state of greatest total happiness. But Axtell's agents are nothing like as community-spirited. They don't give a damn what the collective utility is, so long as their personal utility is as good as they can make it, given the choices open to them.

It's a fair question why agents in this model should wish to join a firm at all. If they like slacking, can they not do that just as happily alone? Conversely, why compromise your hard work by teaming up with potential slackers? This issue was pondered by economics Nobel laureate Ronald Coase in a pivotal paper on firm theory in 1937, in which he wondered why the market did not simply consist of individuals trading with one another.

Coase's answer hinged on transaction costs: the inevitable costs of trading, such as the time, effort and often the financial burden involved in negotiating, drawing up and enforcing contracts, and so forth. Alfred Marshall, on the other hand, suggested in 1920 that the key to organized labour is to be found in the old adage that many hands make light work. That is to say, a certain amount of effort generally produces more returns if exerted collectively rather than individually. Division of labour, sharing of capital costs, specialization – all enable the collective to achieve more than the sum of its parts. Economists call this an *increasing return of scale* – the bigger you are, the more productive you are. Axtell provides an incentive for his agents to form aggregates – firms – by building this property into the mathematical equation that relates an agent's efforts to the productivity of the group in which it belongs.

Standard theories of the firm assume that companies grow because of increasing returns of scale. Crucially, Axtell's model does not. It merely asserts that individual agents get more done for the same effort if they pool their resources. This might lead one to *expect* increasing returns as firms themselves get bigger; but it does not *guarantee* that. (It does not, for example, specify just how hard each agent in a company actually works.) All it says is that it is in the interests of each agent to team up with others, no matter whether they are a workaholic or a slacker. In either event they get more for less, and they all like that.

The agents, each with its own proclivities, are at liberty in the model to join and leave firms (groups of agents) in search of the best deal. Thus firms can grow and decline as agents come and go. Each agent actually has relatively few choices. It cannot see what is happening in the whole market; the rules of the model permit it to know only how it is faring relative to a small number of 'friends', typically two. If these friends are doing better, the agent joins them; otherwise, it stays put.

The key point about the model is that each agent has a choice about how hard to work. The decision varies depending on the situation. If a slacker can find its way into a firm of hard workers, it can do virtually nothing and still profit. Each group of agents constituting a firm gets paid in proportion to the total output, and each gets an equal share. If a single slacker joins a firm of 50 good workers and then puts in no effort, so that the total output does not rise proportionately, it gets 1/51 of the rewards earned by the other 50. Their reward, meanwhile, barely changes: instead of being 1/50 of the total, it is 1/51 of the same total. So if the firm is relatively profitable, its workers might tolerate this slacker in their midst. (No way is provided, within the simplest formulation of the model, for hard workers to gang up on slackers and expel them. All they can do is leave the firm for another, if that will give them better rewards for their efforts.) But if the slacker is on its own, it cannot afford to put in no effort, since then it will have nothing to live on – and even leisure-lovers have to eat. In a small company, the presence of an agent which is not pulling its weight will be more noticeable than it would be in a large company. So the other agents may be more likely to leave, provided they can find pastures greener.

Each agent in the model makes a new decision at randomly chosen time intervals – as if waking up and thinking, 'Do I want to be doing this?' On becoming 'active' in this way, the agent readjusts its levels of effort to the best advantage of its utility. It also weighs up the relative advantages of staying put, joining a friend's firm, or starting a new firm on its own.

One of the big questions for a model like this is whether there are stable states in which every agent is happy with its position. Economists call such states *Nash equilibria*, after the mathematician

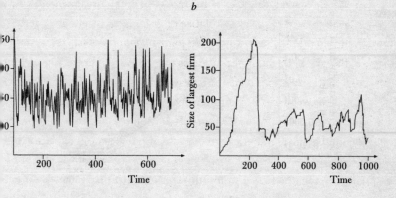

Figure 11.2. There is a constant turnover of businesses in Axtell's model of firm growth. The total number of firms (*a*) never attains a steady value, and even the largest firms (*b*) can collapse catastrophically.

ohn Nash, who proposed them in 1949. They correspond to situtions in which no agent can improve its circumstances by altering its ehaviour. Traditionally, economists are inclined to look for Nash quilibria in their models – these, they think, determine behaviour in he real world. But Axtell's model has no stable Nash equilibria. That is to say, it can never settle down into an unchanging state. There is onstant flux as firms boom and go bust (Figure 11.2). The model is a on-equilibrium one, and this makes it quite different from most microeconomic models of firm growth.

This is not to say, however, that nothing certain can be deduced bout the system. Rather, it means that we are forced to draw statistical nferences. We cannot say, for example, how long it will take for a firm ncompassing 25 per cent of the total labour force to appear. But we an, at any point, determine the probability of such a firm existing. ndeed, a statistical sampling of model 'runs' shows us the full istribution of firm sizes. It is a power law: a (logarithmic) graph of

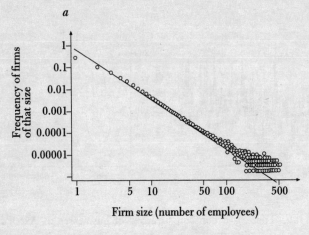

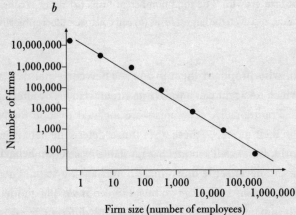

Figure 11.3 (*a*) The statistical distribution of firm sizes in Axtell's model. The relationship is a power law, displaying a straight line on a logarithmic plot. (*b*) The distribution of US firms in 1997, compiled from Census and Compustat data combined with self-employment data. The largest sector consists of 15.5 million firms with 'no' employees – that is, they are self-employed individuals.

firm size against the probability of there being a firm that size is a straight line (Figure 11.3a). And that is just what is observed in practice, as Axtell deduced from the statistics on about 20 million US firms in 1997 (Figure 11.3b). This is a striking result of the model – no other microeconomic theory of the firm has correctly forecast the power-law nature of this distribution.

Another telling test of the model is to ask about firm growth rates. Robert Gibrat, remember, argued that growth rates are randomly distributed. The real economic data show that, on the contrary, they are again distributed according to a power law, producing a tent-shaped plot on logarithmic scales (Figure 11.1). Axtell's model generates this same shape for the distribution of growth rates (Figure 11.4).

The sceptic might suspect that these power laws will arise no matter what we assume about the agents' behaviour – that power laws are somehow built into the model. But if we assume that agents switch firms at random (rather than to improve their utility), or that they choose levels of effort at random (rather than tailoring effort to circumstances), the firm size distribution is no longer a power law. So it seems that the power law is a consequence of purposive behaviour on the part of the individual agents in the market. On the other hand, the power law remains even if we alter many of the details of how the agents make their purposive choices. For example, we can increase the size of each agent's friendship circle (and thus increase its 'know-ledge' of the labour market), or we can boost the factors that promote increasing returns of scale, or we can build in factors such as induce-ments to stay with a firm (loyalty bonuses). These things alter the slope of the straight-line plot, but the line stays straight.

No one is claiming that this is a perfect model of firm growth. It neglects all manner of important things, such as management struc-ture, product specialization, and so on. Yet it seems that the model can

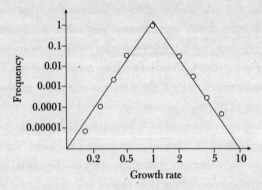

Figure 11.4 The rates of firm growth in Axtell's model show the same double-power-law or 'tent-shaped' distribution as that seen in the real world (Figure 11.1).

make realistic predictions about the statistical properties of firms. How can that be? Axtell suspects that there might be a kind of universality at play here which parallels that in physics, whereby the overall behaviour is insensitive to the details. He sees the origin of this universality in the fact that, regardless of the relative weights of the different factors that influence an individual's decision, the actual options are quite constrained. Basically, an agent can do one of three things: join a firm, leave a firm, or join a rival firm. Tweaking the rules does not alter these choices. As long as these options, and the underlying motivations behind them, remain, the market will acquire certain collective properties regardless of how we specify the details.

RISE AND FALL

As far as one can tell from a comparison with real-world data, this looks like a good description of the way firms develop. But the model doesn't just allow us to predict these broad, cold (if important) statistical measures of firm growth. Because it is built from the 'ground up' through interacting agents, it provides us with a history – in fact, a series of histories, one for each time we run the model on the computer. We can't expect any of these histories to mimic exactly what took place, say, in US manufacturing in the 1950s. But we can search them for typical characteristics of the life cycle of firms, the careers of individual agents, and so forth.

The first thing to notice is that most firms are ephemeral. It is not immediately obvious that the real business world is like this, because we notice the firms that last, such as General Motors or Ford. But most do indeed go under (or get taken over) on a relatively short timescale. Of the largest 5,000 US firms operating in 1982, for example, only 35 per cent still existed as independent entities in 1996. There is a high 'turnover' of companies, which many economic theories of the firm do not acknowledge.

So why do firms fail? In Axtell's model there is a typical trajectory. First, a new firm grows more or less exponentially over time as increasing returns cause workers to flock to it. But at some point the firm reaches its peak, after which collapse is usually sudden and catastrophic. Reduced to a tiny fraction of its former size, the firm struggles on for a while with a handful of determined workers before eventually vanishing (Figure 11.5). This collapse is a consequence of the firm's own success. Once it grows big enough, it becomes a haven for free-riders who capitalize on the efforts of others. So the firm becomes gradually riddled with slackers, until suddenly the other workers decide they have had enough and jump ship. (Note that firms fail in this model because the workers leave for better jobs, not because the market for their products disappears or because there is a terrible warehouse fire, or for any other reason. The failure is self-induced.) Tellingly, just before a firm collapses, the average effort of its workers plummets to zero.

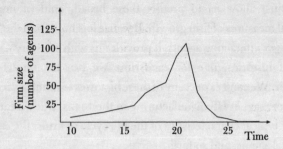

Figure 11.5 The simulated firms have a typical history: exponential growth followed by sudden decline and gradual dwindling. Here is one example.

Agents can't always do as well as they would like – their utility has ups and downs. But on the whole the utility of agents in large companies rarely drops below the average for the whole system; big companies prove to be pretty good buffers against misfortune. Only occasionally, and transiently, do agents in these firms experience below-average utility, and that is because they sometimes get stuck in an ailing firm with no better options in sight.

If we believe that such a simplified model can tell us anything at all about the real world, then we learn some revealing things about firms. First, they are not maximizers. Firms as a whole maximize neither profit nor overall utility (as conventional theories would have us believe). Individual agents *do* try to maximize their utilities, but this does not induce such behaviour in the group as a whole. The firms that do best are not those that aim to make the most profit. Rather, longevity in a company stems from being able to attract and retain productive workers. A firm fails not when its profit margins are eroded but when it is infiltrated by slackers.

The notion that putting profits first does not make a firm successful should not come as a surprise to those in the business community, although they have sometimes been reluctant to acknowledge it. Some market fundamentalists regard profit maximization not just as a principle of sound management but as a social obligation: the notorious 'greed is good' paradigm. But as British economist John Kay points out, it simply doesn't work. If the employees suffer from the profit motive, so does the firm:

The piece-rate systems of car factories were abandoned because they destroyed social relationships in the workplace, provoked endless negotiation and confrontation, and established a working environment in which no one cared about the quality of the product.[9]

Axtell's model does not allow for these complexities of worker interactions; but to the extent that it *does* imbue its agents with a kind of free will (how hard to work, whether to stay or quit), this worker choice comes immediately into play to determine the company's success or failure.

Axtell feels that traditional microeconomic theories of the firm have found themselves in a blind alley because of this neglect of the dynamic, ever-changing face of the labour market. The tendency, he says, has been to develop models that assume uniform intentions on the part of workers, and then to look for stable, equilibrium configurations in which each of them fits happily within one firm or another. These theories, says Axtell,

begin innocuously enough, with purposive agents in strategic environments of one kind or another, notionally similar to some known organization form (e.g., hierarchy). They then go on to derive the performance of the resulting firms in response to strategic rivals, uncertainty, information processing constraints, and so on. But these derivations are almost everywhere characterized by equilibrium theorizing, that is, inter-firm stationarity is seen as the result of intra-firm equilibrium and thus the homogeneity assumption is manifest.[10]

This quest for equilibrium, reminiscent of that of the founders of statistical mechanics, is more transparently misguided in today's era of job-hopping. But even in former times, the idea of a job for life gave a misleading impression of the labour market, afflicted as it was by just as many unpredictable fluctuations and crashes. A theory of the firm must necessarily be a non-equilibrium theory in which the future is never certain.

By looking at real economic data, we can see that there most probably are general laws describing how firms develop and grow. Models like Robert Axtell's give us some hope that we can understand where these laws come from. But ultimately we must ask another question: what do we want? As is ever the case with a statistical social physics, we cannot expect or even hope that it will show us how to predict or control the fine details, such as whether Brian's butcher shop can survive when a big supermarket chain moves into Brecon. This is one of the major limitations of the approach. But we can perhaps hope to identify conditions under which shops like Brian's might have a fighting chance.

12

Join the Club

Alliances in business and politics

The science of society would have attained a very high point of perfection if it enabled us, in any given condition of social affairs – in the condition, for instance, of Europe or of any European country at the present time – to understand by what causes it had, in any and every particular, been made what it was.

John Stuart Mill (1843)

The call for anti-fascist unity was, in some ways, likely to win the most immediate response, since fascism publicly treated liberals of various kinds, socialists and communists, any kind of democratic regimes and soviet regimes as enemies to be equally destroyed. In the old English phrase, they had all to hang together if they did not want to hang separately.

Eric Hobsbawm (1994)

Why do wars and revolutions happen? We do not know. We know only that to produce the one or the other men form themselves into a certain combination in which all take part; and we say that this is the nature of men, that this is a law.

Leo Tolstoy (1869)

*

The truth is, I'm a Mac user. That might mark me out, I suppose, as a radical, an iconoclast, a freethinker. At least, this is what Mac users like to think. Those who choose Microsoft-based PCs might instead see us as mugs who've picked the losing side.

Sometimes the marketplace presents a choice almost too dizzying to contemplate – ordering a coffee in the United States is always a brain-taxing task. Sometimes, in contrast, there seems to be no option at all. For standard mail, for example, there is only one postal service. But sometimes the system bifurcates, and then the decision becomes momentous: will you take the right-hand path or the left, knowing that there is no going back? Life confronts us with many binary choices. We are then like a magnetic atom wondering which way to point our compass-needle spin – and we are likewise inclined to be influenced by the choices of those around us. The result can be a rich and subtle interplay of interactions, leading to outcomes which are impossible to predict without considering the dynamic of the group as a whole.

In this chapter and the next two, I shall consider various social situations in which we are presented with stark and mutually exclusive choices. I begin here with a special class of situations commonly met with in big business and politics: the formation of mergers and alliances. Companies do not always grow in the manner outlined in the previous chapter, by a steady accumulation of employees or a gradual rise in sales. Instead they may merge with a competitor in the hope of gaining a synergistic advantage. Or they might choose to collaborate while retaining separate identities. In the ruthless world of business mergers there may be little room for small fry. Players who do not secure a place as market leaders may have no option but to add their inferior muscle to the strength of the giants: to become engulfed by, or at least conform to the ways of, the big guys.

In multi-party political systems, alliances can decide the balance of power. As the number of parties increases, it may well be that no single party can do without the support of others. Minor players suddenly find themselves with make-or-break power. This has been the case for decades in the German political system, where for example the Social Democratic Party formed (West German) governments in coalition with the Free Democratic Party from 1969 to 1982.

History is alive with the consequences of coalitions between politicians and nations. Would Napoleon have lost at Waterloo without Prussian intervention? Was the United States liberated not by George Washington but by a makeshift union of European powers eager to see British influence diminished in the New World? How would Athens have fared against the Persians without the aid of its long-time foe, Sparta? These may seem idle questions to many historians, who as a rule are interested only in what did happen and not what might have. But we shall see how, with some inspiration from physics, one can enquire further about that 'might have', and start to explore the contingencies of history. While there is always just a single way in which the cookie crumbles, we can begin to systematize, rather than just guess at, the alternatives – and thus make 'counterfactual history' a more concrete and objective pursuit.

SETTING STANDARDS

In the pantheon of businesses there is space at the top of the pile for just a select few, their supremacy defined by size, market share and revenue. But even big firms are sometimes forced to cooperate rather than compete with their rivals, for example when new technologies raise the issue of standards and compatibility. Unless products sold by different companies are technically compatible with one another, each company risks being excluded from some segment of the market.

This is often perceived as a modern concern, exemplified well by the struggle between the VHS and Betamax video systems in the 1980s. But the issue is an old one. It has been claimed that the inefficiency of the railroad system in the southern states of the USA during the 1860s, owing to the existence of three different track gauges, helped to ensure the Union's victory over the Confederacy. (In the North, gauges were already largely standardized.) During the development of the typewriter, companies were faced with the choice of whether to conform to the QWERTY arrangement of keys or to innovate with some new configuration. QWERTY is now a petrified accident of history, and is far from being the optimal design – the accessibility of the keys does not correlate with their frequency of use. Some even suggest that QWERTY was chosen in the nineteenth century *because of* its inefficiency, which avoided the mechanical keys becoming jammed as excessively dextrous typists typed.* When the phonogram was invented, record manufacturers had to agree on the standard rotation speeds of 16, 33, 45 and 78 r.p.m. (and later had to decide whether it was worth persisting with the first and last of these). The conflict between metric and Imperial units of measurement is notorious throughout the engineering industries, and the existence of the two standards still causes difficulties on either side of the Atlantic. (NASA was, to its cost, reminded of this in 1999 when a failure to convert between the two systems of units caused an error that doomed a $200 million space mission). National grid voltages, the size of early compact discs, colour coding for computer graphics – the list goes on.

---

* The standardization of railway gauges and of typewriter keyboards are classic examples of so-called 'path dependence' and 'lock-in' in the evolution of economic markets. In such cases, the eventual outcome is history-dependent. Brian Arthur has developed an agent-based model to investigate how railway gauges become standardized. He argues that path dependence may result in the eventual emergence of a relatively inefficient or less-than-optimal standard, challenging the conventional (neoclassical) idea in economics that market competition always selects the best outcome.

The desire for agreed standards is obvious from the consumer's point of view, but for firms the best option is less clear. If the market is fragmented into many different user groups with functionally equivalent but mutually incompatible hardware, then any firm can capture only a small part of it. (Few consumers will be prepared to buy into more than one system.) If, on the other hand, a firm decides to standardize its product's operational specifications with those of a rival, it can sell to a broader market but at the cost of sharing it. Moreover, the capital costs of altering a manufacturing process to fit a new standard represent a significant disincentive. Yet if one standard begins to get an edge over the others, it might pay to adopt this because, while that may mean compromising with one rival, the company can enjoy an advantage over several others.

In some cases a technical standard is agreed throughout an industry or a nation, perhaps even enforced by legislation. But this is rare. More often, market forces push an industry into adopting a few competing standards, and companies or national industries are obliged to align themselves with one camp or another. Commonly, the competing camps are whittled down to just two. This can be seen as a natural outcome of two opposing tendencies. If a company has to join one alliance or another, it might as well be a big one: so small alliances die out and their total number decreases. At the same time, the company doesn't want to enter into collaboration with a firm with which it is in fierce competition. When there are just two alliances, each company can be part of as big a group as possible while still actively opposing its main rival.

The evolution of technical standards for computer operating systems in the 1980s was a classic example. A computer's operating system is the software package that controls the hardware and the way in which information is processed and passed around; it is, if you like, the computer's native tongue. Anyone with the right technical skills can write an operating system – can invent a language – and in the early days of computer technology, that is just what happened. The computer industry could have become a Tower of Babel.

But in the late 1960s two computer scientists at AT&T's Bell Laboratories in New Jersey devised a sleek and elegant operating system that became known as Unix. It was hugely popular, but AT&T could not profit from it because at the time the company was prohibited by US law from entering the computer business. So it decided to give away Unix at a nominal cost to all who wanted it – which proved to be almost everyone. Yet those who acquired the nearly cost-free licence for Unix were also entitled to alter and improve the language. And that they surely did. By the 1980s there were around 250 different versions of Unix in use, each of them incompatible with the others. It was rather like speciation in nature: populations accumulate small mutations until they become genetically incompatible and can no longer exchange (genetic) information.

The market in computer workstations, most of which ran on Unix, had grown by then to enormous proportions: by 1990 it was worth $10 billion. There was a strong motivation for different companies to standardize their Unix operating systems. And so began a battle of wills. The first move was made in 1987, when Sun Microsystems and AT&T agreed to use the so-called Unix System V, developed at AT&T in the 1970s. AT&T would license this system to others – but now at a price. The prospect of being in thrall to AT&T forced seven of its rivals, including the Digital Equipment Corporation and IBM, to form an alliance in May 1988 called the Open Software Foundation (OSF), which intended to develop a different standardized Unix operating system. In response, AT&T and Sun formalized their own alliance in late 1988 by establishing Unix International Incorporated (UII). The industry became polarized into two camps.

As it turned out, neither was really the winner. UII disbanded in 1993, but OSF had only a short time to savour its apparent victory before it too began to fragment through the inability of its members to resolve their competing interests. Happily, however, a consensus began to emerge in 1994 when a consortium of manufacturers called X/Open Company (which merged with OSF the following year to form the Open Group) agreed on the so-called Single Unix Specification. Supported by governments and commercial suppliers, all eager for a standard operating system, the major computer companies began to ensure that their products conformed to this specification.

The initial growth of OSF and UII forced computer companies to choose one way or the other. How should they decide? Can we ever hope to foresee the outcome of such alliance-forming in industry?

HEAD FOR THE VALLEYS

Political scientist Robert Axelrod at the University of Michigan says yes, we can. And he and his colleagues have a theory to prove it. They call it *landscape theory*, and it is a tailor-made form of statistical physics. The players in this game are like gas particles on the point of condensing – perhaps into one big droplet, but more probably into two or more smaller drops. They are drawn to one another by a kind of attraction, yet are also kept apart by repulsions. Out of this push and pull emerge configurations in which the particle-like agents are aggregated into alliances.

The condensation of particles into discrete clusters is a common phenomenon in physics: think of water vapour in the air gathering into raindrops or the ornate filigree of snowflakes. Clusters also form if a mixture of two different liquids is suddenly 'quenched' – plunged to a temperature at which the two are no longer miscible, so that they begin to separate out. This is common in metallurgy, where two alloyed molten metals segregate as they cool, forming little blobs of the two pure metals. The ideal final configuration of the metal atoms – the one with lowest energy – might be a complete separation of the two types, like a layer of vinegar sitting beneath a layer of oil in salad dressing that has been left standing for too long. But small clusters of each metal may get frozen in place before they can merge into a single layer, leaving the two metals intimately interspersed.

Often the growth of these clusters is self-amplifying: the bigger they get, the faster they grow, because of the increase in the area of the surface on which a cluster accumulates more particles. Small clusters are doomed to disappear or get swallowed up. In physics, this is known as Ostwald ripening. It is, if you like, a case of the rich getting richer and the poor getting poorer; in the business world it could serve as a metaphor for takeovers and globalization.

The picture makes intuitive sense when applied to business alliances: companies are naturally drawn towards the coalitions that look set to emerge as the strongest. But there are other forces at play too. A fundamental impulse is not to align with rival firms but to oppose them and try to squeeze them out of the market. Microsoft's controversial attempts to monopolize the personal computer market in the late 1990s demonstrate this stark Darwinian law of business. Arising out of necessity, alliances between firms can oppose such a one-sided takeover.

On the one hand, then, the force of attraction between two firms might be considered to increase in proportion to their size: the bigger firm A is, the greater the inducement for firm B to aggregate with it (it makes sense, for example, for a tiny computer manufacturer to align itself with a giant such as Sun Microsystems). But the counteracting repulsion between the two firms will depend on how much antipathy exists between them, which is likely to depend on the extent to which their products and markets overlap. Two computer companies which both specialize in PCs for the US domestic market will regard each other as closer rivals than they will a third company which mostly sells software or mainframes to big businesses. And rivals are rarely charitable: firm A might choose to join one alliance not only because it is big but because, by making it bigger, the firm increases the chance that the other alliance, to which rival firm B belongs, will be overwhelmed. Even if B's alliance is bigger, A might join the other side so as not to help B's cause.

In the landscape model of alliance-formation developed by Axelrod and his colleagues, each firm is therefore like a particle with an individually tailored force of interaction towards every other particle. The attractive component of the force that A exerts on B depends on how big A is. The repulsive force depends on whether B is a close or only a distant rival of A. The model is like van der Waals' fluid, except that each particle is unique and there is typically only a handful of them. The principle that governs their final configuration is the same as it is in traditional statistical mechanics: what is the most stable way to arrange them? In other words, what is the equilibrium state?

To find this state, Axelrod and his colleagues defined a kind of 'total energy' for a group of firms, which they calculated by summing all the forces of attraction and repulsion between each pair when the firms are aligned in various coalitions. If close rivals are clustered together in the same alliance, the total energy is relatively high because of the firms' mutual repulsion. A better, more stable arrangement puts such competitors in different camps. In the lowest-energy, equilibrium configuration, no firm can bring about any further stabilization by switching from one camp to another. This is a Nash equilibrium (page 328), and it is the partitioning we should expect to find in reality.

This is not to say that everyone is happy in the lowest-energy configuration. Indeed, almost inevitably some agents will not be. There are usually only two big alliances; yet all firms consider others to be rivals to a greater or lesser degree. So they will all have to share a

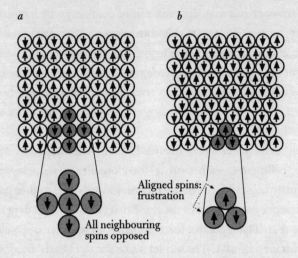

Figure 12.1 (*a*) In an Ising antiferromagnet, the spins of neighbouring atoms point in opposite directions. (*b*) If the atoms are arranged on a triangular rather than a square lattice, it is impossible for each atom to orient its spin to oppose those of all its neighbours: there is *frustration*.

bed with some rivals – perhaps even with close ones. There is no other option, other than the decidedly unwise course of going it alone. Thucydides understood in the fifth century BC how the threat of annihilation makes for unlikely bedfellows: 'mutual fear is the only solid basis of alliance'.[1]

The most stable state, then, inevitably contains some 'frustration' of each firm's desires. I use this term advisedly, for it is drawn from an analogous situation in physics. In the Ising model of magnetism, every magnetic atom sits on a regular lattice and points its needle (spin) in one direction or the other. In a so-called ferromagnet such as iron, the most stable state is the one in which all spins point in the same direction. But in some magnetic materials the interactions are such that neighbouring atoms prefer not to align their spins but to point them in opposite directions. These are called antiferromagnets.

If the spins are arrayed on a square lattice, this requirement of opposed alignments can be satisfied (Figure 12.1a). But if the spins are instead placed on a triangular lattice, complete satisfaction is no longer possible. Arranged thus, the spins can then be regarded as being grouped in threes, with each member of the trio equidistant from the other two (Figure 12.1b). Any two of them can have their spins in opposing directions, but then the third must be aligned with one or other of them. This is called frustration: there is no way to satisfy the conflicting demands all at once. It means that for an Ising antiferromagnet on a triangular lattice there is no unique and clearly defined most stable state – there is always some degree of imperfection or disorder in the orientation of spins, no matter how we align them. Such a system is called a *spin glass.**

Instead of possessing a single equilibrium state, a spin glass has a multitude of different spin arrangements which all have very similar energies. One way of depicting this is to invoke the concept of an *energy landscape*: a map of all possible arrangements and their respective energies. We can see how spin configurations might be represented graphically by analogy with a chess game. The beauty, the art and the enduring appeal of chess lie in the fact that, on its eight-by-eight grid, with sixteen pieces on each side, there are an astronomical number of different possible arrangements and ways of moving between them – so much so that even the world's most powerful computers cannot evaluate the best move by enumerating all possible subsequent outcomes. Yet in principle we could systematically work our way through every configuration of the pieces that can arise in the course of a game, and each of these configurations can be plotted as a point on a graph.

* The word 'glass' is used here by analogy with glassy materials, in which, unlike in crystals, the positions of the atoms are not perfectly ordered and regular.

For example, one graphical axis can represent the position of the white queen, running from 1 to 64, with each square assigned a number. (To be exhaustive, we might go from 0 to 64, with zero denoting absence from the board.) A second axis could represent the position of the left-hand red bishop. Then a configuration in which the white queen is on square 5 and the red bishop on square 42 is denoted as a point on this 64-by-64 graph (Figure 12.2a). Of course, this isn't enough: what about all the other pieces? To fully specify the state of the board, we need a graph with 32 axes, one for each piece. Since we run out of spatial dimensions after three, we can't draw such a graph. But there is nothing in principle to prevent us from thinking in terms of such a multi-dimensional 'cube', as long as we don't worry too much about being able to visualize it. Every grid point on this 'hypergraph' corresponds to a particular arrangement of all the chesspieces.

The same is true for the arrangements of spins in a spin glass, or indeed in any Ising model of a magnet. Each configuration of all those spins – whether each of them points 'up' or 'down' – can be represented as a grid point on a hypergraph with as many dimensions as there are spins. And each of those configurations has a particular total energy, which is calculated by adding up all the different interactions between pairs of spins – some favourable, some not.

Now we can add one further axis to the graph to denote these energy totals. It is asking a lot, admittedly, to visualize this; but the task becomes easier if we pretend that the gamut of different configurations is represented not by a multi-dimensional space but just by a two-dimensional surface (Figure 12.2b). Each grid point on this surface corresponds to a particular arrangement of all the spins. And now we assign each of these points a *height* that corresponds to the energy of that configuration. Join these points up and we arrive at the energy landscape (Figure 12.2b), a kind of topographic map which shows how the energy varies as the configuration of spins is changed. The more

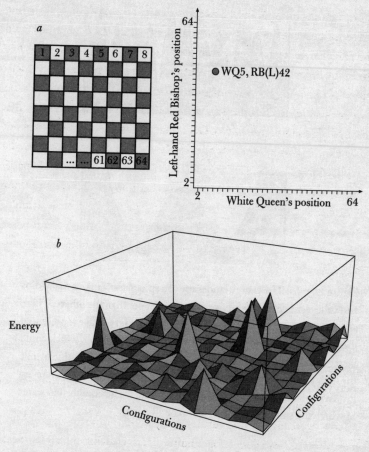

Figure 12.2 (*a*) Configurations of a chess game can be represented by points on a graph. A two-dimensional graph like this one can show the locations of two pieces. To include all 32 pieces we would need a 32-dimensional graph. (*b*) If we reduce the multi-dimensional graph of spin states of a spin glass to a two-dimensional surface, we can imagine an energy landscape in which the height corresponds to the energy of each respective state.

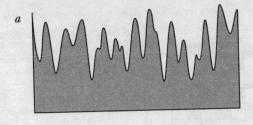

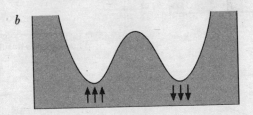

Figure 12.3 (*a*) The energy landscape of a spin glass contains many dips and valleys, none of which is significantly deeper than the others. There are, in other words, many *local energy minima*. Here I have represented these dips and bumps on a simple two-dimensional graph, which could be regarded as a slice through the landscape – like a topographic profile through a mountain range. (*b*) For a ferromagnet, in contrast, there are just two valleys of equal depth, corresponding to the equilibrium states in which the spins are pointing all up or all down.

stable a configuration is, the lower its energy. If a configuration cannot be altered in any minor way (flipping just a few spins) without increasing its overall energy, then it is 'locally' stable and corresponds to a depression in the landscape. Moving in any direction away from such a state means moving uphill.

For a spin glass, the energy landscape is typically very rugged. It is covered in dips, each of them corresponding to a relatively low energy state (Figure 12.3*a*). Some dips are lower than others, but none looks obviously much lower than all the others. For a normal Ising ferromagnet, meanwhile, the situation is quite different. There are just two deep valleys: one corresponding to the configuration in which all spins are 'up', and the other to the configuration in which they are all 'down' (Figure 12.3b). In the absence of anything to bias the choice, these two states are equivalent, with equal energies (equal heights). Small changes to either of these equilibrium configurations – flipping just a few spins to oppose the others – increase the total energy slightly by creating unfavourable alignments. Big changes – flipping a lot of spins – raise the energy a lot. So the landscape rises gradually and smoothly all around the lowest points.

The Ising model, remember, can also represent a fluid that can form a liquid or a gas. So we can imagine an energy landscape for a system of particles in which the grid coordinates correspond to different spatial arrangements of the particles and the contour heights correspond to the energy states determined by the forces between particles. This is the equivalent of Axelrod's landscape model of alliances, in which the 'particles' are gathered into clusters that are forced to endure a certain amount of frustration because of their mutual 'dislike'.

BALANCE OF POWER

To find the most likely configuration of allied agents in their model, Axelrod and his colleagues survey the energy landscape to look for dips. Physicists do something similar when they seek the most stable arrangements of many interacting particles in computer models. Because they are typically dealing with thousands of particles, the number of possible arrangements is immense, forcing them to rely on special computational techniques to feel their way blindly around the landscape and look for the downward slopes. But if the number of agents is small, the search can be done exhaustively simply by calculating the 'energies' of all possible aggregates and picking out the one with the lowest energy. This is what Axelrod's team did.

Because of the frustration in the landscape model, one might expect the energy landscape to be rugged, like that of a spin glass. But the picture is simplified by the fact that the agents are not all identical, as they are in a spin glass. This imbalance means that some alliance configurations are much more stable than others. With two particularly big and powerful rival firms on the scene, for example, configurations in which all the smaller firms aggregate around one or other of these two will tend to be much more stable than any other arrangement. So, contrary to what intuition might suggest, the existence of a few big rivals can help to create stable patterns of alliance. The partitioning of many competing, medium-sized firms, on the other hand, can be very sensitive to small changes in the forces between them, giving rise to a constant shifting of allegiance as the system jumps between one dip and another in a rugged landscape.

The big question is, does it actually work? Can the landscape model predict what happens in reality? To put it to the test, Axelrod and his colleagues used the example of Unix standardization. Would the model provide an accurate 'hindcast' of how the industry split, given the state of play just before the alliances were launched? There were nine principal US computer firms involved in the coalitions of the late 1980s, all of varying size and with different degrees of rivalry. Some, for example, were specialists in making technical workstations controlled by the Unix operating systems; others were computer-products generalists. The researchers assumed that the rivalry would be more intense, and thus the 'repulsion' greater, between two specialists than between a specialist and a generalist.

There are several measures one could use for the 'size' of a firm, such as the market share or the net corporate assets. Axelrod and colleagues chose to represent size by the firms' share of the workstation market in

1987. As for the relative strength of the forces of repulsion between close and distant rivals – well, there was no obvious and unique way of deciding that. But the researchers simply tried out a range of plausible values. They found that, except at extreme ends of this range, the outcome was not greatly affected by the precise strength of the forces.

They assumed that just two alliances would form. But who joins with whom? There are 256 possible ways of dividing nine firms between two camps. Yet the calculations of the energy landscape showed that in general there were just two stable configurations – and the one with the lowest 'energy' was a very close match to the actual split between OSF and UII. In this configuration, only one company (IBM) was placed in the wrong camp ('UII' rather than 'OSF'). As expected, the two alliances had similar sizes in both the stable configurations. Since the probability of getting this close to the historical reality by pure chance is about 1 in 15, it looks as though the landscape model does a good job.* Had it been available to the companies concerned in 1988, they could have used it as a fair guide for predicting how things would turn out – and for helping them to decide which way to jump.

Industrial decisions like this are often made on the basis of all manner of long-term forecasts and cost-benefit analyses. But the landscape model invokes nothing of the sort. Instead, firms act on a decidedly myopic vision: in effect, they simply look at each competitor in turn and

* There is an important proviso, however. French physicists Razvan Florian and Serge Galam have shown that if the model is not constrained to produce just two alliances, but is left to form however many clusters it wants from the nine companies, neither the configurations identified by Axelrod's group nor the one seen in reality emerge as clear contenders for the best arrangement. Instead, the average number of coalitions formed is between six and seven, and those arrangements are not strongly preferred over a situation with no alliances at all. The unresolved question is then whether there are good arguments for insisting that scenarios like this will indeed tend to produce just two coalitions, as I discuss on page 341. Certainly, the model itself does not seem to insist on that.

ask, 'How do I feel about them?' And this feeling is assumed to be uninfluenced by the company that the firm keeps: the interaction between A and B is the same irrespective of whether or not B has joined forces with C.* Moreover, the agents are assumed to converge on their final configuration by a series of small, independent steps, each of which takes the whole group farther down the slope toward equilibrium. In the words of Axelrod and his colleague Scott Bennett,

the idea of descent need not be justified by an appeal to far-sighted rational decision-making, but can easily be the result of a process in which each actor responds to the current situation in a short-sighted attempt to achieve local improvement.[2]

In effect, unless they change their attitudes to their rivals, the firms are at the mercy of an inevitability. It is with this in mind that we turn to a more dramatic and more far-reaching reason to conclude that it is worthwhile searching the landscape to see what the future holds.

EUROPE DIVIDED

If anyone has ever proposed that chess might be played as a game with more than two sets of pieces, it has never caught on. Neither has a version of football evolved in which more than two teams occupy the pitch. We do not play three-way tennis. This must be more than a matter of the limits to physical and mental coordination – for we all

---

* This needs to be stated quite carefully. The incentive for firm A to join an alliance containing firm B depends on the summed interactions of A with each firm in B's alliance. In this sense, whether or not A and B join forces depends on the company B keeps. But the interaction between A and B themselves is not affected by whatever alliance B is in.

know what tends to happen in competitions between three players. Sooner or later two of them will gang up to eliminate the third. There is something inherently unstable about competitions with more than two main players, which is one of the complaints often directed at multi-party politics. Even in the UK, where the Labour and Conservative parties have for decades far outweighed the Liberals, talk of coalition has frequently been in the air.

This takes on an all too deadlier significance when the game is not chess, but war. Cromwell's army represented a complex mixture of different interests, but it disintegrated into fighting factions only after the common enemy, the Royalists, no longer threatened it. Twice in the twentieth century, Europe's many disparate nation states formed up into two big alliances which set about destroying each other. If nerves or luck had not held out, the nations of NATO and the Warsaw Pact would have done the job much more comprehensively in the decades that followed the Second World War. Political scientist Kenneth Waltz sums up the bilateral nature of warfare:

The game of power politics, if really played hard, presses the players into two rival camps, though so complicated is the business of making and maintaining alliances that the game may be played hard enough to produce that result only under the pressure of war.[3]

As I intimated earlier, the Second World War is often portrayed as an inevitable consequence of the First. An embittered Germany, resentful of the indignity of enormous reparation debts, was primed for a leader who would restore Teutonic power and pride. There is surely truth in that; but Germany did not fight alone. Not only Italy but also Hungary and Romania joined Hitler in the Axis alliance, which for a short but dangerous time seemed poised to overwhelm the Allies. Did Europe have to split this way? Did each alliance build up through a series of independent and entirely contingent negotiations? Or were broader forces at work?

There can be few more exacting real-world tests of a model of coalition formation than the onset of the Second World War. Seventeen nations were involved; there are 65,536 ways of dividing them between two camps. Of course, many of these would be historical absurdities (it was never likely that all the rest of Europe would unite against Estonia), but it is by no means trivial to anticipate the alliances that emerged in the late 1930s.

Axelrod and his colleague Scott Bennett were bold enough to confront the landscape model with this challenge. The main difficulty was finding a way of quantifying the interactions between the different nations – the forces that propelled them into one or other camp. Traditionally, political scientists have taken the grimly named 'realist' view that all states view all others as potential enemies: that all 'repel' one another. There is ample historical justification for this rather bleak outlook; but of course states have also tended to respond to or anticipate very particular threats, for example because of economic competition or ethnic or ideological divisions. Similarly, they might feel common cause with other states on the same grounds. Political scientist Glenn Snyder observes that alliances emerge from this mix of 'conflicts and commonalities'.[4]

Translated into the language of physics, conflicts and commonalities become repulsions and attractions. But how does one ascribe numbers to them? Most 'neo-realist' political scientists have no idea. Once the problem is expressed in general terms, however, rather than according to the specifics of historical contingency, one can start to see a way forward. Axelrod and Bennett classified the interactions between each pair of states on the basis of six factors, each derived from the political, economic and demographic situation in 1936: ethnicity, religion, territorial disputes, ideology, economy and past history. For example, some states were predominantly Catholic in 1936, others Christian

Orthodox, others atheist. In terms of ideology, some had democratic governments, others fascist, and so forth. The researchers considered that the existence of border disputes (for example between France and Germany) contributed to repulsion, as did a recent history of armed conflict between two states.

One can tick such boxes easily enough, but it is also necessary to assign a weight to each factor. Axelrod and Bennett made the simplest possible assumption: each factor was weighted equally, set to +1 if there was commonality and to -1 if there was difference or antagonism. This is of course entirely arbitrary, but they had to start somewhere. As for the 'size' of each nation – again, many measures are conceivable, such as population or GNP. The researchers decided to use the 'national capabilities index' assigned to each nation by the Correlates of War project conducted by US political scientists since the 1960s. This index attributes a degree of 'power' to each nation based on six measures of demography and military and industrial strength. Then it was simply a matter of mapping out the landscape of 65,536 points, each of them corresponding to a configuration of all the nations divided into two groups. These groups were taken to represent nations which, at the outbreak of hostilities, had declared war on, or been invaded by, one or more of those in the other group.[*]

On the basis of this criterion, the Allied powers in reality comprised Britain, France, the Soviet Union, Czechoslovakia, Denmark, Greece, Poland and Yugoslavia. The Axis nations were Germany, Italy, Hungary, Estonia, Finland, Latvia, Lithuania and Romania. Portugal

---

[*] In other words, membership in an alliance here does not imply that there was any formal agreement to fight alongside the other nations in that camp. The Baltic states 'joined' the Axis camp by default when the Red Army crossed their borders in 1940, making the Soviet Union the enemy. In fact, the massacres orchestrated by the Stalinist regime reportedly left some of those in Estonia, Latvia and Lithuania praying for a Nazi 'liberation'.

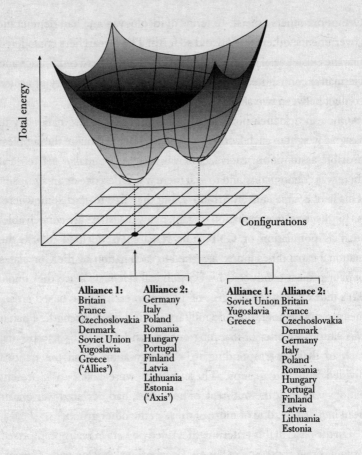

Alliance 1:	**Alliance 2:**	**Alliance 1:**	**Alliance 2:**
Britain	Germany	Soviet Union	Britain
France	Italy	Yugoslavia	France
Czechoslovakia	Poland	Greece	Czechoslovakia
Denmark	Romania		Denmark
Soviet Union	Hungary		Germany
Yugoslavia	Portugal		Italy
Greece	Finland		Poland
('Allies')	Latvia		Romania
	Lithuania		Hungary
	Estonia		Portugal
	('Axis')		Finland
			Latvia
			Lithuania
			Estonia

Figure 12.4 The two basins (energy minima) in the landscape of alliance formation immediately before the Second World War. The deepest basin corresponds closely to the observed split into Allied and Axis powers, with only Portugal and Poland placed in the 'wrong' camp. The other basin predicts a very different history, with Europe united against the Soviet Union.

was an anomaly, formally neutral but with a defence agreement with Britain, which is why it should really be included with the Allies rather than excluded altogether like Switzerland and Sweden. The results generated by the landscape model are remarkable. There are just two broad basins in the terrain (Figure 12.4). One of them (the deepest) corresponds to a configuration in which the alliances match the list above almost exactly: the only wrong assignments are for Portugal and Poland, both placed in the Axis camp. The probability of getting a match as close as this by chance is less than 1 in 200.*

The basin that surrounds this configuration in the energy landscape is more than twice as large as the one that leads to the other low-energy configuration, so it is clearly predicted as the most likely outcome, reached from the majority of 'starting points' in the landscape. But what is the alternative? Strikingly, the second basin speaks of a very different war: one in which almost all of Europe, including Britain, France *and* Germany, is united against the Soviet Union, which is allied only with Yugoslavia and Greece. (These two countries had a recent history of antagonism with Germany, increasing their tendency to join the opposite camp).

This may sound a preposterous prediction. But there is nothing inherently absurd about the notion of a war, in the 1940s, between Western Europe and Stalin's expansionist state. After all, the Nazi–Soviet pact, which lasted until 1941 (a ruse that in no way lessened the two states' aggressive intentions towards each other), all but made Britain and France enemies of Stalin in the war's early years.

---

* Again, relaxing the restriction to just two coalitions changes these conclusions, and degrades the model's ability to match the historical picture. But for the reasons given earlier, the two-alliance condition is defensible. Moreover, without this constraint the model produces three coalitions, with the Soviet Union split from the other allies: a situation which, as I shall argue, was quite likely had it been politically feasible.

There had already been interventionist calls in both Western countries when the Soviet Union invaded Finland in 1939. When Britain declared war on Germany, it was only its lack of military capacity that held it back from fighting the Soviets too. And when Churchill and Stalin did eventually unite, the tension was immense (all the more so when the USA entered the war). The Alliance created, in historian Eric Hobsbawm's words, an 'astonishing unity of opposites, Roosevelt and Stalin, Churchill and the British socialists, de Gaulle and the French communists'.[5] One could see cracks in the Alliance developing well before the war ended.

Thus, if the landscape model is to be believed, the war between Britain and Germany was the most likely but not the only possible outcome of the international situation in 1936. In this picture, the question of which 'valley' Europe entered depended on where it started. Since the Allies/Axis divide has a much bigger basin, it was more probable that history would end up there. But if the balance of relations between nations had been somewhat different, perhaps Britain would have identified Stalin, not Hitler, as its most dangerous enemy.

What about the mistakes in the prediction, minor though they may seem? Portugal is, as we've seen, an unusual case. And it is not hard to understand how Poland could be placed in the 'wrong' camp. It was antagonistic to both of its powerful neighbours – Germany and the Soviet Union – in virtually equal degree, and with good reason: the Soviet Union, after all, invaded Poland just sixteen days after Germany, and Hitler and Stalin carved up the country between them. Arguably Poland, as a twice-conquered nation, belongs in both camps, or neither.

Moreover, even this 'error' is redeemable if one uses the data not from 1936 but from closer to the outbreak of war. The 'size' of the nations changed between 1936 and 1939, most dramatically in the case of Germany as it rapidly expanded its military forces and occupied territory to its south and south-east. Based on data for 1937 there are still two possible configurations, much the same as those above. But by 1938, there is only one: the most realistic one. And by 1939, shortly before the outbreak of war, the increase in Germany's 'size' has increased Poland's repulsion towards it to such an extent that Poland falls instead into the Allied camp in this single equilibrium configuration. Thus, just before war was declared (and so before the alliances were cemented), the model predicts precisely the way things will go – a result that has a less than 1 in 3,000 chance of arising fortuitously.

The sceptic might well ask whether the Allied/Axis division was not a foregone conclusion, insensitive to the details of what we assume about international relations. Does the landscape model, for all its apparent success, actually do any better than a 'realist' approach which assumes that everyone regards everyone else with equal suspicion? It is easy enough to run the landscape model under this set of conditions instead – whereupon the uniformity of interaction forces makes the landscape resemble that of a spin glass, with no fewer than 209 stable configurations instead of just two. None of them is an accurate picture of the historical truth.

REWRITING HISTORY

The landscape model does something more profound than provide a retrospective prediction of the course of history, impressive though that is. It gives us a picture of the historical landscape: a map of possibilities. What are we to make of this? Many historians baulk at the idea of discussing might-have-beens, what they call 'counterfactual histories'. The task of the historian, they say, is to interpret what happened, not to construct stories about what might have but didn't.

The philosopher Michael Oakeshott has pointed out, for example, the absurdity of attributing the spread of Christianity to St Paul's escape from Damascus, as we seem forced to do the moment we start asking what would have happened if Paul had been caught. 'When events are treated in this manner,' says Oakeshott, 'they cease at once to be historical events. The result is not merely bad or doubtful history, but the complete rejection of history.'[6]

In his book *Virtual History*, British historian Niall Ferguson defends counterfactual history against such criticism. Some historians have argued that the past is an ever-branching tree, an 'infinitude of Pasts, all equally valid'[7] in the words of André Maurois – like Borges' Garden of Forking Paths. That may be true, opponents will say, but once each choice is made the other branches stop short, and we can say no more about them.

Some historians favour a determinist view which relegates the role of chance in directing the course of world events. Modern determinists have often been inclined towards Marxism, with its tendency to invoke inevitability in human affairs. They and other 'materialist' historians, taking their lead from Kant and Comte, regard history as being guided by laws analogous to those that govern the natural sciences, and so are generally opposed to counterfactual scenarios on the grounds that these laws, not fickle fortune, banish them. It might seem strange, then, to use analogies from physics to argue the case for counterfactual history. But the physics we are using here is statistical – it is concerned not with how one thing led to another, like a succession of colliding billiard balls, but with the range of possibilities and the likelihood of each.

The landscape model promises something more concrete than flights of historical fancy. Not only might it offer an objective rational-ization of the path that events took, but it can potentially draw the map of the terrain. This can help counterfactual historians to ground their discussions in the realm of the possible, and not to succumb to vague speculation. It gives concrete form to the range of influences that any historian knows must operate in shaping the course of real events. As historian Hugh Trevor-Roper has said, 'History is not merely what happened: it is what happened in the context of what might have happened.'[8]

If we concede, then, that the successes of the theory are not flukes, we must grant that the landscape that provides those successes has some broader validity. We can talk in (somewhat) quantitative terms about worlds which might have been, and identify the factors that helped things turn out this way and not that. An ability to do this, says Ferguson, can validate counterfactual history as a way of understanding the past. We must be concerned, he says, not with all things that *could* have happened, but with those that were most likely – 'with possibilities which seem probable in the past':

By narrowing down the historical alternatives we consider to those which are *plausible* – and hence by replacing the enigma of 'chance' with the calculation of *probabilities* – we solve the dilemma of choosing between a single deterministic past and an unmanageably infinite number of possible pasts.[9]

Here Ferguson seems to be implying that these 'calculations' rely largely on the good judgement of the historian. Counterfactual scenarios, he says, should also be cognizant of 'those alternatives which we can show on the basis of contemporary evidence that contemporaries actually considered'[10] (such as a British assault on the Soviet Union in the 1930s). Fair enough – but how are we to assign quantitative values to any of this? Landscape theory shows how, in the particular case of alliance formation, it is possible to make a real calculation that produces numbers whose values, although highly approximate (and how could they be otherwise?), hold some degree of validity.

THE EDGE OF HISTORY

As war approaches in the late 1930s, according to the landscape model, the European political terrain divides into two realms of possibility: one in which the strongest democratic nations unite against Germany, the other in which they oppose the Soviet Union. That one scenario occurred and not the other is due to the conditions from which history began its 'walk through the hills'. This starting point, of course, itself follows from the history of the earlier twentieth century, which in turn is shaped by the conflicts of the Victorian era. That is in one sense nothing more than a banal statement that the present depends on the past – except that we can, using this theory, imagine watching that development take place in the form of a fluctuating landscape.

As we have seen, the Ising model of a ferromagnet or a many-particle fluid also has a twin-valleyed landscape, corresponding to the stable configurations of spin-up and spin-down states, or of liquid and gas. Which state prevails depends on which is lower in energy under the prevailing conditions. At the boiling or condensation temperature of a fluid, the liquid and gas 'valleys' are equally deep, and a transition can occur from one to the other. This is exactly what van der Waals described; couching the phase transition in terms of an energy landscape is just another way of looking at it.

It follows that something like a phase transition is possible between 'alliance energy' valleys – that it is possible, if two valleys become equal in depth, to jump from one to the other, from one alliance configuration to a totally different one. And just as, close to the liquid–gas transition, a small change in pressure or temperature can bring about a profound change in the state of the entire system, so too might small changes in political attitudes or in circumstances markedly alter the configuration of players, if alternative 'energy' valleys are nearly matched in depth. Thus it becomes important to know not only which valley is deepest (since that is the one corresponding to the predicted outcome of pact-making), but whether there are nearby valleys that are almost as deep – in which case the alliances may be prone to changing abruptly into a quite different configuration.

For a fluid of interacting particles, altering the pressure while keeping the temperature constant changes the energy landscape so as to alter the relative depths of the 'liquid' and 'gas' valleys. At the transition point these depths are equal. At a slightly higher pressure the liquid becomes more stable than the gas: the gas is a *metastable* configuration (see page 201). In the same sense, the anti-Soviet alliance was metastable in 1936. We saw earlier that metastable states can exist if the system is prepared in the right way, even though they are constantly at risk of collapsing into the more stable state. But if we continue to increase the pressure, the metastable gas state becomes ever less stable than the liquid state. Not only does the valley rise, but it becomes shallower. At some point the metastable dip vanishes altogether: a gas-like configuration is no longer possible, even in theory. This is called a *spinodal point*, and it represents the limit of metastability (Figure 12.5*a*).

In Axelrod's landscape model, altering the 'pressure' might be equated with changes in the attitudes or the 'sizes' of the various agents – that is, changes in the strengths of the attractive and repulsive forces between them. We find that such changes alter the historical landscape between 1936 and 1939 from a two-valleyed terrain to one with a single stable state. This implies the occurrence of a kind of historical spinodal point: a set of circumstances under which the less stable alliance ceases to be viable. Between 1937 and 1938 the political landscape apparently passed through such a point (Figure 12.5*b*); after that, a war would inevitably involve the Axis and Allied coalitions.

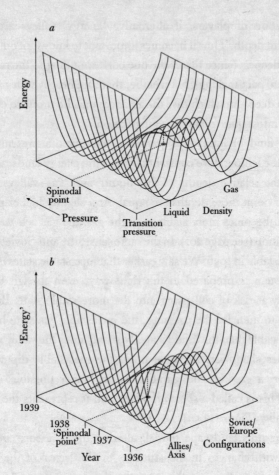

a

Energy

Spinodal
point

Gas

Liquid   Density

Pressure

Transition
pressure

b

'Energy'

1939

1938

'Spinodal
point'   1937

Soviet/
Europe

Allies/   Configurations
Axis

Year   1936

Figure 12.5 (*a*) At a spinodal point, the least stable of the two possible
configurations of a fluid (the metastable state) vanishes. (*b*) Something
analogous happens in the landscape model of national alliances
between 1936 and 1939: history seems to pass through a kind of
spinodal point where the anti-Soviet alliance ceases to be viable.

It is not hard to find contemporary political scenarios on which to exercise the landscape theory. Indeed, Axelrod and his colleagues have used it to retrospectively 'predict' that, given the balance of power and international relations in 1989, just after German unification, the desertion of the Soviet Union by its Eastern European allies was already on the cards. They predicted that only Bulgaria would stay with the former Warsaw Pact, while the others would align themselves with NATO. By the time the Soviet Union began to disintegrate, only Romania in fact remained in contradiction of this hindcast. Others have applied the landscape model to the break-up of the former Yugoslavia in the early 1990s.

Looking back is all very well. But the model might be of greatest value for looking forward. What can it teach us to expect of relations in the volatile Middle East, where for example Israel, Syria, Iran and Jordan are locked into a frustrated mutual antipathy? Might religious similarities and fear of Western interference outweigh political differences in creating an alliance of Islamic states? Where would that leave Turkey? How is this picture affected by the new (and currently anarchic) political situation in Iraq?

It is neither likely nor desirable that anyone will decide international policy on the basis of a single and, let's face it, simple model. But there seems ample reason to believe that such an approach can help us foresee the general lie of the land. It might even assist in identifying the best points of leverage for engineering favourable alliances and avoiding conflict, although that is a challenge of another order: dishearteningly, most international alliances are forged in the run-up to or in the midst of war, not in order to avert it.

There are endless opportunities to explore the factors that affect the emergence of groups, coalitions and organizations in many areas of society. It would be valuable to know whether characteristics such as race, class, religion and ideology lend predictability to the way in which social groups (both small, such as office friendship circles, and large, such as national groups and communities) develop. Which of these characteristics tends to dominate the clustering process, for example? In organizations, it would certainly be helpful to know how best to subdivide employees. It is not always clear, for example, to which university faculty a particular department should be assigned (is geology an engineering discipline or a natural science?). If one had a means of determining how things should fall 'naturally', such dilemmas could be more reliably resolved.

That we can even speak about such questions in the same breath as wondering about the fate of nations is a hint that there is something like a fundamental principle operating here. For a single model successfully to predict the allegiances of computer companies and the alliances during the descent into global war suggests that we have moved beyond the compartmentalized, case-dependent perspective of much traditional social and political science and hit a deeper seam in the order of things.

13

Multitudes in the Valley
of Decision

*Collective influence and
social change*

Only in a quite limited sense does the single individual create out of
himself the mode of speech and of thought we attribute to him. He
speaks the language of his group; he thinks in the manner in which
his group thinks.

   Karl Mannheim (1936)

When I'd had my coffee this morning and went upstairs to get
dressed for work, I never considered being a nudist for the day.
When I got in my car to drive to work, it never crossed my mind to
drive on the left. And when I joined my colleagues at lunch, I did
not consider eating my salad bare-handed; without a thought, I
used a fork.

   Joshua Epstein (1999)

To look closely at complex behaviors like smoking or suicide or
crime is to appreciate how suggestible we are in the face of what we
see and hear, and how acutely sensitive we are to even the smallest

details of everyday life. That's why social change is so volatile and
so often inexplicable.

Malcolm Gladwell (2000)

*

Perhaps the best reporting on the US presidential election in late 2000
was to be found in the satirical magazine *The Onion*, 'America's finest
news source':

In one of the narrowest presidential votes in U.S. history, either George W.
Bush or Al Gore was elected the 43rd president of the United States
Tuesday, proclaiming the win 'a victory for the American people and the
dawn of a bold new era in this great nation.'

'My fellow Americans,' a triumphant Bush or Gore told throngs of
jubilant, flag-waving supporters at his campaign headquarters, 'tonight, we
as a nation stand on the brink of many exciting new challenges. And I stand
here before you to say that I am ready to meet those challenges.'

'The people have spoken,' Bush or Gore continued, 'and with their vote
they have sent the message, loud and clear, that we are the true party of the
people.'

With these words, the crowd of Republicans or Democrats erupted.[1]

This is no more farcical than the events that followed, as the man
who would later lead the USA into two wars (and counting) came to
owe his seat in the White House not to the nation's constitution or to
a majority of popular votes, but to an escalation of the legal process.
While the media grappled with the hitherto unknown concept of the
'hanging chad' (an imperfectly punched hole on voting cards in
Florida's notoriously error-prone 'butterfly ballot'), the Supreme
Court of Florida ordered a recount of its pivotal votes. In overruling

this decision by a single vote, the US Supreme Court then decided their country's leader.

In December 2000, the USA's treasured democracy fell prey to statistical variability. While some methods of casting and counting votes are clearly better than others, all incur inevitable statistical uncertainties. The result of the 2000 election was simply too close to enable any of those uncertainties to be discounted. When the presidency hangs on a hundred votes or so, there is no way of judging who has won that does not involve a strong element of the arbitrary. Trapped within the margin of error, we find that all outcomes are possible.

Democracy is supposed to be about making choices. The Bush v. Gore election has highlighted how much contingency can be involved in the expression of those choices (for example, how clearly you make your mark at the polling station, and how voting ambiguities are resolved); but the principles at least are clear. Every eligible voter participates in the selection of their leaders, and no person's vote counts for more than any other person's. In practice, of course, each voter's power depends considerably on the details of how voting districts are defined and how the votes are counted; but the popular image of a democratic process is of millions of ballot papers accumulating on two (or more) great piles until the candidate with the highest pile emerges victorious.

But, leaving aside the technicalities of a balloting process, it is still not that simple. For one thing, not even the most ardent supporter of individualism could reasonably claim that our choices are truly independent. How, in a society flooded with mass advertising, can we hope to make decisions free from the influences of our environment? Today this applies as much to politics as it does to soap powders. In the 1999/2000 election campaign the Republicans and Democrats between them spent $300 million – more than the marketing budget of many multinational companies.

Few stones have been left unturned in studies of how we make our decisions for, after all, our political institutions hinge on it. And yet there is no consensus view on this question, let alone any reliable means of prediction. Opinion polls led British voters to believe that the 1992 UK general election would be a close call: that Labour's candidate Neil Kinnock might narrowly end the 13-year reign of the Conservatives. In the event, there was no photo finish at all: the Conservative Party won a big majority, leaving the pollsters on that occasion thoroughly discredited.

There is a name for the business of election prediction: psephology. Behind that respectable-sounding word lurks a black art compared with which economic theory looks like a science of infallible accuracy. This is not necessarily the fault of psephologists, for they face human nature at its most slippery. People announce one intention in opinion polls and then go and do the opposite come polling day. Substantial changes in voters' affiliations may be triggered by last-minute events, such as a speech by a party leader. Single and sometimes quite trivial issues can mobilize large proportions of public opinion.

In the face of such psychological volatility, can we expect models of social behaviour to reveal anything useful about the political decision-making that characterizes any society with a degree of democracy? Economic traders may take risks, but they do not make their decisions lightly. Irresponsible choices by vehicle drivers are strongly suppressed by heavy penalties – financial, legal and medical. But some voters choose their leader because they like his hairstyle. Can science cope with *that*?

What physics can bring to the social science of decision-making is not a fully fledged and precise theory but a deeper understanding of an ingredient that has often been ignored or clumsily handled in the past. By now we should know this ingredient well. It is the effect that one person has on another: the influence of interaction. One of the features of collective behaviour arising from local interactions is that it becomes impossible to deduce the global state of a system purely by inspecting the characteristics of its individual components. This is physical science's most important message to social science: do not be tempted too readily into extrapolating from the psychology of the individual to the behaviour of the group.

Throughout the twentieth century, sociologists have stressed that their studies are concerned not with the individual but with the group. Yet frequently they have been able to do little more than embed their subjects within a pre-existing set of cultural norms: group behaviour is postulated a priori. Now, instead, we can begin to examine how, through the interplay of personal choice and interpersonal exchange, such norms arise and change. We can start to understand how a society creates its leaders, its customs, its fashions and its problems through a mass of mutually interdependent decisions.

CAST YOUR VOTE

One of the issues that the US election debacle has highlighted is the absurdity of wrangling over a few hundred votes in an election in which over half the enfranchised population was silent. And that is hardly the worst of it: US county and city officials are typically elected by less than 20 per cent of the eligible electorate.

Brazil allows no such apathy. It enforces compulsory voting. In October 1998, over 100 million voters determined the Brazilian president, Senate and Congress. (The country's political structure is built essentially on the US model.) In addition, the voters selected a governor and state deputies for each state in the country. Physicist Raimundo Costa Filho and his colleagues at the Federal University of Ceará in Brazil analysed the election results. Since state deputies made up the biggest group of candidates, Costa Filho and colleagues focused on the voting statistics for these 10,535 prospective minor officials, looking at the proportion of the total vote that each candidate received.

What might one expect to find from such an exercise? If voting were a purely random process – say, if the choice of each voter was determined by the throw of a die – then we should see a gaussian distribution. That is to say, most candidates would receive some average proportion of votes, a few would receive rather less, and a few rather more. The probability distribution – the number of candidates receiving some proportion p of the votes, plotted against p – would then be a bell-shaped curve. Of course, voters don't, on the whole, make their choices by rolling dice. But we might imagine that the factors determining how individuals vote are so many and so varied that the end result for a large number of voters is indistinguishable from them all having chosen at random.

But the researchers did not see a gaussian distribution. Instead they found that that the voting statistics fitted a power law (Figure 13.1). What is more, this particular power law seemed to have an exponent close to –1, like that seen in self-organized criticality (page 297): this means that the number of candidates receiving a fraction p of the votes is inversely proportional to p. The same power-law relationship also turns up in a state-by-state analysis of the voting patterns – it seems to be reliably reproduced in subgroups of the total electorate.

This tells us at once that the voting process does not consist of millions of independent decisions being made essentially at random.*

* The conclusions I draw here about non-randomness due to the interdependence of choices are in fact not at all airtight. Costa Filho and his colleagues pointed out that, under certain conditions, a concatenation of independent, random decisions by each voter can give rise to behaviour which looks, within a certain window of parameters, like a power law. This emphasizes the point that any one data set is generally insufficient to allow us to uniquely identify the mechanism that produces it.

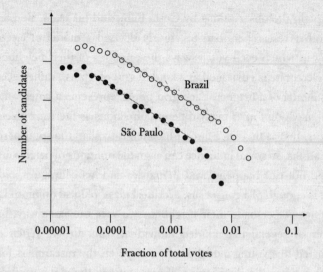

Figure 13.1 Voting statistics in the Brazilian elections of 1998. The number of candidates who gathered various fractions of the total votes follows a power law. The black circles show the results for São Paulo only, and the white circles are for Brazil as a whole. Both straight lines have essentially the same slope.

As we saw earlier, a power law is typically a signature of some process governed by strong interactions between the 'decision-making' agents in the system. At the critical point of a magnet, a power-law distribution in the sizes of islands of oppositely oriented atomic spins arises from the influence of each spin on its neighbours. In a hypothetical sand pile, each grain 'chooses' whether to participate in an avalanche on the basis of its interactions (via collisions and friction) with other grains. We might suspect, then, that a power law in voting statistics is telling us that voters are influenced by one another's decisions.

That is what Americo Tristao Bernardes of the Federal University of Ouro Preto in Brazil and his colleagues believed when they set out to explain the observations by Costa Filho and his team. Bernardes and other researchers had previously devised a model of electoral voting in which each voter sways the opinions of their neighbours. The electorate is envisaged as a kind of grid of 'spins', rather like the Ising model of a magnet. Each grid point represents a voter, and its 'spin' may point in as many different directions as there are electoral choices. Just as there is a force which tends to align the spins of magnetic atoms, so social influence can align the opinions of neighbouring voters. But this happens, said Bernardes and his colleagues, only if there is enough of a consensus: a 'critical mass' of local opinion. Lone voters can't persuade their neighbours to vote the same way, but a cluster of several like-minded individuals can do so. When they simulated this voting model on the computer, the researchers found that it produced a distribution of votes among the many candidates with precisely the power-law form seen in the real election results: even the slope of the power-law graph was the same.* Voting is, it seems, very much a group decision.

The idea that we are influenced by our friends, colleagues and neighbours is, needless to say, hardly a revelation. Like most people, I have been on holiday to places my friends have recommended, and I have gone to see films and plays they have enjoyed. And we don't even need to know other people to feel a mutual influence. The reason organic produce has become popular, for example, is surely not

* To obtain the right slope, however, Bernardes and his team had to make a particular assumption about the 'persuasiveness' of each candidate – that is, the way in which the ability of voters to persuade their neighbours depended on which candidate they were voting for. But they were able to do away with any such assumption in the model when they represented the network of social interactions not as a grid but as a branching, 'scale-free' network of the kind discussed in Chapter 16, which seems to be a better description of real social interconnections. In this case, the model agreed with the real voting statistics without the researchers having had to make any assumptions about how convincing each candidate is.

because a significant fraction of the population independently decided that they prefer it. (There is an additional positive feedback at play here, since the law of supply and demand pushes prices down as more people buy organic goods, which makes them more attractive to others with marginal preferences.) There is no reason to expect our choice of political affiliation to be immune to such influences.

All the same, the notion sits uncomfortably with our sense of democracy and freedom of choice. Whatever else a democratic election is, it is apparently not a summation of so many million independent votes. The interdependence of voters' decisions means that it is far from easy to predict how public opinion will respond to particular events or inducements. There is, for instance, ample reason to suspect that small imbalances in the visibility of candidates (a result of differences in campaign funding, say) may not lead to correspondingly small differences in voter choices. I have a strong suspicion that the outcome of the 1992 UK general election was at least in part a consequence of the same collective herding behaviour that occurs in economic markets (page 269). Whether people confessed to one another that they could not after all bring themselves to trust Labour's Neil Kinnock, or whether they simply sensed this in the prevailing mood of the nation, the large discrepancy between poll predictions and results does not seem a likely outcome of so many random, independent decisions.*

Some political scientists have explored a model of party-political electioneering which has strong resemblences to models used in physics. It is a kind of landscape model, akin to that developed by Robert Axelrod (see the previous chapter), in which voters give their allegiance to one of several parties. The topography of the landscape is

* The concerted scaremongering in the right-wing British press no doubt played a strong part too, though perhaps only in tilting the political landscape to favour one-way herding.

defined by the voters' preferences on a range of issues, and political parties are considered to rove across this landscape looking for a suitably attractive pinnacle on which to plant their flag: to erect their 'platform' position on the various issues of the day. Thus, the parties aim rather nakedly to tailor their policies so as to attract as many votes as possible, although they are at the same time conditioned by the different positions in the spectrum of opinions from which they start.

This so-called 'spatial' model of voting was first developed in the 1950s by political scientists Anthony Downs and Duncan Black, who adapted it from a similar model of how businesses make economic decisions. The central idea, which aims to bring decision-making within the embrace of quantitative, 'scientific' analysis, is that each voter takes a particular position within a spectrum of views about each issue. That is to say, if we represent the range of possible opinions as a line between two extreme stances, each voter can be placed somewhere along that line.

We conventionally think about political opinions in such spatial terms. We speak of views that are on the left, on the right, or centrist. This terminology derives from a genuinely spatial division of the National Assembly (later the National Convention) of the French Republic just after the Revolution. The Assembly was split into two main camps: the Jacobins, supporters of Robespierre who favoured radical political and social change, and the somewhat more powerful Girondins, who were concerned to preserve the status quo. These two groups sat in the French parliament as far apart from one another as they could: the Jacobins on the left as one entered the chamber, the Girondins on the right. Those with more moderate views took their seats in the middle. Ever since then, the political left has been associated with change, the right with conservatism.

But political factions can rarely be neatly divided along just a single axis. Elections are fought over a range of issues, many of which can be characterized by two mutually incompatible extreme views. The spatial model of voting broadens the one-dimensional picture of a political left and right into a multi-dimensional one in which there is an axis of opinion for each important issue. Every voter, in this model, can be placed somewhere along each of these axes. In other words, the political 'space' has as many dimensions as there are significant issues, and each voter can be represented by a point somewhere in this space to indicate his or her position on each issue.

This is, of course, a rudimentary way of looking at how people define their political stance. For one thing, not everyone has a precisely articulated preference on every particular electioneering topic. But the approach can be expected to capture some of the basic features of the election process. Certainly, it seems fair to suggest that political parties try to gauge the shape of the landscape of public opinion and to adopt policies they think will attract the greatest number of voters. This is not to suggest that all politicians are so cynical as to take up *any* viewpoint just because it will win votes: there are typically large regions of the landscape which are characteristic of one persuasion or another (liberal or conservative, say), and boundaries that politicians might prefer not to cross. But political parties manifestly do try to mould themselves to public opinion, a point emphasized by the recent fad for 'focus groups' in liberal politics in the West. In many European countries, parties across the political spectrum have had to adjust their policies in response to strong public opinion on issues such as immigration and crime that have traditionally been the preserve of the political right.

The spatial model of voting has become one of the key tools for a political science that aims to understand how democracy works – and how it breaks down. It has been used, for example, to explore the factors that promote polarization or convergence of political stances, or to contrast the differences between two-party and multi-party systems, or to investigate the consequences of single-issue politics. One of the most striking findings is that there is not necessarily a 'best' winner in democratic elections. One might imagine that, in an ideal voting system, the winner will be the party whose policies lie closest to the preferences of the majority. But often it is impossible to determine just where that point lies in the space of opinions. Aristotle realized as much. Even in the simplest case, where public opinion can be represented on a single axis, problems arise when the distribution of opinion has more than a single peak. Aristotle imagined a society in which there are two dominant constituencies with very different preferences: the rich and the poor. Is the ideal political leadership the one that adopts the policies of whichever of those two groups happens to have the loudest voice at any moment? Or is it better to elect a government whose policies lie in the middle between these two polarities – policies which therefore match no one's preferences precisely? Aristotle concluded that a society like this can never be stable. For true stability, he said, it would be best to have a large 'middle class' with centrist views.

Condorcet, in his pioneering treatise on the statistics of voting and public choice in 1785 (page 66), identified another problem with democracy. What, he asked, if the preference profiles are not single-peaked even for individuals? This might sound odd at first – surely we always have a single preference? If, for example, we are considering how much we think the government should spend on public services, we are likely to have a certain ballpark figure in mind, and look progressively less favourably on policies that propose spending increasingly more or less than that figure. But in some situations, if we can't have exactly what we want then our second choice might be something quite different. When the USA was debating whether to intervene in the ethnic conflicts in former Yugoslavia in the 1990s, some Americans felt that their country should either make a massive intervention that would end the fighting definitively, or stay out altogether. Something in between – a peace-keeping force in which US soldiers were exposed to risk, but which could not end the conflict – was deemed less desirable than either extreme. Many had felt the same way about the Vietnam War two decades earlier.

Under such conditions, Condorcet concluded, majority rule – which is surely what we regard democracy as being about – may not be stable. He considered elections in which each candidate faces each of the others in turn in a one-to-one contest. In modern theories of voting, the contestant who beats (or at least ties with) all others in a series of such head-to-heads is called the *Condorcet winner*. It is possible to show that in some elections the Condorcet winner is not necessarily the best compromise for reflecting the views of the electorate. Indeed, there are cases where the Condorcet *loser* (the candidate who loses each of those one-to-one bouts) is arguably the best overall winner.

Condorcet's voting procedure is by no means the only alternative to the principle of simple majority rule. The discipline called *choice theory* explores the consequences of different voting schemes. For example, the French political theorist Jean Charles de Borda proposed in the late eighteenth century that elections to the French Academy of Sciences should be made by the voters giving each of the candidates a score denoting the number of candidates considered to be *less* preferable. The higher the score, the more desirable the candidate is. So the winner would be the candidate who accrues the highest total score from all the voters. This is a form of proportional representation, but it is far from ideal since the 'Borda winner' can depend on the rankings of candidates that nobody wants.

An ideal voting system would not be plagued by apparent logical absurdities such as this. Choice theorists look for a certain logical consistency in methods of voting. For example, the outcome should be *transitive*: if candidate A is preferred by the electorate over candidate B, and candidate B over candidate C, then candidate A should also be preferred over candidate C. Another desirable feature is that the outcome should not change under a contraction of the choices: if A is picked from A, B and C, then A should also be picked from A and B alone. And so forth.

The trouble is, no voting mechanism satisfies all the logical criteria that one might reasonably demand from it. This is not because we haven't yet found the right system, but because it simply cannot be done. That rather shocking result is proved in the 'impossibility theorem' of economist Kenneth Arrow. He argued that any truly democratic decision-making process should possess a particular set of logical features, and went on to show that the only collective-choice mechanism that satisfies all these characteristics *and* allows one set of preferences (that is, one candidate) to be unambiguously selected over the others is dictatorship: rule by a single person. Clearly, this is not a collective process at all! Arrow seemed to find himself forced, by a very different route, to the same conclusion that Thomas Hobbes had reached three centuries earlier.

The implication of Arrow's paradox is that there is no perfect alternative to dictatorship. Either we need to accept that majority rule has some undesirable consequences (including instability), or we need to find an alternative to simple majority rule. Such alternatives have been thoroughly explored in theory, but it is very hard to find ones that are transparently fair and immune to manipulation. In other words, democracy is a very slippery concept. Western politicians and public alike frequently fail to acknowledge this crucial point: the argument for adopting a democratic system of government is not that it is the perfect or even in some vague sense the 'fairest' system, but that it is (probably) the least susceptible to corruption.

And all of this, let's remember, follows from models in which each voter makes an *independent* choice. The statistical analyses of voting that physicists have conducted undermine this simplifying assumption, and make the path to a workable and equitable democracy even murkier. It all drives home the message delivered by Otto von Bismarck to the Prussian Herrenhaus in 1863: 'Politics is not an exact science.'[2]

WHICH WAY TO TURN?

Joshua Epstein of the Brookings Institution points out that, whereas many psychologists and sociologists have exercised themselves over how we make decisions, society is geared largely towards removing that need. Many social norms exist simply so that we no longer have to think about other options. If each day we had to choose afresh every aspect of our appearance, conduct and activity, we'd never get anything done.

Thus, for example, the inhabitants of every nation have agreed to drive solely on the left-hand or the right-hand side of the road.* This agreement is enforced by law, of course; in general, though, laws simply consecrate pre-existing social norms. In Britain, for example, driving on the left may be a legacy from the preference of passing an approaching horseman or carriage on the left-hand side, so that one could wield one's sword right-handed against any sudden attack. In Continental Europe, in contrast, postillions (mounted riders guiding a

---

* Beware of exceptions! There is a 10–20 km stretch of Autoroute 20 in south-western Montreal, Canada, for example, on which one must drive on the left. And it was decreed around 1929 that vehicles in Savoy Court in London, a short stretch of road which gives access to the Savoy Theatre, should drive on the right so that vehicles queuing to drop people at the theatre would not block access to the Savoy Hotel.

wagon team) were mounted on the rearmost left horse and thus preferred to pass left side to left side (that is, to drive on the right). Today these conventions avoid accidents, of course, but they also free us from the need to guess, each time we get into the car, how everyone else might be driving today.

If this seems like trivial common sense, consider how the same kind of mass conventions apply to many seemingly unimportant aspects of our lives. In Baroque England you would not think of appearing at the royal court without your wig; in the 1920s only ill-mannered and disreputable men walked the streets of New York City without a hat. There is no law (as far as I am aware) against shopping in your underwear, or shaking hands with your left hand, but few of us consider doing these things. Epstein suggests that we structure society so as to keep the need for thought to a minimum. We are compelled to make a decision only when no social convention prescribes it. The stronger the norm, the less we as individuals need to think about it. This urge to conform exists not just in society as a whole, but within particular subgroups whose conventions, however alien to outsiders, are observed without thought by group members.

Few if any social rules are engraved by nature; they have to be learnt. That is why children and tourists are more likely than others to bump into people in the street (page 165): they haven't yet learnt on which side it is usual to pass people. As people familiarize themselves with the rules, says Epstein, they also learn *how much* to think about how to behave. Unless we are born into the aristocracy, we generally know that we have to think carefully about what to wear to a royal occasion. Epstein proposes that this learning process involves checking what others are doing. If we find that they are all doing the same thing, it is clear what we should do. If their behaviour varies, we may need to watch a larger sample before deciding what seems to be 'normal'. But in any event, our sample is usually limited: we don't ask the entire population of Manhattan whether it is safe to wander around the Lower East Side or how much to tip cab drivers.

To investigate how norms emerge from such a system, Epstein imagines a line of agents (individuals) arranged in a circle. Each agent makes a dual-option (binary) choice – this or that – based on a census of its neighbours. The circle isn't meant to mimic any particular social situation. It is just the simplest arrangement possible: a line, or what physicists would call a one-dimensional array. Joining the line at each end avoids anomalies at the ends, where agents would otherwise have fewer neighbours.

Individual decisions in this model are reached according to simple rules. The underlying philosophy is that agents want to do as little thinking as possible: they want to conform, but with minimum effort. This means that they look for the smallest feasible sample that will tell them how to make up their minds. Each agent updates its decision once every 'round'. It adds up the choices of all its neighbours within a given distance to each side, and conforms with the majority. But then it checks this decision by extending the sample by one agent in each direction, to see whether this produces the same recommendation. If it doesn't, the agent concludes that the initial sample wasn't representative. It continues to expand its field of vision until the consensus from one sample matches that of the next-largest sample.

Because these agents are lazy, they also look for ways to make the sample size smaller. Once they find a consensus view, they look to see whether they get the same consensus from a slight contraction of their field of view. If they do, they contract it further, and keep doing so until the sample is as small as possible while still reflecting the view of the 'wider world'. As Epstein puts it, the idea is that 'When in Rome, do as the majority of Romans do – but with the smallest feasible definition of Rome.'[3] When a particular norm becomes entrenched across a wide span of the ring, the agents become small-minded: they

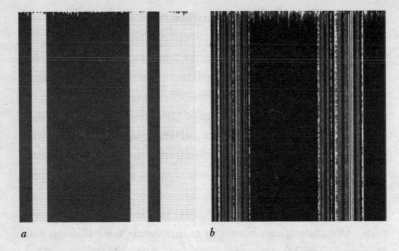

a *b*

Figure 13.2 Conformity to a norm eliminates the need to think too hard about decisions. In Joshua Epstein's model, agents placed around a ring consult their neighbours, out to a certain distance, in order to choose between two courses of action: whether, for example, to become 'grey' or 'white'. They continue to extend their opinion-sampling range until they are confident of a local consensus. In these figures each horizontal slice shows the affiliations and activity of each agent around the ring at each moment (so that in reality the left and right edges of the plots lie adjacent). Time advances from top to bottom of the plots. An initial random distribution of 'grey' and 'white' voters (barely visible here along the top line of *a*) develops quickly into a stable global pattern of wide grey and white domains. Within each domain there is a clear consensus, and the sampling range for each agent dwindles to zero. This is shown in *b*, where the sampling range is indicated by a grey scale. Black regions denote that this range has fallen to the minimum value (only immediate neighbours are consulted) – the agents here need no longer 'think' about their choices. Only those agents near the boundaries of the grey and white domains need check much beyond their immediate neighbours in order to decide which camp to join.

look no further than those right next to them. You could say that they stop thinking about their choices.

Epstein finds that, when preferences are initially assigned at random to a large number of agents arranged in a ring, the ring becomes rapidly segmented into regions where one or other preference prevails. This is shown in Figure 13.2*a*, where the two preferences are marked as white and grey. To make the results easy to display, the ring is straightened out here into a horizontal line, and successive time steps run down the page. Vertical stripes in the diagram indicate regions of the ring where the agents' preferences stay fixed over time. In Figure 13.2*b* a grey scale denotes the corresponding 'search distances' of the agents. Away from the boundaries of the domains, 'thinking' quickly contracts.

If this system is 'shocked' by briefly introducing a random element (noise) into the decision-making so that all agents make their choices purely at random for one round, the ring settles into a new configuration (Figure 13.3). We might imagine that some new revelation has momentarily shaken up people's preconceptions and forced them to rethink their position. In a sense the outcome is not surprising, but it is a valuable reminder of the fickleness of human nature: some people who were unthinkingly in the 'grey' camp before the shake-up end up equally unthinkingly in the 'white' camp. Epstein suspects that many social norms and beliefs are determined this way. Who would doubt that the virulent opponents of the male fashion for long hair in the 1960s would have been any less ardently opposed to short hair in the seventeenth century, and for much the same reason? Epstein suggests that most American citizens believe the Earth is round not for any reasons they can articulate but because it is as much the received wisdom as a flat Earth was in older times.

Injecting a little, constant random noise into this model (a little 'confusion' or arbitrariness in the choices agents are making) doesn't

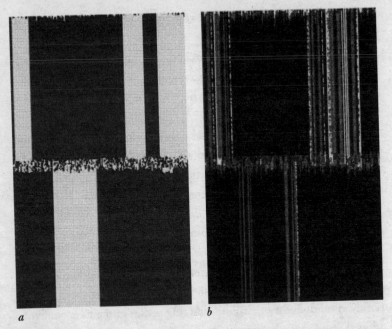

a *b*

Figure 13.3 The pattern of domains in Epstein's model is arbitrary, yet stable. As in Figure 13.2, *a* shows the domains of 'grey' and 'white' agents, while *b* shows the corresponding variations in sampling range. When given a 'kick' by randomizing the affiliations of agents for several time steps (middle of plots), the model settles down into a new configuration. Agents that were confidently grey before the kick (in the sense that they didn't need to 'think' about it) may be equally confident in their whiteness afterwards.

prevent it from becoming resolved into distinct blocks of one persuasion or the other, but it makes the boundaries between the blocks more mobile. Majority views can alter over time in any region of the ring; but at any moment there are still broad regions encompassing many 'minimal-thinking' conformists, with the hardest thinkers prevaricating at their edges, where they sample other points of view. As a metaphor for social decision-making, the model is absurdly simplistic – and irresistibly attractive.

WORLDS APART

This kind of polarization among ostensibly open-minded individuals was explored over two decades ago by Thomas Schelling, a political scientist at the University of Maryland. In 1978 Schelling published a groundbreaking work in which he demonstrated how often the collective consequences of many individual decisions can defy intuition. This book, called *Micromotives and Macrobehavior*, is one of the cornerstones of the kind of social physics that I am discussing. Schelling lacked access to the tools of modern statistical physics, but nevertheless saw clearly how physical laws find analogies in the social sciences.

It is common in physics, for example, to speak of non-sentient systems as though they exhibit purposive behaviour – to say that a soap bubble *tries* to minimize its surface area or light *tries* to find the quickest path. What physicists mean is that the laws that govern such processes are determined by the tendency of the world to maximize or minimize some quantity. There is nothing teleological about this; it is simply akin to (indeed, formally allied to) the way in which, as we saw earlier, entropy increases in processes of change. Of course, vehicle drivers in contrast really do *try* to find the quickest route through town (although they lack the apparent omniscience of light). Social behaviour may also thus be guided by an impulse to minimize (effort, time, distance, . . .) or maximize (profit, happiness, . . .) – in short, to optimize. This was the basis of George Kingsley Zipf's 'principle of least effort' (page 305).

In economics, Adam Smith's 'hidden hand' is generally considered to optimize the market: to make it efficient at providing and shifting goods. 'Somehow', said Schelling,

all of the activities seem to get coordinated. There's a taxi to take you to the airport. There's butter and cheese for lunch on the airplane. There are refineries to make the airplane fuel and trucks to transport it, cement for the runways, electricity for the escalators, and, most important of all, passengers who want to fly where the airplanes are going.[4]

And if economics works this way, why shouldn't similar principles apply to other human activities? Might society have some spontaneous organizational ability, unguided by laws or coercion, that allows it to run its affairs?

If economists have studied the matter for two hundred years and many of them have concluded that a comparatively unrestricted free market is often an advantageous way of letting individuals interact with each other, should we suppose that the same is true in all the rest of those social activities, the ones that do not fall under the heading of economics, in which people impinge upon one another as they go about pursuing their own interests?[5]

Here, Schelling had in mind anything from decisions about when to turn on the car's lights to choosing whether to vaccinate children, whether to wear safety helmets in sport, whether to carry tow ropes in cars, whether to break the law, where to sit in an auditorium, and whether to play golf on Fridays. The crucial aspect of all these activities is that in conducting them we are influenced and affected by the choices other people make. Furthermore, we have to make our own decisions on the basis of imperfect knowledge about what others are doing or intend to do. This interdependence is what makes group behaviour different from a trivial extrapolation of individual behaviour.

Schelling's approach cried out for the tools and insights of statistical physics. In fact, he used physics-based concepts in a qualitative manner without knowing that he was doing so: he presented models which show phase transitions between different types of behaviour, although he never explicitly used the term. But perhaps the most memorable and influential scenario that Schelling handed down to later generations of 'micromotive' modellers is that of racial segregation in demographics.

The 'multiculturalism' of Western nations was once celebrated, but increasingly it is becoming clear that behind this optimistic label often lies not integration but segregation. Neighbourhoods in the big cities are often demarcated by race and culture. The non-white inner cities become the ghettos of Baltimore, Chicago and Los Angeles, while 'white flight' creates the well-to-do suburbs. The result is all too well rehearsed: racial tension, mutual suspicion and a recipe for explosive unrest.

It is not simply a black and white issue, nor white and brown or Hispanic. There are Chinatowns in North American cities from Toronto to Boston to San Francisco; London has its Greek and Irish neighbourhoods; Grenoble has a Jewish quarter, Berlin a Turkish quarter. These ethnic neighbourhoods enrich the cultures in which they are embedded and generally coexist amicably. But as Britain discovered in 2001 when riots exploded in the Asian Muslim communities of some northern cities, racism and resentment often smoulder below the surface. It is worth remembering that there is nothing 'modern' about this kind of segregation. Before the lines were drawn according to race, they were defined in Britain by class or, less commonly, religion. (Arguably, racial segregation is still in part a separation of rich from poor.) 'A preference to mix with people similar to oneself is as old as humanity',[6] says Paul Ormerod.

Schelling wanted to know how segregation comes about. It is

 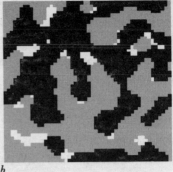

a   b

Figure 13.4 In Thomas Schelling's model of neighbourhood inter-actions, each agent prefers to have a majority of neighbours of the same colour (here grey or black), but only marginally so: they will move to free squares (white) only if more than 45 per cent of their neighbours are of a different colour. Thus the agents have only a mild degree of prejudice. ('Colour' here might represent a difference in class, race, religion, or whatever.) A random initial distribution of the two kinds of agent (*a*) develops quickly into a highly segregated arrangement (*b*) – one which, at face value, might lead us to infer a much higher degree of prejudice. The configuration in *b* here is the result of an average of only two moves of each agent from the configuration in *a*. Notice that empty squares tend to appear at the boundaries, alleviating the 'tension' between adjacent domains. This reflects the fact that the interfaces are relatively unstable – they have a kind of 'surface tension'. (These results, based on Schelling's original model, were produced by Paul Ormerod.)

sometimes assumed that it reflects a high degree of racial intolerance. But it is certainly unfair to impute to all American citizens, say, an unwillingness to live next door to someone of a different race or colour. On the other hand we can imagine that, however regrettable it might seem, many people will not wish to remain in a neighbourhood in which they are in a racial or cultural minority (assuming that they could easily move elsewhere).

Yet people move house for all sorts of reasons, and one might suppose that the cultural stirring this generates, coupled with a reasonable degree of tolerance, should keep the population relatively homogeneous. Schelling devised a model – we'd now regard it as an interacting-agents model, like the ones we encountered in earlier chapters – which showed that, on the contrary, there is an unexpectedly strong *collective* pull towards racial segregation. The model contained two different types of agent – two 'colours', which could represent race or ethnicity or some other difference. Schelling specified the rule that in a mixed neighbourhood, a family will move if more than one-third of its neighbours are of a different 'colour'. Note that this is consistent with a desire to escape bigotry as well as to express it. But it nevertheless allows a degree of tolerance too: people don't mind if, for example, a quarter of their neighbours are of a different 'colour'.

Schelling found that such a society quickly devolved into segregated enclaves, despite an initially uniform mixing of two types of agent (Figure 13.4). This situation has an analogy in physics: two well-mixed substances whose molecules have only a slight preference for others of their kind will gradually 'phase separate' into distinct regions, like oil and vinegar separating in a salad dressing.* The segregation process is a collective effect because if one agent leaves a

---

* Recent models of segregation, taking an approach similar to Schelling's, show more to the trained eye of a physicist. The Sugarscape model of Robert Axtell and Joshua Epstein (see the next chapter) generates segregation distributions which suggest a clear relation to critical points. Under some conditions, segregation produces domains with a 'typical' size, as it does in physical phase separation. Under other conditions the segregated regions have a range of sizes, with no dominant scale, implying that they are nearer to a critical point. In this case the mixing is more intimate – the majority of the domains are very small, although a small proportion are rather large. So integration might be encouraged if one can find a way to make the demographic system 'closer to criticality'.

neighbourhood, it becomes proportionately less likely that others of the same 'colour' will stay.

There are few better illustrations of how difficult and contentious it can be to separate the often straightforward predictions of a physical description of society from the interpretations and implications for policy-makers. One way of interpreting these results is to say that we should not worry too much about racial (or cultural or class) segregation, since it is virtually inevitable. Or to look at it another way, we cannot infer that a segregated society is a highly prejudiced one (and therefore prone to confrontations). Such conclusions would certainly suit those who would rather worry about other things – US Senator Daniel Patrick Moynihan notoriously advised Richard Nixon in the early 1970s that race relations should be treated with 'benign neglect'. But would it not be more useful to ask how, if there is a strong likelihood of segregation arising, we might want to respond? For example, it seems entirely possible that a separation of cultures will promote an increasing ignorance of, and thus fear of and hostility to, other ways of living – that mild preferences could become transformed into entrenched prejudices. (Schelling's model does not include such feedbacks on the agents' behaviour, but one can imagine ways in which this could be done.) So it might be profitable to focus available resources not on trying to suppress segregation but on fostering close interactions between the distinct communities that result from it. By the same token, we might benefit from knowing that the introduction of choice and freedom of movement into social environments (such as schools) which have previously had cultural mixing imposed on them is likely to lead to rapid and extreme segregation. In policies on public services, 'consumer choice' has become a mantra for some Western governments. Would they be so keen to promote it if they knew that unfettered choice begets segregation of, say, rich from poor, or of more able from less able? In other words, while this kind of modelling could be used to defend a laissez-faire attitude, a bowing to the inevitable, it need not carry that implication. Rather, it may force us to think more carefully about what kind of society we consider desirable, and help us to identify realistic (as opposed to idealistic or simply naive) means of achieving it.

Perhaps the most important general lesson of Schelling's work is that pronounced segregation does not necessarily imply a high degree of intolerance. In other words, individual tendencies *do not necessarily extrapolate to group behaviour*. This cannot be emphasized enough, particularly to scientists seeking to make sociology more scientific. The Harvard biologist Edward O. Wilson argues that such efforts should take more account of the evolutionarily determined predispositions of human nature: rather than simply positing modes of individual behaviour, one should look for the fundamental tendencies for which the brain has been hard-wired by natural selection. This is an important objective – too many models, including many of those described in this book, begin with apparently commonsensical but ultimately rather arbitrary assumptions (or preconceptions) about human behaviour. But Wilson fails adequately to acknowledge that group behaviour may not be simply a scaled-up version of individual behaviour. Rather, characteristics of the group appear which cannot be predicted from the nature of the brain's instincts alone. Schelling's mildly tolerant agents do not organize themselves into moderately mixed neighbourhoods, but instead display a high degree of segregation. Looked at from the top down, we might misread people's natures if we judged them on the basis of how they distribute themselves.

Michael Lind, a political scientist at the New America Foundation, puts it very aptly:

A friend of mine who raises dogs tells me that you cannot understand them unless you have half a dozen or more. The behavior of dogs, when assembled in sufficient numbers, undergoes an astonishing change. They instinctively form a disciplined pack. Traditional political philosophers have been in the position of students of canine behavior who have observed only individual pet dogs.[7]

Today it is time to study the pack.

CRIME AND PUNISHMENT

Another reason why people change homes is to escape crime. This can instigate an all-too-familiar downward spiral, for crime is indisputably linked to social deprivation. No one who is not a criminal wishes to live in a high-crime neighbourhood (in fact, even criminals probably don't), but only those who can afford to move to more genteel areas will do so. This makes high-crime areas poorer, bringing with it more crime. The equation is not at all simple, however. Some poor rural areas have lower crime rates than richer urban areas. Periods of high unemployment and lower standards of living in the past were not necessarily accompanied by higher crime rates than is the case today. We cannot say with confidence that there is a predictable relationship between social deprivation and the incidence of crime.

This makes it hard to identify the best solutions. The prevailing political fashion on both sides of the Atlantic is to be 'tough on crime' – with harsher penalties, including longer prison sentences. This pleases certain sectors of the voting population, but there is no unambiguous evidence that it reduces crime rates. Rather, we can see it as a backlash against the more liberal attitudes of the 1970s, when crime was regarded as a disease whose cure was understanding and compassion rather than punishment. There is no reason, however, to believe that this earlier approach did any better either.

The hard-line attitude to crime control is supported by an 'economic' model of criminality advanced in the 1960s by the American economist Gary Becker. He suggested that one can speak of a market for crime just as there exists a market for aluminium or bananas. The traders are the criminals, who enter into transactions (commit crimes) by making a cost-benefit analysis. If you get away with it, crime pays. If you don't, crime costs – either financially, for a minor offence, or through the curtailment of personal liberty at Her Majesty's pleasure, as we say in Britain. In Becker's model, criminals make a rational assessment of these pros and cons, and decide accordingly whether or not to commit an offence. If this is how crime happens, it stands to reason that it can be reduced by increasing the cost – by imposing harsher punishments. That is a very Hobbesian outlook: man's will to power can be subverted only by a greater power, which he dare not provoke.

But many conventional criminologists regard this rational decision process as pure fantasy. Crime, they say, is rarely undertaken after a cool weighing-up of advantages and disadvantages. Most criminals don't consider seriously the consequences of being caught. If that is so, tougher penalties have next to no deterrent effect, and one must look elsewhere for cures. Dostoevsky's Raskolnikoff knew the inclemency of the penal system he faced; he even conceded that 'almost all crimes are so easily discovered'.[8] But this did not deter him from committing a frenzied murder – and for reasons that even he could barely articulate.

Dostoevsky has given us some of the most compelling insights into the criminal mind. If his portraits come anywhere close to reality (and as a convicted criminal who endured a mock execution in Siberia, he arguably knew whereof he spoke), there is no easy identification of the factors that deter and spur on the offender. But one issue which has perhaps not received the attention it deserves is social pressure: the effect of one's fellow people. Japan is by no means a country free of crime, but its streets and parks must surely derive some of their remarkable safety from the shame and social exclusion heaped upon criminals. In a culture steeped in social etiquette and peer pressure, this is perhaps a more powerful deterrent than any draconian penalties imposed by law. The same applies in some small communities in the West. If criminals know their neighbours, not only do they feel less able to depersonalize their victims, which seems to be a necessary part of most random crime, but they risk the disapprobation of everyone they subsequently encounter.

The economists Michael Campbell and Paul Ormerod have proposed a model of criminal activity which owes much to Schelling's approach, including as it does this element of interaction between its protagonists. They assume that, simply put, there is a 'cost' to crime which depends on how much it deviates from social norms. If nearly everyone in your street is a burglar, you aren't likely to risk social exclusion if you join them. Indeed, you'd be better off doing so, since otherwise you'd be in a minority of losers. If, on the other hand, everyone is a pillar of society, there is more pressure on you to abide by the law.

The choices here might be better regarded as a matter of calculated risk. One could argue that criminally inclined individuals have everything to gain by living among lawful citizens, since then they benefit from their crimes without suffering from those of their neighbours. On the other hand, that same temptation is open to everyone else too. This is the kind of scenario for which game theory, described in Chapter 17, was devised. But Campbell and Ormerod take the view that social pressures make a potential criminal more likely to conform in a lawful neighbourhood.

There is no reason to suppose a priori that these pressures are the major influence on the incidence of crime. Campbell and Ormerod simply ask what effect they have in societies subject to particular penal regimes and social conditions. They divide the population into three groups. The first is immune to the temptations of crime, come what may. Most women and pensioners are likely to belong to this group. The second consists of active criminals. The third represents the 'floating voters', who are susceptible to becoming criminals under the right circumstances but who might instead decide to live lawfully. Individuals can switch from one group to another, and do so on the basis of a kind of peer pressure: the higher the relative proportions of each group in the population, the more likely it is that others will join it.*

Campbell and Ormerod assume a straightforward relationship between social deprivation and the propensity to become an active criminal. Hardship can be considered as a kind of overall 'driving force' for criminality which subsumes a variety of social factors, such as wage rates and unemployment levels – one could imagine it being linked to a standard economic index. On the face of it, increasing social deprivation would be expected simply to increase the proportion of active criminals in the population. But because of the interactions between individuals – the group pressure to conform – the relationship between social hardship and actual incidence of criminality is not simple. For a wide range of levels of deprivation, the model produces two possible levels of criminality: one high, one low (Figure 13.5a). In other words, there is no unique stable state for a wide range of conditions; it depends on the history of the system.

If we start with a low level of crime and little deprivation (the lower curve, to the left of the graph), worsening social conditions lead to only a small increase in the proportion of criminals. Only at the point where this curve terminates with a sudden upturn *must* there be a shift to the

---

* Changes from the non-susceptible to the susceptible group are caused not by individuals changing their mind, which would contradict the very notion of non-susceptibility, but by demographic shifts in the population.

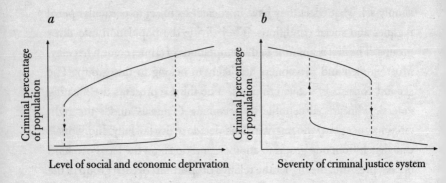

Figure 13.5 The incidence of criminality in a society can depend on social factors in ways that are not obvious or easy to predict. Ormerod and Campbell's model shows sudden jumps between 'high-crime' and low-crime' states that can be triggered by only small changes in social conditions. Changes in criminality are shown in relation to (*a*) the degree of social deprivation and (*b*) the severity of punishment. In both cases there are abrupt phase transitions between two different states, which must happen at the end of each branch (if they don't do so before then).

higher branch. Thus, at this point only a slight worsening of deprivation can lead to an abrupt and large increase in crime. Conversely, if we begin on the upper branch (to the right of the graph), even substantial improvements in social conditions (moving to the left) lead to only a small decrease in crime – until the end of the upper branch is reached. Here there is a jump down to the lower curve: improvements that previously seemed ineffectual suddenly bring about a tremendous change for the better.

Ormerod points out that, despite the evident simplicity of the model, this result has an important implication which is likely to be general: a particular social policy can have different effects on crime depending on the initial conditions. The liberal view maintains that crime can be reduced by combating deprivation. If a neighbourhood is on the high-crime branch and towards the high-deprivation end of the scale, anti-deprivation measures alone are unlikely (according to the model) to have much effect, and conservatives would then denounce such policies as foolish. But if one were near the end of the upper branch, even small improvements could tip the system down onto the lower branch, creating a dramatic drop in crime.

The same kind of behaviour emerges if one considers how, for fixed social conditions, the proportion of criminals changes if the strength of the deterrent is altered. In the model, this amounts to simply changing the probability that susceptible individuals will turn to crime, on the assumption that tougher sentences have a proportionate effect on this probability. Again, a two-branched set of solutions emerges. Increasing the deterrent could have either a small or a large effect depending on where you start from (Figure 13.5b). So again, depending on the circumstances, the observations could substantiate arguments either for or against the case for being tough on crime.

There is more to be gleaned from Ormerod's results. I once spent three years generating graphs which look exactly like Figure 13.5. But I was not researching crime – I was investigating the conditions under which a liquid and a gas interconvert. As the pressure of the fluid is altered, one finds exactly the same kind of dualism (Figure 13.6). That is to say, there are some pressures at which the fluid can adopt either the liquid or the gaseous state. The two branches show exactly the same gentle slope, ending in a little upturn or downturn, as is generated by Ormerod's model. It is clear that the crime model is displaying a first-order phase transition. A phase transition, remember, happens abruptly as the driving force is changed. For a first-order transition, such as the freezing or evaporation of a liquid, the transition is marked by a sudden jump in some property of the system, such as volume or density. Yet Ormerod's model seems, at face value, to be implying that there are two such jumps, occurring at different values of the 'driving

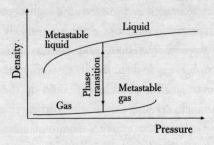

Figure 13.6 The first-order phase transition between a liquid and a gas (evaporation and condensation) can be induced suddenly by only a slight change in pressure. Each state can persist metastably beyond the transition point, until the branch ends at a spinodal point.

force' (the severity of the justice system, or the depth of social deprivation). One jump happens when we reach the end of the upper branch; the other at the end of the lower branch.

For real first-order phase transitions, however, there is only ever one equilibrium state for the system under a given set of conditions, and only one point at which there can be a jump between these equilibrium states. I have marked this point on Figure 13.6, and it lies roughly midway along the region of overlap. Beyond this point – to the right of the lower (gas) branch – the gas is no longer the most stable state: it is not an equilibrium state. Instead, it represents a *metastable* state: provisionally stable, but only for as long as nothing happens to trigger the formation of the liquid. In theory, a metastable state will always convert to the more stable state if we wait long enough (see page 202). How long we have to wait depends on how near we are to the end of the branch. In practice, metastable states can persist for an awfully long time. For pressures beyond the end of the gas branch, a metastable state is no longer tenable. The end of the line, as we saw in the previous chapter, is called a spinodal point.

I am quite sure that all of this pertains to the crime model too. That is to say, the 'equilibrium' switching point between the low-crime and high-crime states will lie somewhere near the middle of the region where the two branches overlap. Once we have passed this point *without* the switch having occurred, there is always the potential that it will. In practice, what destroys a metastable state is the phenomenon of *nucleation*. If a sufficiently large region of the more stable state happens to form by chance in the metastable state, it can expand rapidly to engulf the whole system. Thus, for example, in a liquid cooled below freezing point (supercooled), a tiny crystal of the solid will provide the seed from which freezing spreads throughout the liquid (see page 202). By analogy, we would expect for example that, towards the end of the low-crime branch, a local neighbourhood with a high crime rate could provide a nucleus from which criminality spreads throughout the surrounding population. Such behaviour makes it even more difficult to discern cause and effect in crime policy. It can be interpreted correctly, however, once the analogy with phase transitions is recognized.

Do crime rates really show such sudden jumps? Malcolm Gladwell's book *The Tipping Point* cites the case history of New York City: a dangerous, fearsome Gotham in the 1980s, a proud, safe and neighbourly place in the late 1990s. The ghettos of Brownsville and East New York were no-go areas after sundown in the early 1990s. But within the space of five years, murders in New York City had declined by about 64 per cent and total crimes fell by half. The change is commonly attributed to Mayor Rudolph Giuliani's celebrated 'zero-tolerance' policy, but there is no agreement over precisely what this entailed or how its effects were manifested. Was it all down to better policing? A decrease in drugs trading? An increase in the average age of the population? An upturn in the economy? The determined removal of graffiti from subway trains?

Gladwell argues that this abrupt improvement in crime rates, like social phenomena such as the spread of fashions and ideas or the popularity of successful books, should be regarded as a kind of epidemic. In other words, such behavioural trends spread exponentially in a society. This might be true – although, as we shall see in Chapter 16, there is more to the spread of epidemics than biologists and epidemiologists have traditionally realized. But exponential epidemics are not the only scientific model for sudden change. Phase transitions provide another, somewhat more subtle mechanism. A first-order phase transition represents a switch between two alternative global states of a system. Physics shows us that such switches do not always happen under a unique set of conditions (that is, metastable states can persist), and that it is the past history of the system that really matters. Like Gladwell's epidemics, phase transitions are big changes caused by small effects: 'All of the possible reasons for why New York's crime rate dropped are changes that happened at the margin; they were incremental changes.'[9]

Critical phase transitions, meanwhile, provide another prescription for rapid change – one in which two choices coalesce into one. And spinodal points create yet another scenario, one in which a first-order-like jump in the state of a system is prompted by the disappearance of a metastable state. Gladwell's 'tipping point' can therefore be reached in several different ways. And the distinctions matter, because they imply different things about how the change unfolds and what might be done to induce or suppress it.

THE TIES THAT BIND

Until relatively recently, marriage in Western society was not so much a choice as a social obligation. George Bernard Shaw once said,

'However much we may all suffer through marriage, most of us think so little about it that we regard it as a fixed part of the order of nature, like gravitation.'[10] But no longer so. The average age of marrying couples, the proportion of people who never marry, and the number of divorces all increased in Western society through the twentieth century. More and more people seem to be asking themselves, why marry?

Shaw went on to claim that 'The common notion that the existing forms of marriage are not political contrivances, but sacred obligations . . . influences, or is believed to influence, so many votes, that no Government will touch the marriage question if it can possibly help it.'[11] Well, times change. Today some governments, alarmed at the declining figures and, in supposed consequence, the diminution of 'family values', feel a duty to try to engineer a revitalization of marriage. But how do you do that? It is possible in effect to pay people to marry, through the machinery of tax incentives. It might even be possible to engineer a social climate that smiles upon families welded by marriage (though generally at the cost of implying disapproval of families which are not). Tampering with employment regulations might make families easier to rear – for example, enforcing the right to generous maternity and paternity leave – in the hope that raising a family, or the contemplation of it, increases the likelihood of marriage.

But in the end one will not get far without some understanding of *why* people choose to get married. Needless to say, there is no criterion that fits every couple – but presumably some reasons are commoner than others. If so, what are they? Gary Becker has brought an economist's eye to bear on this issue. In the late 1970s he proposed that a cohabiting couple gains in efficiency over single households by specialization, just as in the pin factory that Adam Smith invoked to advertise the benefits of division of labour. In the traditional arrangement, one partner fulfils the domestic duties while the other goes out to earn a wage. In this way the couple maximize their utility, like dutifully rational market traders.

This may seem an awfully cold-blooded way to prise open the mystery that brings two people together. In Becker's now-classic analysis of the economics of the family (which helped to win him a Nobel prize in 1992), romance and familial love take a back seat while people coolly evaluate what they stand to gain from their spouses (potential or actual), their children and their parents. 'Participants in marriage markets', argues Becker, face a difficult choice because they 'have limited information about the utility they can expect with potential mates.'[12] People are compelled to marry across boundaries of race, religion and class when 'they do not expect to do better by further search and waiting'.[13] Let us be thankful that Shakespeare did not have Romeo and Juliet put it that way.

It is easy to glibly deride Becker's analysis as a case of rationalism gone mad, with family decisions of all sorts hingeing on complex differential equations that describe cost-benefit analyses. Applying economic ideas to family life might look like a caricature of scientism, giving rise as it does to statements like this one:

Consequently, with no costs of supervision and no fixed costs of allocating time between different sectors, the output of a multi-person household would depend only on the aggregate inputs of goods and effective time.[14]

But with this no-nonsense language Becker nevertheless does grasp the essence of many issues surrounding family life. Customs and social norms exist, after all, for a reason. They may sometimes be unfair, out of date or oppressive, but they become established initially because a society finds (or considers) that it needs them. A cool assessment of the 'utility' of potential marriage partners is quite consistent with the common practice in Japan of exchanging curricula vitae for inspection by the partner's family, or with the use of marriage brokers in some Jewish and Eastern European communities. The marriageable age has often been used by societies as a means to adjust birth rates, as for example in the practice in traditional Irish Catholic communities to marry later than is typical in England.

Moreover, Becker's analysis is able to throw considerable light on customs and social traditions whose explanation is by no means obvious. Polygyny – the taking of many wives – is, he argues, commoner than polyandry (multiple husbands) in cultures throughout the world because 'the marginal contribution of women to output has significantly exceeded that of men'.[15] In other words, women are more useful – not just to men but to one another. In some social circumstances a household with several wives makes all the individuals better off than they would be on their own, whereas a woman typically stands to gain less by taking many husbands. This is not to defend polygyny (or indeed to condemn it); rather, it focuses attention on the socio-economic conditions that support it. Certain aspects of patriarchy, according to this model, may not be the preconditions but rather the consequences of a particular kind of society.

The same may be true for the inequalities that continue to exist between levels of earnings for working men and women. Men collectively have always earned more than women, and this would be the case today even if they were paid the same for doing the same job – because on average, women tend (or are, perhaps, compelled) to take lower-paid jobs. This disparity may now be narrowing, but in the past. it has been huge. The natural assumption is that this is because there has been tremendous discrimination, with men being awarded the higher-paid jobs in preference to women. That, at least, might be the concern of the liberal; the conservative might be tempted to claim that this is simply a matter of biology, with men being better adapted to take on such jobs.

Becker shows that neither need be the case, provided mixed male–female households seek to maximize their joint utility: to get things done as efficiently as possible, and with the maximum profit. Under such circumstances a division of labour is always mutually beneficial. The couple are bound to split the duties; and only minor distinctions, such as a slight biological bias towards maternal childcare when children are young, can tip the balance. 'Even small amounts of market discrimination against women or small biological differences between men and women can cause huge differences in the activities of husbands and wives.'[16] We should not expect the effects to follow in the same proportions as their cause.

Thus, says Becker, social inequality can result from purely rational behaviour: 'An efficient division of labour is perfectly consistent with the exploitation of women by husbands and parents – a 'patrimony' system – that reduces their well-being and their command of their lives.'[17] He has been criticized for seeming to imply a kind of determinism – that's just the way things are and we had better put up with it – or worse still, a justification for inequity. But Becker's work does not in fact do this at all. Rather, it is very much in the spirit of a 'physics of society' in asserting that, to understand the reasons why things are the way they are, we are often ill advised to rely on intuition and preconception. Rather, we should seek models which illustrate how certain circumstances can arise if certain rules are followed. Far from demonstrating inevitability, such an analysis may then show us what needs to be changed if we wish to arrange things otherwise, and may save us from empirical tinkering motivated by little more than wishful thinking, the consequences of which might well prove the opposite of what we intended.

Some of Becker's conclusions make uncomfortable reading for liberals like me, such as his suggestion that welfare payments can encumber single women who have larger childcare responsibilities, because it then becomes in their 'rational' interest both to have more children and not to marry. But again, the implication is not that welfare payments are a bad thing but that their impact cannot be assessed in isolation from the wider socio-economic picture.

The great value of Becker's work is to identify a fallacy in conventional neoclassical economic modelling, which regards birth and marriage rates largely as 'given'. Economists have not traditionally been interested in such statistics other than as background features of the economic landscape. In contrast, Becker has shown, economics both affects and is affected by these social factors. Inequality in society is not just the result of human greed but

depends on the relation between fertility and family income;* on the underinvestment by poorer families in their children's human capital; on the tendency for mating to be determined by education, family background, and other characteristics; on divorce rates and the amount of child support to divorced women; and on any inequity in the distribution of bequests among children.[18]

On the other hand, his analysis is itself neoclassical in its foundations, being rooted in the assumption of rational agents independently maximizing their utility. No one would seek to deny that many decisions about conception, marriage and divorce fly in the face of what, objectively weighed, would be in the person's 'best interests'. There is – mercifully – no precise calculus of falling in and out of love. And most significantly, what we decide about our partners or our children depends not just on our relationship with them as individuals

---

* Becker uses 'fertility', somewhat unusually but consistently, to denote not the biological capacity to produce children but the actual (often consciously determined) number of children a woman or couple has – their 'child productivity', if you will.

but on the patterns that society in general has established. A neoclassical approach can only ever tell a partial truth, because – yet again – it neglects the crucial factor: interaction.

THE CONJUGAL CHOICE

When considered from an economic point of view, marriage is destabilized by conditions which make it less 'profitable'. For example, as more women find well-paid jobs available to them, they find their personal utility compromised rather than enhanced by marriage. Changes in the gender demographics of employment have indeed been cited as a cause of the decline of marriage. Yet the corollary – that an economic boom with high employment and high wages disrupts marriage – does not seem to be consistently observed. Some European countries (such as Denmark and the UK) have much higher divorce rates than others (such as France and Germany), despite their comparable economies. It appears that social attitudes to marriage also play a large part in determining how likely a couple is to opt for it, and to stick with it.

Thomas Schelling recognized marriage as the prototype of many social situations in which we make interdependent, binary choices. Either we marry or we don't. (Or rather, at any one time we are either married or not.) A married person affects the options of non-married people by depleting the pool of choices, and also by exerting an influence on social norms. 'The social consequences of marriage', said Schelling, 'make this activity one of the central phenomena in the landscape of social science.'[19]

Paul Ormerod and Michael Campbell saw this as another opportunity for modelling with interacting agents. As with crime, individuals can be regarded as more likely to make a certain choice if their fellow citizens do the same. One can then investigate how the various possible driving forces for marriage alter the proportions of married and unmarried people in an interactive society, while other factors are maintained at a constant level. In Ormerod and Campbell's model the population is again apportioned into three groups: single, married and divorced. Being single is, according to this definition, rather like virginity: once you've left this state, there's no going back. But one can be as Liz-Taylor-esque as one likes in switching between marriage and divorce. Two general factors are assumed to influence these choices: economic incentives (be they wage-earning potential, tax breaks, job opportunities, or whatever) and social attitudes (public disapproval of unmarried cohabitation, unfashionableness of marriage).

If the strength of social attitudes is weak, then the model produces a proportion of married people that simply increases as the economic inventive to be and stay married increases. But if social attitudes are stronger, the outcome is different (Figure 13.7a). Again, we find two branches: a high-marriage and a low-marriage state of the population. In fact, Ormerod makes the point that the two branches are linked by a continuous curve (shown by the dotted line). This seems to imply that for a range of strengths of the economic incentive, there are *three* possible states – a vertical line in the figure intersects the curve at three points. However, one can show that beyond the turning points of the upper and lower curves, the states represented by the dotted curve are neither stable nor metastable, but unstable – they transform instantly into something else. This is exactly what emerges from van der Waals' theory of the liquid–gas phase transition: a single, kinked curve divided into stable, metastable and unstable portions (Figure 13.7b).

We can therefore expect all the same features in the marriage model as we saw in the crime model: an equilibrium transition between the upper and lower branches, beyond which nucleation of the more stable state within a metastable state can trigger the switch at any time.

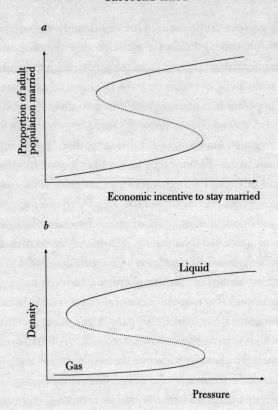

Figure 13.7 In a model that describes how the proportion of married people in a population depends on social pressures and economic incentives, identical social and economic conditions can lead to two possible outcomes: a 'high-marriage' and 'low-marriage' society (*a*). These two states are connected by a continuous, looping curve (the dotted line); but between the two turning points of the loop, the states that emerge from the model are not stable. This is entirely analogous to the loop that connects liquid and gas states in van der Waals' theory of phase transitions (*b*). Van der Waals showed that there is no stable state in the dotted portion of the loop.

And there is again a dependence on history: economic incentives might not significantly increase the proportion of married individuals if this proportion is initially low, even though a different social system with the same level of incentive might have a high incidence of marriage. The loop in the curve appears only if the influence of social attitudes is strong enough. A three-dimensional graph which shows the dependence of marriage both on this and on economic factors therefore looks like Figure 13.8a. The lines in Figure 13.8 trace out a *surface* which develops a kink. Ormerod compares it with a folded mattress.

You may not be surprised to learn that this too is a familiar image to statistical physicists. Josiah Willard Gibbs owned a plaster cast of this strange wave, made for him by James Clerk Maxwell. It is the curve that shows the first-order liquid–gas transition vanishing into the critical point. The critical point itself corresponds to the place where the kink first starts (or ceases) to appear – where the upper surface is so twisted that it overhangs the lower surface. On Gibbs's surface, 'strength of social attitudes' is replaced by temperature, 'economic incentives' by pressure, and 'proportion of married population' by density (Figure 13.8b). In other words, Ormerod and Campbell's marriage model has a critical point. At this point there are no longer two possible states of society, with a high and low proportion of marriage: these states merge into one. So these interacting agents, influenced by the choices one another makes, display the whole gamut of behaviours that characterize the particles of a fluid, influenced by their mutual forces of attraction and repulsion.

When wilful human acts such as marriage and crime first entered the roster of social phenomena governed by statistical laws and regularities, the response was a mixture of amazement, delight and dismay. Gary Becker's attempts to make these things accessible to economic modelling drew much the same reactions. We'd be wise to

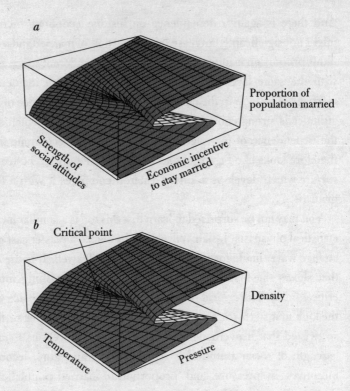

Figure 13.8 The dependence of marriage rate on *both* of the factors in the model – social pressures and economic incentives – can be represented as a surface in a three-dimensional plot (*a*). This surface has a fold: outside the fold, a given set of social conditions allows for only one state of the system, while inside the fold there are two possible states. The fold ends at a kind of critical point. Exactly the same behaviour is found in the dependence of the density of a fluid (liquid or gas) on its temperature and pressure (*b*).

anticipate something similar as physics makes its mark in social science. William Newmarch, addressing the Statistical Society of London in 1860, sketched a positivist perspective no doubt buoyed by the first flowering of statistical science. Yet his words remind us that, at the very least, sound policy must, if it is to be just and effective, draw on something more than preconception and intuition:

The rain and the sun have long passed from under the administration of magicians and fortune-tellers; religion has mostly reduced its pontiffs and priests into simple ministers with very circumscribed functions . . . and now, men are gradually finding out that all attempts at making or administering laws which do not rest upon an accurate view of the social circumstances of the case, are neither more nor less than imposture in one of its most gigantic and perilous forms.

'Crime', Newmarch added, 'is no longer to be repressed by mere severity.'[20] And indeed we can now see that there are reasons to expect that the effects (on average and not individual behaviour) of 'mere severity' may be hard to predict or assess – if we fail to recognize that the choices we make, about this as about so much else, are consequent on the relations and dealings we have with our fellow people.

MINORITY RULE

Voters, clearly, hope to end up in the majority. Similarly, the kinds of social interactions we have considered for crime and marriage tend to lead to normative modes of behaviour: if marriage is considered the 'normal' thing to do, more people will do it. But there are many situations in which we would prefer to do what everyone else *doesn't* – we want to be in the minority, and the smaller the better. Driving to work, we hope to choose a route which most others will not take, and thereby to avoid traffic jams. If we are selling our house, we might prefer to wait until it is a seller's market – when there are more prospective buyers than sellers. That way, we can demand a higher price.

Economist Brian Arthur discovered a version of this 'minority problem' when he went to study complex systems at the Santa Fe Institute in New Mexico. He dubbed it the El Farol problem, and it was all about how to have a good night out. El Farol was a bar in Santa Fe, close to where the Institute originally stood. On Thursday nights it offered Irish music – a great attraction to the Irish-born Arthur. But not to him alone. Researchers from the Institute started to flock there in such numbers that the place often became uncomfortably over-crowded. Eventually people began to stay away on Thursday nights – the music was great, but it wasn't worth the crush. And so some Irish nights were relatively quiet, allowing a lucky few to enjoy the bar like they did in the good old days. And there's the dilemma. Do you risk going to the El Farol in the hope that it will be a 'good' (that is, uncrowded) night, or do you stay at home? Either way, the minority does best. If those at the bar are in the minority, they have a great night out. If the minority is instead the stay-at-homes, they have a relaxing night in while the rest are jostled and swamped in the overcrowded bar.

How, Arthur asked rhetorically when he gave a talk on the El Farol problem in 1994, would you expect attendance at the bar to vary over time? 'Will it converge,' he asked, 'and if so, to what? Will it become chaotic? How might predictions be arrived at?'[21] The problem, he pointed out, is that there is no 'right' choice: no 'deductively rational solution'. No individual has any way of knowing what is best, since they cannot know what others will choose to do. All they can do is guess, based on whatever hunches or intuitions they may have. Their choice is made on the basis of *beliefs* about the likely actions of others.

To an economist like Arthur, this situation was familiar. Market traders buy and sell on the basis of beliefs about the market: whether the prevailing feeling is optimistic (so that others will buy) or pessimistic (when everyone wants to sell). But whether prices actually rise or fall depends on what everyone decides, largely in ignorance of what everyone else intends to do. Arthur pointed out that while economists have traditionally assumed that traders use *deductive* rationality – that they make decisions on the basis of solutions to well-defined problems – in fact they are in general facing a problem which is ill-defined. There is no right answer – except retrospectively, which is not much use. In this situation one can only apply *inductive* reasoning, based on subjectivity and experience.

'As economists we need to pay great attention to inductive reasoning',[22] Arthur argued. And he suggested that the El Farol problem provided the perfect model system for studying it. He devised a simple, idealized description of the situation in which agents use a wide range of different rules of thumb for predicting what the attendance at the bar would be, and thus for deciding whether or not to go there themselves. These rules make use of past attendance figures: one agent's belief, for example, might be that 'the attendance will be the same as last week', or that 'the attendance will be an average of the past four weeks'. Each agent applies several such rules, and makes its decision on the basis of the one that experience has shown to be the most accurate.

Arthur found that this model produced constantly fluctuating attendance figures, with no obvious pattern to them. Some nights, only 30 per cent of the agents would go; on other nights, 90 per cent. But the attendance averaged about 60 per cent, and seldom deviated from this by more than 20 per cent. In other words, while the attendance neither settled down to a regular number, nor showed any sign of regular rises and falls, it maintained a constant average. Arthur compared it to a forest whose edges stay fixed while individual trees grow and die. Why 60 per cent, though? Because this is the number that Arthur chose arbitrarily to designate the tolerable limit. If attendance was greater than 60 per cent, the bar was overcrowded and it was better to stay at home. Thus the agents automatically 'find' the optimal attendance level, on average, even though there is no rule that guarantees this.

In 1997, physicists Damien Challet and Yi-Cheng Zhang at the University of Fribourg in Switzerland formulated a better-defined model for the El Farol problem, which became known as the 'minority game': a game that an agent 'wins' by ending up in the minority. In Arthur's model the strategies used by each agent to reach a decision were more or less arbitrary. In the minority game they are more systematically defined. Each agent keeps a record of which of the two available choices (to stay in or to go to the bar, say) produced the majority in the previous 'rounds' of the game (that is, on previous nights). This record – which can be written as a string of binary digits, like the 1's and 0's used for computer logic – is then plugged into a particular strategy in order to reach a decision about the next round. For example, a strategy might say that if bar attenders were in the minority over the previous three rounds, the agent should go to El Farol in the next round. Again, each agent has several strategies to choose from, and at any time it uses the strategy that has so far proved the most successful.

Challet and Zhang found that the average attendance was 50 per cent: half of the agents elect to stay away, on average, and the other half go to the bar.* So at first glance it looks as though the agents do a pretty good job of organizing themselves. Even though they cannot devise any collective plan, the 'minority' is as big as it possibly can be on average – which is to say, essentially the same size as the 'majority'. But how 'efficient' is the game really? Again, the attendance figures

---

* In this model, the two options have equal weighting: in essence, you just want to end up in the minority, that is, the group containing less than 50 per cent of the agents.

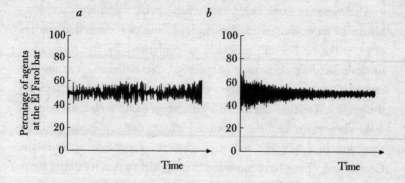

Figure 13.9 In the minority game, the proportion of agents which make each of the two available choices (going to the El Farol or staying at home) fluctuates around 50 per cent (*a*). In an ideal world, the minority would always be as big as possible – that is, as close to 50 per cent as it can be. But because of the fluctuations, the minority is often smaller; the game is 'inefficient'. If, however, agents with poor strategies are weeded out over time in an evolutionary process, the fluctuations get progressively smaller: the efficiency of the game improves (*b*).

fluctuate constantly either side of the average (Figure 13.9*a*). Each over- or undershoot means that more players *could* have 'won' (been in the minority) than was actually the case. The larger the fluctuations, the more 'inefficiently' the game is played.

The researchers found that the fluctuations get smaller as the strategies take into account a greater number of past rounds – that is, as the agents' memories improve, or, you might say, as they become smarter. Moreover, the efficiency of the game increases – the size of the fluctuations decreases – if agents are allowed to *evolve* in a Darwinian fashion such that the most successful ones multiply and the worst players (those that end up most often in the majority) die out (Figure 13.9*b*). In this case, the population as a whole is capable of 'learning' to play well.

The minority game bears only a superficial, metaphorical resemblance to an economic market. Indeed, one can argue that traders often strive to be the *majority*, as evidenced by the inclination towards herding behaviour (page 269). Nevertheless, it embodies enough features of real markets – competition and 'selfish' behaviour, the empirical testing and use of a wide range of strategies, the need to make choices on the basis of limited information – to make it highly tempting to apply the minority game to economic phenomena. Challet and Zhang have modified their model so that it mimics more closely the conditions of the marketplace, and have used it to investigate the effects of, for example, 'noise traders' (who make decisions based on perceived trends in market indices) and insider trading. They find, for instance, that the range of chosen strategies narrows when there is 'exploitable information' (privileged or 'inside' information) available: there then become 'best' ways to play the game, which do not exist when everyone has to use the same information.

One of the more remarkable discoveries of the physics of society is that behaviour which looks strangely 'human' can emerge among agents which are in effect nothing but robot-like automata. We saw in Chapter 6 how such agents can appear to panic when trying to leave a crowded room. Shahar Hod and Ehud Nakar in Israel have demonstrated that the minority game can mimic another human attribute: indecision. They considered a version of the game in which, having weighed up the outcomes of past rounds, all agents use the same calculation to select its next move. But the agents don't necessarily heed this calculation: they do so with a probability that can vary between 1 (the calculation is always heeded) and 0 (the agent always does the opposite of what the calculation suggests). An agent's personal 'strategy', therefore, amounts to the probability with which it follows the apparent lessons of past history. In successive rounds, each agent learns from experience to increase or decrease this probability.*

In the usual form of the game, this process leads to very decisive behaviour: the agents end up polarized into extreme groups, which invariably choose either to heed or to ignore history. In other words, the strategies become either 'always heed what the calculation recommends' or 'always do the opposite.' But Hod and Nakar found that the situation changes if the payoffs for 'good' and 'bad' choices are unequal – if, say, it hurts more to lose (to be in the majority) than it pays to win. This might be the case, for example, when selecting the best route to work. A bad choice, which dumps you in a traffic jam and makes you arrive late, could mean that you miss a crucial meeting or even that you get sacked, while a good choice means only that the day unfurls as normal. Similarly, traders in a depressed market stand to lose more in a bad deal than they will gain in a good one.

When the outcomes have this kind of inequality – when the stakes are loaded to our disadvantage – the best strategy is to make cautious, safe choices rather than extreme ones. Hod and Nakar found that under such circumstances, agents in the minority game tend to become indecisive: rather than selecting a particular strategy invari-ably (that is, with probability 1 or 0), they pick one or the other option with equal probability. This generally means that the rewards over time are small, but so are the dangers. Agents who stick to extreme behaviours, on the other hand, lose more than they win on average.

This indecisive, hedging behaviour may be best for each individual, but it makes the population as a whole inefficient: the average size of

---

* That an agent should choose to ignore the lessons of history might strike us as odd. But there is no guarantee that the procedure an agent uses to guide future choices by looking at past experiences will prove to be beneficial. If this procedure seems at any moment to be leading to bad recommendations, the agent will attribute less validity to it.

the winning minority is considerably smaller than it could be. It would mean, for example, that the average number of attendees at the El Farol bar is consistently lower than its ideal capacity. Indeed, the population as a whole typically achieves a worse average than it would if each agent made its decision purely at random: attempts to make decisions by inductive reasoning become destructive to the group's overall efficiency.

The minority game is certainly psychologically naive; in fact, it assumes virtually no individual psychology at all, beyond a crude ability to reason from experience. Yet the range of group behaviours which emerge from it are extraordinary, subtle and often impossible to predict. We do not need to believe that we really make our decisions this way if we are to acknowledge the value of the model's central message, which is surely this: that we should not be too ready to impute complex psychological motives to the decisions made by a group without first appreciating how much complexity can arise even from the most basic, stripped-down description of the process. There is little that is obvious about how we make our choices.

14

The Colonization of Culture

Globalization, diversity and synthetic societies

There are today no longer Frenchmen, Germans, Spaniards, Englishmen . . . ; there are only Europeans.

Jean-Jacques Rousseau (1772)

They hate being Indians. They want to be Canadians and Americans . . . Nobody ever told us that all this would be coming in with TV. It's like some kind of invasion from outer space or something. First it was the government coming in here, then those oil companies and now it's TV.

Cindy Gilday
Communications director of the Dene Nation, Northwest Territories, Canada

Too many people still believe Margaret Thatcher's TINA, There Is No Alternative, whereas we should say TATA, There Are Thousands of Alternatives.

Susan George (2002)

*

Among my many vivid memories of travelling in Japan is that of Tokyo subway advertisements for lavish Western-style marriages in full white bridal outfits. The modern Japanese convention, money permitting, is to have both a traditional wedding in kimono and a Western-style white wedding. 'More cultural imperialism', I muttered at the time.

In the past, the spread of cultural beliefs and values has certainly been bound up with dreams of empire. Today it follows a commercial imperative. The Victorian conviction that 'primitive' cultures simply had not yet found the 'right' way to behave, to speak and to worship is, one hopes, dead and buried, but it is being replaced by the multi-national conviction that there are few parts of the world yet to learn the benefits of Coca-Cola and hamburgers. But the transmission of cultural values from one society to another has surely not always been a bad thing. The Moorish architecture of southern Spain is a beautiful monument to the flow of learning from the Islamic world to the early medieval West. From the blight of slavery in the New World arose the sublime consolation of jazz music that enriched American culture. The paintings of van Gogh and Matisse betray the influence of Japanese woodblock prints; modern European dance floors throb to the rhythms of India and Latin America.

When one culture subsumes another, or the two merge into a monoculture, the world loses a degree of diversity and richness of experience. On the other hand, commonality of belief and tradition lowers the potential for conflict. A shared language lubricates communication. The relative roles of choice and enforcement in cultural transmission are hard to distinguish. Even conquering nations often adopt some of the traditions of their subjects; and did not Rome eventually abandon its gods for a Judaeo-Christian monotheism? Whenever and however two cultures come together, there will be exchange of values, of arts, sciences and technologies, of customs and beliefs, of languages. Within these processes lies the history of the human world, from the siege of Troy to the eastward expansion of the European Union. History can never be solely about the impact of great men and women, but must always also examine the interactions between vast groups of people.

Making models of how people pick up criminal habits from their neighbours or succumb to the social influences of friends and family is simply to attempt to describe a tiny microcosm of this ebb and flow of cultural and social values. Some scientists are seeking to explore a wider vista: to uncover the way in which cultural transmission shapes the demographic, political and linguistic boundaries of our world. It is, needless to say, an awesome objective, but its beginnings can be simple. The questions these researchers ask are broad ones. What determines the diversity of a culture? Why do some minority cultures or traits survive (like the Basque and Welsh languages) while others vanish? Indeed, why do any two cultures not generally converge completely when they interact?

At the beginning of the twenty-first century, such questions have seldom been more urgent. The latest conflicts in the Balkans reflect a persistent lack of convergence in values and beliefs that made an ad hoc nation like Yugoslavia fatally fissiparous. A comparable process proved bloodless in the break-up of Czechoslovakia, but the collapse of the Soviet Union continues to be a painful affair in Chechnya and Georgia. A nation cannot survive without a sense of national identity, and that identity cannot arise without at least some degree of uniformity of culture.

Most modern civil wars are at root no different from the one that impelled Hobbes to seek a calculus of social order. They explode out of unresolved conflicts whose fault lines go so deep as to resist decades or even generations of effort to heal them over. For England in the seventeenth century the line was, on the surface, drawn between those who felt that the sovereign should wield the ultimate power and those who would rather see the king as the servant of the people and their parliament. But below this dispute was the old and festering religious divide – the same wound that had bled Europe for a century. It is a wound that claims lives today on the streets of Northern Ireland.

There seems to be nothing inevitable about convergences of cultural, social or political values. The cohesion of the United Nations has been severely tested by the differing attitudes of its member nations towards the wars in Kosovo and Iraq – if the bombing of Belgrade highlighted these divergences, the conquest of Baghdad by US- and British-led forces stretched them almost to breaking point. Even the unprecedented degree of international unity over the war in Afghanistan in late 2001 may well disappear before a stable nation arises from Kabul's ashes.

Yet as fast as the world map seems to be fragmenting, we also see efforts to weave new unions. The European Union is a uniquely ambitious experiment which searches for shared economic and technical values while attempting to preserve and respect social and cultural differences. In January 2002 it embarked on its grandest experiment with a near-unification of hitherto national currencies. And concerns about the global environment, world trade and world health have stimulated the emergence of international groups and organizations that share beliefs quite unrelated to matters of nationality.

The stability of such unions may depend in the long term on the extent to which their members can align their views and goals – on their ability to modify behaviour through interaction. The resulting push and pull poses difficult questions for liberal advocates of diversity. Many of us may feel that too much homogeneity is a regrettable thing, yet we argue for 'universal' human rights and expect nations to conform to certain codes of conduct – a trend that has been dubbed 'liberal imperialism'. We may rejoice in cultural diversity but lament the segregation that some ethnic minorities believe to be necessary to avoid erosion of their traditions and identity. If there is too sharp a division between cultures, the result is the persecution of minority groups, alienation and unrest. All these issues point to a burning question: can distinct cultures and beliefs coexist stably, or must one inevitably engulf the others?

CULTURE SHOCKS

A long list of factors has been advanced by social scientists to explain why some cultural differences persist while others dissolve. There seems to be a natural human tendency to seek group identity, which depends on emphasizing differences with the 'outside' and can sometimes encourage the adoption of extreme views. The spread of fads and fashions may, in contrast, gain momentum from a desire to conform. Languages evolve and become differentiated partly by a seemingly inevitable process of random drift resulting from the accumulation of small changes – the equivalent of Darwinian random mutations. Linguistic and technological compatibility, meanwhile, greatly facilitate the exchange of ideas, while geographical isolation can inhibit it.

One thing many of these considerations suggest is that the transmission of cultural ideas depends on how much common ground already exists. A shared language is a particularly strong conduit for exchange, although that does not in itself guarantee convergence – one is immediately reminded of George Bernard Shaw's apocryphal remark that Britain and America are 'divided by a common language'.*

Robert Axelrod has developed an interaction-based model that seeks to sketch in broad outline how cultures and customs are

---

* Oscar Wilde perhaps deserves the credit: 'We have really everything in common with America nowadays except, of course, language.'¹

disseminated. The central assumption in this model is that likeness encourages more likeness: Axelrod does not worry about the mechanistic details of *how* cultural exchange happens but merely says that it becomes more likely the more it has already happened. There is, in other words, positive feedback in the cultural convergence of the interacting agents.

Axelrod imagines a map divided into a regular grid with an 'agent' on every site. These agents are not individual people but sub-populations inhabiting a geographical region – a village perhaps, or a district. Each agent possesses a number of cultural 'features': a style of pottery, for example, or a preferred method of agriculture, or a local dialect. There can be any number of such features, and in reality we would expect them to be both rather numerous and difficult to define and disentangle. Nevertheless, it seems reasonable to suppose that in some sense a culture can be identified by a carefully compiled list of such attributes.

To each cultural feature Axelrod ascribes a certain number of variations. For example, his model world might include five different languages: five different 'values' of the 'language feature'. I shall call these different versions of a single feature 'traits'. Axelrod assumes that any two traits of a given feature are equally different. This is clearly a simplification. For example, the technological bases of Japanese and British societies are different, but are clearly more similar than either of them is to that of Bhutan. Italians and Spaniards can almost understand each other's language; both are at a considerable and comparable disadvantage with German. But this is, after all, meant to be a simple, minimal model.

Thus each grid site in the model is labelled with a set of values for each cultural feature. Figure 14.1 shows a small part of a grid for a version of the model with five cultural features, each of which may take one of ten values (labelled 0–9; but remember that the numbers have

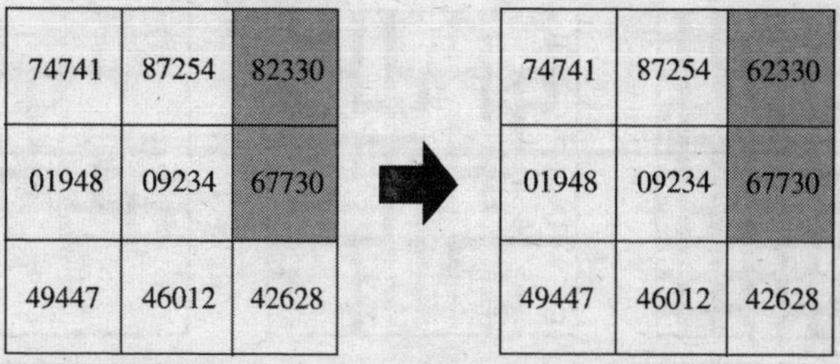

74741	87254	82330
01948	09234	67730
49447	46012	42628

74741	87254	62330
01948	09234	67730
49447	46012	42628

Figure 14.1 In Robert Axelrod's model of cultural dissemination, each site on a regular grid represents a local culture with a specific number of features (here five). Each feature may take one of ten 'values' (0–9). If two adjacent sites share one or more values, they can converge further by aligning one of the existing dissimilar values. One such step is shown here for the two sites indicated in grey.

no quantitative significance). Transmission of cultural values then takes place by iterating the following procedure:

1. Pick a site at random, and select one of its adjoining neighbours at random. (There are four neighbours on a square grid.)
2. The probability of interaction between these two sites depends on their cultural similarity. If they share no features with the same value (that is, no traits), they will not interact. If they share one feature, they interact with a probability of, say, 1 in 10. If two features are the same, the probability might be 1 in 5.
3. 'Interaction' consists of setting one other – randomly selected – feature on the chosen site equal to that of the neighbouring site. In other words, the cultures become even more alike.

A single step like this is depicted in Figure 14.1. Its effect, obviously, is to make the sites increasingly more similar. The end result, then, might appear to be a foregone conclusion: gradual erosion of

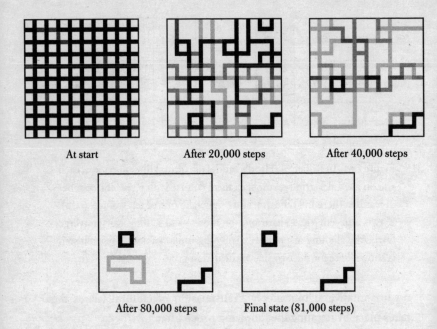

At start   After 20,000 steps   After 40,000 steps

After 80,000 steps   Final state (81,000 steps)

Figure 14.2 A simulation of cultural spread on a 10 × 10 grid. Each grid point is shown in white, and the heavy black/grey lines separate regions whose cultural features differ in at least one respect. The degree of difference is reflected in the shade of the line, with black representing complete difference of all five features. As the simulation advances, differences are eroded until only three cultural 'islands' exist: two small ones surrounded by one large one. Further change is then impossible because the wholly different cultures can no longer interact.

difference and spread of a monoculture. Sometimes this is indeed observed – but not always. The model can reach stable states in which culturally different regions persist until no further change is possible. If, for example, a region of sites of one culture were to appear in the middle of another culture with which it has no traits in common, the two cannot interact – so the 'island' culture remains, even if the culture that surrounds it grows to dominate the rest of the grid.

The emergence of two such islands of diversity in a 'global' monoculture is shown in Figure 14.2. Here the grid consists of 100 sites, each of which (as in Figure 14.1) has five cultural features that may each take ten different values. The number of possible states for even this 'simple' system is astronomical: 10 raised to the power of 500, which vastly exceeds the number of atoms in the universe. The cultural traits are initially chosen at random, so most adjacent sites share no traits; a few share just one. The similarity between neighbours is indicated by the shading of the lines dividing them: black denotes one or no common traits, white denotes all five values the same.

After thousands of steps the grid has developed several small patches in which the sites all share the same culture. Note that although all the grid sites appear as white in the figure, this does not mean that they all have the *same* culture – each separate white region bordered by black or grey lines represents a different culture. As 'time' progresses in its step-by-step manner, some cultures grow while others are enveloped. Eventually a single culture covers most – but not all – of the grid.

What does this exercise tell us? We would, after all, *expect* the grid to develop greater homogeneity purely from the very way the rules are set up. But we see that this need not lead to the complete eradication of cultural diversity. Moreover, we can use the model to deduce something about the dependence of persistent diversity on initial diversity. Suppose, for example, we were to increase the number of different values a cultural feature can take – to allow, for each cultural feature, fifteen different traits instead of ten. What effect does this have on the number of stable regions (the number of 'countries', local dialects, or whatever) in the final, unchanging form of the grid?

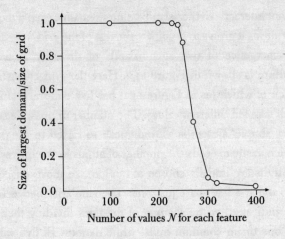

Figure 14.3 The geographical extent of the largest culture in the final, stable state of Axelrod's model depends on how many 'values' each cultural feature can take. If this number (N) is small, a monoculture develops: the size of the largest culture is more or less equal to the size of the whole grid. If N is large, even the largest culture remains of insignificant size relative to the whole grid – there is a lot of cultural diversity. The changeover between these two extremes is not gradual but sudden – it has the characteristics of a phase transition. In this example, the phase transition happens at around $N = 240$.

Well, what would we expect? It seems reasonable to imagine that, by increasing the potential for diversity in this way, we will increase the actual diversity in the final map – that there will be more stable regions. And that's just what we find. With ten values per feature, the average number of stable regions on the 10 × 10 grid, for many model runs with different starting configurations, is 3.2. With fifteen values per feature, this average rises to 20. If, on the other hand, we reduce the number of traits to five, the average number of stable regions is just 1: we create the conditions for a monoculture.

So far, so predictable. But Alessandro Vespignani of the Abdus Salam

International Centre for Theoretical Physics in Trieste, Italy, and his colleagues looked more closely at how the final diversity depends on the number of traits (call it N) in Axelrod's model. They discovered something unexpected: the switch from a monoculture (for small N) to a polyculture (for large N) is not gradual but abrupt. To illustrate this, they measured the size of the largest region as N varies. For small N, this region encompasses more or less the whole grid. At some critical value of N, the largest region becomes dramatically smaller until it occupies an insignificant proportion of the grid and is really no different from all the other regions (Figure 14.3). This sudden change has the characteristics of a genuine phase transition between a uniform and a fragmented state, rather like the melting transition between an orderly solid and a disorderly liquid. Close to the phase transition, slight differences in the cultural diversity of a territory (that is, in N) can have a substantial effect on the number of different cultures it can support.*

The researchers also found that within the range of N where the rapid change happens – that is, within the transition region – the size distribution of the stable regions follows a power law. This is just the same behaviour as is found for the size distribution of different domains (gas-like and liquid-like, say) at a critical point (page 289). Thus Axelrod's model has several characteristics in common with simple physical systems of interacting particles. And these characteristics are not obvious: we wouldn't be able to predict them just by looking at the rules of the game.

* Konstantin Klemm at the Niels Bohr Institute in Copenhagen and his colleagues have looked closely at this transition between a polyculture and a monoculture, and find that it is modified by randomness. If there is some chance that any cultural trait will change unprovoked – an effect anthropologists call 'cultural drift' – then the polycultural patchwork is at risk of collapsing into a monoculture, even for large N. That's to say, the polyculture can become metastable, and a monoculture is the most stable state. So the behaviour of the model depends rather subtly on the conditions under which it unfolds.

There are more surprises. Suppose we introduce greater diversity in another way, by increasing the number of different cultural features. Thus, each grid site will be labelled not with a string of five numbers, but with a string of, say, ten. Will this increase the cultural variation of the final state? Not at all. For a 10×10 grid with ten features, each of which can take ten values, the average number of stable regions is 1. Remember that for five such features, it is 3.2. For fifteen traits per feature, a five-feature model gives an average of 20 regions whereas a ten-feature model gives only 1.4 on average, and a fifteen-feature model gives just 1.2. Contrary to intuition, increasing the potential diversity in this way actually *reduces* the model's resistance to monocultural takeover.

How can this be? The answer is not hard to find. Remember that adjacent sites can interact only if they share one or more traits. These interactions promote homogenization. The greater the number of *features*, the greater the chance that one of them will match those of a neighbour, and so the greater the potential for interaction. In contrast, as the number of allowed *traits* increases for each feature, the chance of a match is diminished – just as two people are less likely to pick the same number between 1 and 15 than they are between 1 and 5. So increasing the number of versions of each cultural feature diminishes the chance of interaction. Verbal communication can be harder in India, for example, where there are many local languages, than in the USA, where just one predominates.

The actual numbers here have no significance for the real world. But the model, despite its simplicity, offers two general insights. First, it is not obvious exactly what we mean when we talk about the 'complexity' of a culture. The technologically sophisticated cultures of the West arguably have greater diversity than the cultures of developing countries, in terms of number of *traits* – we identify ourselves according to a vast number of highly specialized jobs, the kind of cars we drive, the many types of music we listen to, the wide spread of incomes, and so forth. But it is not clear whether – or how – real-life cultures differ in the number of *features* that characterize them: most features in Western culture, such as music, transport, art, religion and language, have direct analogues in Bangladesh or Cambodia. If a breadth of traits has different implications from a breadth of features, this distinction matters. And second, Axelrod's model tells us that we cannot necessarily rely on intuition when evaluating how easy it is for a culture to spread.

The model can also help us to explore the influence of the size of the overall territory. What happens if we expand the 10×10 grid to 100×100, or shrink it to 5×5? Intuition suggests that the number of stable regions will increase as the grid expands: if we can fit 3.2 regions into a 10×10 grid, should we not expect a hundred times that number in a 100×100 grid? Again, this common-sense thinking trips us up. In fact the number of stable regions decreases as the grid gets bigger. For a 12×12 grid with five features and fifteen traits per feature, there are on average 23 stable regions. For a 50×50 grid the number falls to about 6, and for 100×100 there are just 2. This contrasts, for example, with size effects on biodiversity: island species are less numerous than those on large continents because the number of evolutionary niches is smaller.

We can see in Figure 14.2 how the stable regions emerge from a kind of gradual 'freezing-out' of the model. The two small cultures that remain in this example are frozen in place by the impossibility of interaction with their neighbours across the border. But what if there were different, non-frozen regions with which to interact, just beyond the edges of the grid? These might extend their influence into the 10×10 grid, rearranging its component sites and resurrecting the possibility of interaction with the 'island' regions. This is just what happens in large grids: interactions go on for longer, so there is less likelihood that the map will become frozen into a configuration with many remaining regions. But the smallest grids also fall prey to monocultures, simply because, as on ocean islands, there is not enough space to accommodate much diversity. There seems to be an *optimal* territory size that will support a large number of stable regions – as though a continent has an ideal number of nation states it can accommodate.

A map of the world is a peculiar, uneven patchwork. Most African nations are larger than most European nations; is it a coincidence that two of the smallest, Rwanda and Burundi, have recently seen some of the greatest turmoil? (Likewise the larger nations of Western Europe have enjoyed greater stability than the smaller Eastern European countries.) And why are the West African nations smaller than those in the east and south? While one can offer contingent historical answers to these questions, Axelrod's model seems on the brink of telling us something more general about how geographically distinct regions or continents are likely to draw their national borders, and what consequences this will have.

THE LAND OF SUGAR AND SPICE

We might object that this model is too absurdly simplistic to bear any relation to things that happen in the real world. Think of what it neglects: geographical influences, the effect of technological change, mass media, different organizational and governing systems, tourism . . . One can imagine how to include some of these influences in the model (and this will probably be done in time), but the broader question remains: given that there will always be some arbitrariness in the assumptions made, why should we have any faith at all in the conclusions reached? It is a fair objection, and to address it Axelrod has looked at whether the general predictions of the model transfer to other cultural models erected on a different set of assumptions. To this end he has ventured into the world of Sugarscape.

Since its release in the early 1990s, the computer game SimCity has acquired cult status. The aim is to build and maintain an entire city from the ground up. You give it a power system, you feed its inhabitants, you conduct repairs and maintenance. You get to play God, only to discover what a mundane, maddeningly intricate and yet wholly compulsive business that is. Sugarscape is like an eighteenth-century SimCity. It was devised by Robert Axtell and Joshua Epstein of the Brookings Institution as a general-purpose testing ground for social theories. Suppose that someone claims that this law or that constraint applied to a society will produce a particular result; you can simulate the process in Sugarscape, and see if the prediction holds. 'Our rules create the cliffs we drive off', say the two researchers. 'Computational systems such as Sugarscape can offer "headlights", if you will, by permitting us to project, however crudely, the evolutionary consequences of certain rules.'[2]

Sugarscape is another grid world – but this grid is imprinted on a doughnut-shaped object, a torus. This seemingly peculiar shape is chosen so as to eliminate boundaries, which makes computation easier and removes 'edge' effects.* The grid sites are populated by agents, but usually sparsely – there is plenty of open space in Sugarscape. The agents have one key impulse: to seek sugar. Sugar cane is distributed in some chosen manner over the grid, and the agents move towards whichever grid point in their field of vision has the most of it. There they chop it down and consume it; eventually, more sugar cane will grow. The agents also interact with one another – and do they ever interact! Specified rules enable them to fight, trade, collaborate, exchange cultural traits and have sex (what else do you do with your

* Sometimes one might wish to retain edge effects in models like this, for example to gain some understanding of how cultures change at a frontier.

life, after all?). Cultural interactions are somewhat similar to those in Axelrod's model. The agents have a set of cultural attributes which can become equalized during interaction through a process of random selection. Often agents will form tribal alliances ('nations') in which cultural attributes coalesce according to the rule that the commonest attribute is adopted by all.

This is an extremely complex model, and it offers a framework for simulating all manner of scenarios. One can observe the effects of a sugar shortage: will it precipitate war or 'international' cooperation? Epstein says, 'We think of our model as a laboratory for social science . . . We can examine population growth and migration, famine, epidemics, economic development, trade, conflict, and other social issues.'[3] The researchers can even examine evolution and how social forces shape mating patterns. Sugarscape is like a fantastic elaboration of chess, where the pieces make their own decisions and have their own desires and fears.

One scenario, for example, concerns the development of trade. Here Epstein and Axtell introduce a second resource: spice. Agents can have different appetites for sugar and spice, so some trade for one commodity and some for the other. They are programmed to obey certain rules for bargaining and exchange, from which prices emerge according to the interplay of supply and demand. If the agents are 'humanized' by being given finite lifespans and evolving preferences, the markets never reach equilibrium but fluctuate constantly.

Trade does some interesting things to the cultural landscape. It enables more agents to survive – individuals can buy what they need where, in the absence of trade, they would have starved. But it also skews the distribution of wealth so that the majority of agents are poor while a small number of them accumulate great riches. Here, then, is Pareto's law of wealth (page 307), conjured up from a computer game. Some economists have argued that this emergent power-law distribution represents a kind of immutable property of a trading society. Sugarscape can help to explore just how immutable it is in the face of changing social and economic conditions. The model could, for example, be used as a testing ground for ways of redistributing wealth more equitably.

Sugarscape can arguably embody the brutish world that Thomas Hobbes envisaged: it can approximate to his fearsome State of Nature. Axtell and Epstein simulate warfare by allowing one agent to take over the position occupied by another – basically to kill its opponent and reap the spoils. This might happen, for example, if the victorious agent is somehow 'bigger' or more powerful than the vanquished agent. Typically that superiority is manifested in Sugarscape by a greater store of sugar, which means that the stronger agent is better fed and provisioned. In Hobbes's world each man has to look out for himself, since each is pitched against all his neighbours. Sugarscape's creators have been more concerned to simulate wars between rival groups, in which entire armies of agents do battle. In such circumstances there is safety in numbers, so warfare is conducted through group co-operation. Depending on the rules of combat, the researchers find they can generate both rapid 'blitzkrieg' takeovers of territory and grinding 'trench wars' of attrition.

Axtell and Epstein don't claim to have produced a model of today's industrialized society; they suggest instead that this is a new way of doing social science. The behaviour that emerges from Sugarscape is sometimes extraordinarily complex, yet the generative rules guiding the behaviour of agents are rather simple. Might that, they ask, also be the case in the real world? Might the enormous diversity in behaviour that social science seeks to study turn out to be based on simple foundations? 'Imagine', they say,

that we had begun the entire discussion by simply running [the model] which shows a buzz of agents 'hiving' the sugar mountains, and that we had then bluntly asked, 'What's happening here?' Would you have guessed that the agents are all following [a simple rule]? We do not think we would have been able to divine it. But that really is all that is happening. Isn't it just possible that something comparably simple is 'all that is happening' in other complex systems, such as stock markets or political systems?[4]

In 1996 Axelrod and his colleague Michael Cohen at Michigan approached Axtell and Epstein to see whether they could 'align' their model of cultural spread with Sugarscape. They wondered whether, if the two quite different models were made as equivalent as possible, they would generate the same results. This entailed, for example, setting the number of cultural features and traits to be the same in the two models, and conducting them on grids of the same size and shape. This levels the playing field; it does not by any means make the models identical, since the interaction rules differ. Yet the result of the exercise was reassuring: Sugarscape showed the same number of stable regions as Axelrod's model, as the cultural diversity (the number of features and traits) and grid size was varied.

Ambitious things are planned for Sugarscape. In a collaboration called Project 2050 involving the Brookings Institute, the Santa Fe Institute and the World Resources Institute in Washington, the model will be used to develop policy recommendations on population growth, use of resources, migration, economic development and other social issues, in order to identify pathways towards sustainable world development. Such difficult matters call for a complex, multi-factored model. Once we reach this level of sophistication, any direct analogy with physical models of interacting particles becomes harder to sustain. Nonetheless one can expect – and Axelrod's alignment study supports the notion – that even in the most densely spun artificial society there will remain a core of what we might regard as inevitable physics: robust, collective modes of behaviour, such as phase transitions and power laws, which emerge over a wide range of plausible conditions. Simple models may see these things first and most clearly.

Many social scientists are now pinning their hopes on that being the case. Sugarscape is just one example of a broad range of computer-based model systems that aim to provide an understanding of how social structures, institutions, behaviours and traditions arise from the bottom up: from a consideration of how individuals interact locally with one another. This is called agent-based modelling, and in earlier chapters we have seen it applied to phenomena such as economic trading, the growth of firms and the movements of pedestrians. It has even been used to predict the progress of rafting trips on the Colorado River through the Grand Canyon National Park, so that more efficient and rational scheduling can be developed.

Thomas Schelling was one of the modern pioneers of this approach, and Herbert Simon believes it may hold the key to a more rigorous science of society. In principle, agent-based modelling should make some of the grandest social and political questions of our time accessible to rational experiment, such as whether the globalization of the economy is likely to lead to greater cultural harmony or to cultural conflict. But some social scientists remain uneasy, suspecting that any particular agent-based model of a social phenomenon risks coming to conclusions that depend on the underlying assumptions of the model. How do we know whether any one set of rules or assumptions will lead to truly representative behaviour, and not to an excessively crude caricature of the real situation? In short, such models can hardly be expected to provide a sound basis for policy until we can distinguish what is contingent from what is robust: what a particular model will produce as opposed to what all good models will produce. This is something physics knows about. It is the difference between a model of particles in motion that freeze at 0°C and a knowledge that there are liquid and solid states of matter separated by a phase transition. It is the distinction between a model of society and a physics of society.

15

Small Worlds

Networks that bring us together

I consider extremely fruitful this idea that social life should be explained, not by the notions of those who participate in it, but by more profound causes which are unperceived by consciousness, and I think also that these causes are to be sought mainly in the manner according to which the associated individuals are grouped.

Emile Durkheim (1879)

There are a number of morals to this story. Perhaps the most important is that your friends just aren't normal. No one's friends are.

Mark Newman (2001)

*

In 1941 the American movie actor Eddie Albert appeared in *The Wagons Roll at Night*, which starred Humphrey Bogart. Little known today outside the United States, Albert was never a major star, although he was nominated for an Oscar for his role in *Roman Holiday* (1953), alongside Audrey Hepburn and Gregory Peck. To American viewers he is perhaps best known for the sitcom *Green Acres*, an episode of which reputedly plays every day somewhere in the world. But Albert was still acting in Hollywood films fifty years after *The Wagons* rolled. In 1989 he appeared in *The Big Picture*, with Kevin Bacon.

Albert is thus the crucial link which gives several famous actors of yesteryear a low 'Bacon Number'. The Bacon Number is the smallest number of movies that link the actor in question with Kevin Bacon, where each movie brings together two of the actors along the path. Albert has a Bacon Number of 1, because he has actually appeared in a movie with Bacon. Bogart has a Bacon Number of 2: he appeared in a movie with Albert, who appeared in a movie with Bacon. James Dean and Ronald Reagan also have a Bacon Number of 2 thanks to Albert, who starred with them both in the TV drama *I'm a Fool* (1953). Errol Flynn is linked to Bacon via Albert, who shared the screen with Flynn in *The Sun Also Rises* (1957).

The Kevin Bacon game became a favourite among movie buffs and college students in the 1990s. The aim is to find the shortest route to Kevin Bacon – the lowest Bacon Number – for any movie star. But why Bacon? Like Eddie Albert, he became renowned for appearing in a large number of films in which he was *not* the star. In this way he links many bigger and lesser names in a network of movie relationships. Kevin Bacon, it was ironically suggested, was the real centre of the Hollywood industry.

The surprising aspect of this game is just how densely connected the network is. Something like 150,000 distributed films have been made since motion pictures began, together featuring about 300,000 different actors as named cast members. And yet nearly all of these can be assigned a Bacon Number (BN) of 3 or less. The last time I checked,[*] there were 1,686 actors with a BN of 1 (that is, those who have appeared in a film with Bacon), 133,856 with a BN of 2 (those who have appeared with someone who has appeared with Bacon), and a phenomenal

[*] The latest figures are all on The Oracle of Bacon web site at the University of Virginia: http://www.cs.virginia.edu/oracle/

364,066 with a BN of 3. As of October 2003, the average BN for all movie actors (at least, all those listed on the Internet Movie Database at www.us.imdb.com) is 2.946. These numbers certainly seem to support the contention that there is something special about Kevin Bacon – that he connects actors like no other. But is this really so?

The Kevin Bacon game is not a new invention. Mathematicians, not surprisingly, played their own rarefied version first, which features so-called Erdös Numbers. Paul Erdös (1913–96) was an extraordinary, influential and prolific Hungarian mathematician who wrote hundreds of papers. Because of this unusual productivity, he seems to be the centre of the mathematical universe.* Other mathematicians and scientists can be assigned an Erdös Number (EN) according to whether they wrote a paper with Erdös himself (EN = 1), with a co-author of Erdös (EN = 2), with a co-author of a co-author, and so on.

This construction links a large number not only of mathematicians but also of physical and social scientists into a vast network of connections. Albert Einstein has an EN of 2. Werner Heisenberg's EN is 4, and that of the German physicist Erwin Schrödinger is a surprisingly large 8. I have an Erdös Number of similar magnitude.† This does not, of course, mean that I am a scientist comparable in status to Schrödinger, any more than it makes Heisenberg twice as good as him or Einstein four times better. I have written very few scientific papers, none of them of significance, yet I can trace a link via

---

* Actually, productivity in itself does not guarantee this – the crucial point is that Erdös had many *coauthors*, and was thus well connected in the network. A lone mathematician could churn out a hundred papers a year and still be irrelevant to the network as a whole. The eighteenth-century Swiss mathematician Leonhard Euler was a prime example – he was the most prolific mathematician in history, but wrote only by himself.

† I'm grateful to Mark Newman for establishing that I have an Erdös Number of 8 at most – I might find a shorter link to the great mathematician if I were to search more exhaustively.

my co-authors to any of these great names. This is no particular source of pride, since the vast majority of other scientists could do the same. All it is saying is that the network formed by co-authorship in the scientific community is a highly connected one, so that for any scientist an eminent name is only a few steps away.

We all know the phenomenon behind these games. A colleague of mine finds himself talking at a party to a friend of the best man at his sister's wedding. The pianist of a singer I know is an old school-friend of my own old school-friend's wife. I discover in conversation that someone I've never met before knows my mother. These social networks of friends and acquaintances are continually reminding us that it is a small world.

The highly interconnected nature of social interactions has tantalized social scientists for decades. But they have, by and large, had little more to go on than anecdotes. In the past few years, however, some physicists have focused their attention on networks which display this 'small-world' behaviour, and have come to understand some of their defining features. These studies reach far beyond the sphere of social dynamics, revealing that the same features can be found in a wide range of networks in systems as diverse as neural wiring in the brain, interdependent metabolic biochemical reactions in the body, and power-line grids. There is, in other words, a kind of universality to certain types of network. By studying one of them, even on the basis of a simplistic model, we can gain insights into apparently unrelated systems and processes. We can even figure out whether there is really anything special about Kevin Bacon.

THE SIX DEGREES

The networks formed by friendships are complex and fascinating, but extremely hard to pin down in a way that makes them amenable to mathematical or physical analysis. A central difficulty becomes evident each Christmas. Do we know Amy well enough to send her a card? She sent us one last year, but she's not so prominent in our friendship circle as we seem to be in hers. And then there's Roger – we've known him for years, but we don't really keep in touch any more.

In the movie network, either you've been in a film with Kevin Bacon or you haven't (although one might argue about whether TV films such as *I'm a Fool* really count). Questions of friendship are more ambiguous. In particular, they may not be reciprocal in equal degree: I see Harry as a friend; he sees me as an acquaintance. And if one tries to establish the structure and extent of such a network, as social scientists have, one soon discovers how bad people are at describing them accurately. We forget to include people, and we are very bad at estimating the size of our friendship circle. And of course the network changes over time as we let old friendships slip away and form new ones.

In the 1970s, social scientist Mark Granovetter at Johns Hopkins University in Baltimore highlighted the importance of different 'strengths' of friendship ties. Strong ties are relatively easy to identify and record, but they may not be the most significant in the overall architecture of the network. Many of our close friends know one another, so their own 'strong-tie' network looks similar to ours. This property binds groups of individuals together in clusters. But Granovetter suggested that it is the weaker, 'acquaintance' ties that reach between these cliques and so bind the global network together (Figure 15.1). He called this characteristic 'the strength of weak ties'. But these more tenuous links are precisely the ones that are hard to identify reliably.

Sociologists began to take a close interest in social ties and networks even earlier than this. In the 1950s, political scientist Ithiel de Sola Pool and mathematician Manfred Kochen at the Massachusetts

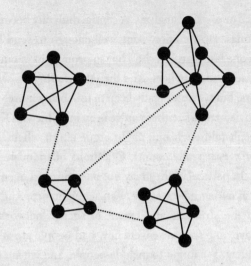

Figure 15.1 A friendship network. Clusters of close friends are typically linked by weaker acquaintances (dotted lines).

Institute of Technology wondered how political power gets mobilized in society: how do individuals acquire political power? Friends of the president may whisper in his ear and influence national policy. But do friends of friends also exert an influence? If so, how much? How close do you have to be to power in order to affect it?

Pool and Kochen made an attempt to formulate a theory of social networks, and they predicted that people in a population might be more closely connected to one another than they imagine. The two researchers wrote a paper but never got round to publishing it (it appeared in print retrospectively in 1978). The paper was circulated widely, however, and one of the people who saw it was psychologist Stanley Milgram, down the road from MIT at Harvard University. In 1967 Milgram, inspired by Pool and Kochen, conducted one of the most famous and most elegant studies into the structure of social networks. He sent packages to 196 people in Omaha, Nebraska, with a request that they forward them to the intended recipient, a stockbroker living in Sharon, near Boston.* But he provided no address. The stockbroker was identified only by his name and profession, along with the fact that he lived in the Boston area. The Nebraskans, chosen at random, were asked to send the packages to friends, acquaintances or colleagues who they felt might be better placed to channel the package to the right destination – along with a request that they do the same thing.

Someone receiving the package might, for example, send it to a stockbroker they knew, in the hope that he'd have a better idea of where to send it next. No one was to make any systematic attempt to track down the addressee; he or she was simply to pass the package along to someone else they knew, chosen for their possible greater proximity (in geographical, social or professional terms) to the addressee. At each stage, participants were asked to include in the package the same kind of details about themselves as had been provided about the addressee. In this way, if and when the packages arrived at the right destination their route could be reconstructed.

'When I asked an intelligent friend of mine how many steps he thought it would take,' Milgram recalled, 'he estimated that it would require 100 intermediate persons or more to move from Nebraska to Sharon.'[1] It didn't. Milgram found that on average just five intermediaries (that is, six separate journeys) were needed before the packages found their way to the right people. The implication was that any person in the United States could be linked to any other chosen at random via an average of five acquaintances. The exact number can be debated – for example, it is unlikely that each intermediary will make the ideal choice of who to pass the package on to, and the relevance of packages that never get delivered owing to

* Milgram apparently chose Omaha because that was the furthest he could imagine from East Coast 'civilization'.

apathy has to be factored in. But there is a consensus that between five and seven steps is enough to link everyone – a conclusion supported by Milgram's later experiments in 1970 with parcels sent from Los Angeles to New York.

These studies were restricted to the United States, but their message was expanded onto the global stage in John Guare's 1990 play *Six Degrees of Separation.** Guare's character Ouisa says:

Everybody on this planet is separated by only six other people. Six degrees of separation. Between us and everybody else on this planet. The president of the United States. A gondolier in Venice . . . A native in a rain forest. A Tierra del Fuegan. An eskimo . . . It's a profound thought . . . How every person is a new door, opening up into other worlds.[2]

But is it really a profound thought? The world is full of things which surprise us not so much because they are strange or perplexing but because our intuition is askew. It is natural to assume that people who are geographically very distant will also be socially distant. But to a corresponding degree? Is there any reason why I should be socially more remote from people in the Shetland Isles than from people in Bromley, a short bus ride from where I live? Will people in Nebraska be proportionately further removed from my social sphere?

These may seem like whimsical, even trivial questions.† But they have ramifications that go far deeper. The people with whom I come into close physical proximity are the pool from which I pick up many

---

* Milgram's 'six degrees' seem to have been predicted as early as 1929 by the Hungarian poet Frigyes Karinthy, who made the same claim – without any scientific justification for it.

† For a healthy dose of whimsical triviality, see the op-ed in the *New York Times* by D. Kirby and P. Sahry entitled 'Six degrees of Monica' (21 February 1998).

of the diseases I shall suffer during my life. More tangible still is the network of sexual and intimate contacts through which, for example, Aids spreads throughout a population. There is little fuzziness about this network – either you have a sexual encounter that puts you at risk of contracting Aids or you do not. (Infection by blood transfusion is harder to trace, but is ultimately an equally concrete link in the chain.) Yet when epidemiologists began to try to understand and forecast the expansion of the disease, they soon encountered the practical difficulties of reconstructing such a social network – even though it was in principle relatively clearly defined. In a few cases, crucial connections have nevertheless been identified. A single, promiscuous Norwegian sailor appears to have been instrumental in some of the earliest cases of Aids in Europe, having contracted the disease in West Africa in 1960 and then transmitted it to people in Cologne and Reims during his later life as a truck driver in the 1970s. A French-Canadian flight attendant named Gaetan Dugas, later dubbed 'Patient Zero', was linked to at least forty of the earliest known cases of Aids in California and New York.

Some movies, books, plays and music become huge hits not because of major advertising campaigns but by word of mouth. Gratifyingly, the same process can have the reverse effect – a much-hyped movie can flop when word spreads that it's a turkey. In this age of mass communication and global information systems, the process of dispersal of cultural products and ideas becomes very hard to identify precisely. But there is no doubt that social networks play their part, and only a few decades ago person-to-person connections were paramount. For anyone concerned with the dynamics of globalization, networks of encounters and information exchange between people are of central importance.

BETWEEN ORDER AND CHAOS

Paul Erdös is the ideal subject for the mathematicians' game of connections because of his productivity. But that is not the only thing that makes him appropriate. In the 1950s and 60s, Erdös pioneered the study of networks, and until recently his work provided the starting point for nearly all social-science studies in this area. With his colleague Alfred Rényi (who has an Erdös Number of 1, naturally) he clarified the properties of networks called *random graphs*.

A graph, we learn at school, has a vertical axis, a horizontal axis, and a scattering of data points with a line drawn through them. But Erdös and Rényi were not talking about this kind of graph. For them, a graph is simply a series of points connected by lines (Figure 15.2*a*). The points are called *vertices* and the lines are *edges*. This abstract entity can represent all manner of things. The vertices could be cities, and the edges the roads that join them: the graph then represents a road transportation network. Or the vertices could be movie actors, and the edges show links between pairs of actors who have appeared in a film together. An actor's Bacon Number is the number of separate edges one has to traverse to get to the relevant actor from the vertex representing Kevin Bacon (Figure 15.2*b*). A graph like this displays the relationships between the entities represented by its vertices.

When the vertices are cities and the edges are roads it is easy to comprehend the rules of the graph: it is like a map. The distances and compass directions between the vertices representing cities generally reflect those in geographical reality. But for the movie-actor graph, the rules are not so plain. How close to Kevin Bacon do we place Eddie Albert, and in which direction? Does Jack Nicholson (BN = 1: *A Few Good Men*, 1992) get placed at the same distance? We might, say, decide to make all the edges the same length, and not to attribute any

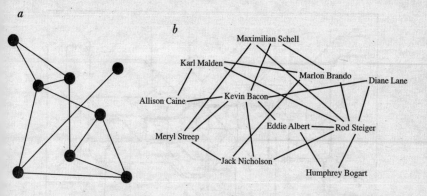

Figure 15.2 (*a*) A typical network graph – a series of points (vertices) connected by lines (edges). (*b*) A small part of the graph that connects movie actors, showing some of the connections in the vicinity of Kevin Bacon. The respective movies are listed in the endnotes.

significance to the directions. But we quickly find that this won't work: soon, we are unable to place two specific co-starring actors close enough to link them with a single edge of fixed length, because their links to other parts of the graph keep them apart. In fact, why should we draw the graph in two dimensions on the flat page? Why not a three-dimensional web, like the timber frame of a building? Or, now that mathematicians are comfortable negotiating their way around spaces with more than three dimensions, why not a ten-dimensional web?

The answer to these dilemmas is that it really doesn't matter how we choose to draw the movie-actor graph, provided the connections between vertices are correct, linking only those actors who have appeared in movies together. To understand the important features of the network, all that matters is this system of connections, which mathematicians call the *topology*. Graphs that look different when drawn on paper can nevertheless be topologically identical. The distances and directions don't mean anything.

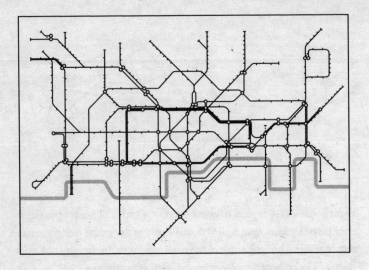

Figure 15.3 The London Underground is a relational graph, showing how vertices (stations) are connected but giving only approximate indications of their relative geographical positions and the distances between them. The thick grey line is the River Thames.

We can call this kind of graph a *relational* graph, which concerns itself only with showing the relations between its vertices. In contrast, the city network is a *spatial* graph, where distances and positions correlate with something real. Of course, we don't have to make the city graph a spatial one. The iconic map of London's Underground railway lines (Figure 15.3), devised by Harry Beck in 1931, is a wonderful example of a relational graph; but it retains some element of a spatial graph because that makes it easier to draw the links and it helps people to find their bearings. The stations appear roughly in their geographical arrangement, but the distances aren't to scale and the compass directions aren't precise. It can be quicker to walk between Covent Garden and Holborn stations, one edge apart, than to take the tube; but you wouldn't want to walk from Hatton Cross to

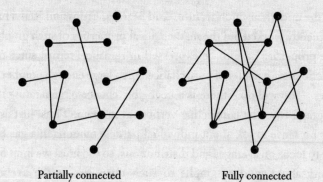

Partially connected   Fully connected

Figure 15.4 A random graph becomes fully connected when all the vertices are linked into the network – when there are no isolated vertices or clusters.

Boston Manor, even though they are the same distance apart on Beck's graph-like map.

In the random graphs studied by Erdös and Rényi, a scattering of vertices is interconnected at random. You could make a six-vertex random graph by numbering the vertices and then throwing two dice. An edge is added between the two vertices whose numbers come up. When each of the vertices is linked to at least one other, the graph is said to be fully connected (Figure 15.4). On a fully connected graph it is always possible to travel by some route from any one vertex to any other. The London Underground is fully connected in this sense. In general there will be several possible routes between destinations, and we'll often be interested in finding the shortest.

Several investigators of social networks, beginning with Anatol Rapoport and colleagues at the University of Chicago in the 1950s, assumed that they have the topology of random graphs. This is almost certainly not the case, but random graphs seemed like a good starting point for such studies because they make only a neutral assumption about the underlying architecture. And besides, Erdös and Rényi had conveniently worked out the mathematical properties of such graphs. These properties have to be expressed in statistical terms, since the connections are a matter of pure chance. For a large enough number of vertices – 100 will do – there is a negligible chance of generating the same graph twice by connecting vertices at random. Thus, just as it makes no sense to ask about individual particle motions in a gas but rather to focus on averages and distributions, so too must we limit our curiosity about random graphs to such quantities as the average number of connections per vertex, and the probability distribution of this number. Erdös and Rényi showed that the probability distribution is the familiar bell-shaped gaussian curve. The average number of connections corresponds to the peak of the curve. This average obviously depends on how long we go on adding edges to the graph, but for any specific random graph it has a well-defined value.

Another kind of graph that has been used to study networks lies at the opposite extreme to the random graph. It is the regular grid: an array of identical vertices connected by identical edges (Figure 15.5). There is no need to talk of averages here – every vertex (except those at the edges and corners) has exactly the same number of connections (four in the example shown). These 'ordered' graphs are easy to describe mathematically. Random and ordered graphs have very different properties. To get from one vertex to a distant one on an ordered grid, you have no choice but to proceed in a series of little hops between neighbours. The path length, measured as the number of edges traversed, is long. For a random graph, on the other hand, there is a good chance that a vertex near your starting point will be connected 'long-distance' to one close to the target vertex. There are lots of short cuts, in other words. This means that the path length between apparently widely separated vertices may be small.

One way of making this difference precise is to ask about the average

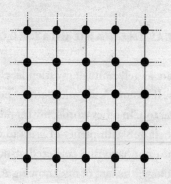

Figure 15.5 On a regular grid, all vertices (excluding edges) have the same number of connections.

path length between any two vertices chosen at random. This is called the *characteristic path length*, and it is a statistical property exactly analogous to the average number of journeys Stanley Milgram's packages had to make to reach their destination. For an ordered graph the characteristic path length is long, and gets proportionately longer as the number of vertices increases. For a random graph the characteristic path length is small. Moreover, it doesn't increase much as more vertices are added, because there is a good chance that a random connection will be made between a new, outer vertex and one deep in the middle of the graph.

Judging from this, you might expect that social networks like the movie-actor graph will be like random graphs. Isn't that the point of the small-world concept – that you're always socially closer to another random individual than you think? But in 1998 two scientists at Cornell University showed that social networks aren't like random graphs. Instead they fall into a class of their own, lying somewhere between the perfect disorder of random graphs and the perfect order of regular grids. They are called, suitably enough, small-world networks.

OF CAVEMEN AND CHAT ROOMS

Friendship circles are largely mutual: my friends' close friends tend to be my close friends. If I know Andy and Betty, Andy and Betty probably know each other too. In other words, my friends and I form a kind of social cluster, albeit one with plenty of links to other clusters. But random networks do not have this clustered structure. In networks with a random topology, the fact of my knowing Andy and Betty has no bearing at all on the probability of them knowing each other. Ordered grids do have an element of clustering, because each vertex is connected only to its immediate neighbours – there are no big leaps.*
Many of the neighbours of one vertex are also neighbours of a neighbouring vertex. So there is a good chance that close friends connected by an ordered grid share a mutual close friend.

When Steven Strogatz at Cornell and his graduate student Duncan Watts began to investigate the problem of social networks in the late 1990s, they found a way to represent this clustering phenomenon. They weren't initially interested in human social structures; they were thinking instead about how some animals synchronize their behaviour, such as the way a field of crickets will synchronize their chirping. But the question soon led them into the tangled web that social scientists had begun to weave.

Strogatz and Watts imagined two possible extreme societies. The first is fragmented into clusters of people who socialize with one another but with next to no one outside the group. We can think of examples that approximate this situation – for example businesses which dominate their employees' lives, providing them with almost their sole source of social contacts, as is the case in some Japanese

* One can create ordered grids with links that reach beyond near neighbours, but that doesn't alter the tendency towards clustering.

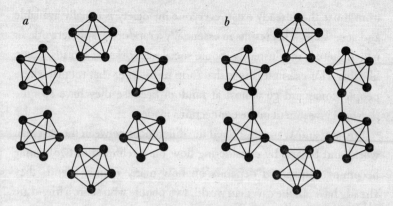

Figure 15.6 In 'caveman world' (*a*) people form social clusters that are closely linked internally but are isolated from one another. Even if this graph is made fully connected by linking the 'caves' (*b*), the characteristic path length is long.

firms. But Strogatz and Watts found a more basic analogy: the world of the caveman, consisting of sparsely scattered groups, each of which share a cave, do everything together and shun outsiders. In Thomas Hobbes's time many peasants in villages were still living in this sort of world. It can be represented as a series of small graphs which are highly connected internally but disconnected from, or only sparsely connected to, one another (Figure 15.6). This is not an ordered graph, but it shares with such graphs the properties of high clustering and long characteristic path length.

The two researchers drew the other extreme not from the past but from the future. Isaac Asimov's 1957 novel *The Naked Sun* portrays a future society in which people interact almost entirely through robots and computers. In Solaria it is as easy – and as likely – for you to forge a relationship with someone on the other side of the world as with your neighbour. These virtual friendships have become so weak and superficial that new ones are established completely independently from those that already exist: everyone in society is equally available and accessible. This results in essentially a random-graph network, in which there is no clustering. Some social networks are already a little like this – for example, those that form in Internet chat rooms, where people come and go almost at random because they have so little personal investment in the connections made.

Strogatz and Watts described the differences between the caveman world and Solaria by considering how the likelihood of two people becoming acquainted depends on how many mutual friends they already have. In the caveman world, two people who share a friend are almost certain to become friends themselves, because it means they must be in the same cluster. In Solaria, two people who have even a large number of mutual friends have no greater chance of becoming friends with each other than either of them has with any other random person.

Our real world presumably lies somewhere between these two extremes. But where? The mathematics demands that any graphs one uses to mimic social networks must be fully connected. If they are not, then some vertices remain completely inaccessible to other vertices, and this immediately pushes up the characteristic path length to infinity – which isn't realistic at all. So Strogatz and Watts devised a method for converting a fully connected ordered graph (comparable to a fully connected caveman world) into a fully connected random graph (like Solaria) in steps that always left the intermediate graphs fully connected too. This is called random rewiring. You start with an ordered grid and select a vertex at random. Then you select one of its connecting edges at random, disconnect it from its destination, and reconnect (rewire) it to another vertex in the graph, also selected at random (Figure 15.7). As the rewiring progresses, more and more short cuts are created that link distant parts of the graph in a single jump, and the graph becomes increasingly random.

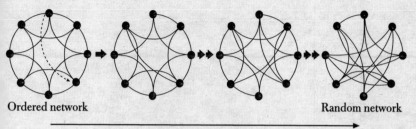

Ordered network

Random network

Increasing randomness

Figure 15.7 Random rewiring on a ring of vertices converts an ordered graph gradually into a random one. Somewhere between these two extremes lies the small world.

The researchers chose to begin with the simplest ordered graph they could think of: a ring of vertices. (A ring is better than a row of vertices, because it eliminates anomalies at the ends.) They began rewiring these graphs and watched what happened to the topologically determined features: the characteristic path length L and the amount of clustering, which can be measured as a numerical quantity denoted by C.* The amount of rewiring can be quantified by the probability that a vertex chosen at random will have been rewired from its initial configuration. When this probability is zero, the graph is a perfectly ordered (ring) network. When the probability is one, the graph is fully random.

As would be expected, both L and C decrease as the amount of rewiring increases: the ordered grid loses its clustering but acquires

* C is defined as the average, over all vertices, of the number of edges connecting a vertex v divided by the total number of possible edges in the neighbourhood of v. It is a quantitative answer to the question: of all the possible ways of connecting vertices in the neighbourhood of v, how many are actually realized? If the clustering coefficient C is large, this means that most potential connections within the neighbourhood of any vertex are indeed made: there is a lot of clustering. In a friendship network with large C, one's friends are likely to know one another.

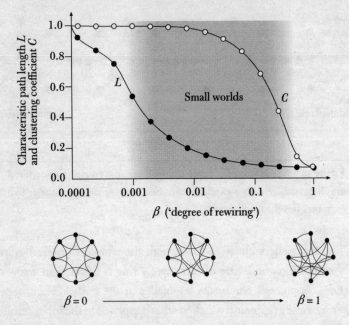

Figure 15.8 The characteristic path length L and the clustering coefficient C for a rewired ring graph change rather abruptly as rewiring proceeds. The quantity β here is essentially a measure of the extent of rewiring. Notice that the horizontal axis has a logarithmic scale, which means that small values of β are 'stretched out'. This is so that it is easier to follow the changes in L, which happen mostly in the first few rewiring steps. L and C are both large in the ordered ring graph and small in the random graph. But they do not both change at the same stage. Graphs with large C but small L are small-world graphs.

more short cuts. But there are three surprises. First, nearly all the action happens in the first few rewirings. By the time one in ten vertices are rewired, the graph typically has properties indistinguishable from those of a random graph. Second, the changeover from ordered-like to random-like is quite abrupt – Watts compares it to a phase transition in statistical physics. It is as if the ordered, 'solid-like' graph melts into the disordered 'liquid-like' graph. And finally, L and C do not change together but at different stages of the rewiring process (Figure 15.8).

The last observation seems truly weird. The characteristic path length L plummets to a value typical of a random graph while the graph still retains a high degree of clustering (large C). Then, after a little more rewiring, C plummets too. So for a small range of rewiring extent, the graph has small L but large C. The coexistence of these two characteristics is what makes a network a small world. The small world of friendship circles, for example, incorporates a high degree of clustering but possesses numerous short cuts between clusters, creating the short average path lengths responsible for our 'six degrees of separation'. So Strogatz and Watts dubbed these transitional graphs *small-world graphs*.

DO I KNOW YOU?

Do social networks in the real world actually look like the small-world graphs created by random rewiring? In other words, could Strogatz and Watts *prove* that indeed 'it's a small world', and thereby perhaps deter us from trotting out this truism every time we discover previously unsuspected connections at drinks parties?

For all its apparent frivolity, the movie-actor network remains one of the best test cases, since the data are unambiguous, abundant and readily available. It is clear enough from the Kevin Bacon game that there is a short characteristic path length L in this network (the average Bacon Number is low), and it doesn't take much inspection to see that there is a high degree of clustering. Actors of the same nationality tend to form clusters, for instance, which are interlinked by a few cross-cultural linchpins such as Bruce Lee (Hong Kong), Gérard Depardieu (France) and Gong Li (China).

To this extent the movie-actor network has properties compatible with a small world. But what about the network topology? Some kind of comparison can be made by defining a quantity called the vertex contraction parameter, which is a measure of how many short cuts exist in the network between widely separated vertices. This number can be calculated for any network, and it increases for the ring graphs as they are progressively rewired. Strogatz and Watts constructed a small-world rewiring network with the same value of the vertex contraction parameter as the movie-actor network, thereby ensuring that it was somewhat comparable topologically. The question then is how the two networks compare in terms of their characteristic path length L and clustering parameter C.

This comparison is quite good – certainly, the rewiring model can mimic the parameters of the movie-actor network more closely than can either a caveman network or a Solaria network. But we must nevertheless ask whether these are the only alternatives. That is, is the rewiring scheme the only way to make a network with small-world properties (low L, high C)? We shall explore this question in the next chapter.

And what of the most burning issue – is Kevin Bacon really the centre of the movie universe? To answer this, one must calculate the average Bacon Number for the entire network and see how it compares with the equivalent measures for other actors: the Elvis Number, the Bogart Number, the Brando Number, and so on. If Kevin Bacon is really the most important linchpin in the network, all other actors will, on average, be closer to him than to anyone else.

It turns out that not only is Kevin Bacon not the most important hub of the network, but he is not even in the top one thousand (the list of course changes daily as new films are made). Currently up at the top is Rod Steiger (the average Steiger Number is 2.652), followed by Christopher Lee, Dennis Hopper, Donald Pleasence and Donald

Sutherland (who appears in the film version of *Six Degrees of Separation*). Marlon Brando is number 202, Frank Sinatra number 443. By the time we get to Kevin Bacon's level, the differences in the average Actor Number that separate successive actors in the list are tiny, about 0.0001.

So why did Kevin Bacon get picked out for this game? The answer contains the entire essence of a small world: for in such a network, *everyone* appears to be at the centre. Some are more 'central' than others – but not by very much. Even relatively minor actors like Eddie Albert have a comparable network status to major stars. (Donald Pleasence was a capable actor, but hardly a superstar.) There is something refreshingly egalitarian about this message. If social networks bear any resemblance to the movie-actor graph – and this is a reasonable guess, as we shall see – we need no longer gaze with envy upon those few who seem to represent the hub of our social circles, nor lament our paucity of friends. By and large, it is simply a matter of getting the right perspective on the web that links us all together.

There is a postscript to this story. In 2003 Duncan Watts (now at Columbia University) and colleagues staged a rerun of Milgram's classic experiment, but using email instead of the postal service. They asked for volunteers to take part in this electronic project, and were inundated with offers, finally registering 61,168 individuals in 166 countries. To each of these participants they assigned a target person, to whom an email message was to be relayed. There were 18 targets in 13 different countries, ranging from an Ivy League US professor to an inspector of archives in Estonia and a Norwegian military veterinary surgeon. The rules were much the same as Milgram's: participants were to pass on the message to a friend, colleague or associate who they considered likely to be 'closer' to the designated target.

Well, it was not a good way to deliver the mail. There was an initial drop-out rate of about 63 per cent, but that wasn't bad for an email survey, and it meant that 24,163 chains got started. Yet only 384 reached their target. These completed chains contained an average of just four steps – but that number was biased by the fact that the shorter chains were more likely to run to completion. By allowing for the attrition rate, Watts and colleagues estimated that a typical chain length between any of the participants and their target was five to seven steps – just the same as Milgram had found, although his chains were confined only to the USA. That bolsters the notion of six degrees of separation, but it does not exactly imply that our small world makes for efficient person-to-person contact. The researchers established that few of the people who dropped out of a chain did so because they couldn't think of anyone appropriate to pass the message on to. They simply lost interest. What was in it for them?

Yet for one target the attrition rate was significantly lower than for the others: the American professor. Did that mean he was better connected and so more easily found? Probably not, say Watts and colleagues – they estimated that his average degree of separation was not so different from that of the other targets. But over half of the project's participants were middle-class, professional, college-educated Americans. Those in the chains directed at the professor probably believed that it would be easier to get the message to him than did participants faced with relaying a message to, say, a technology consultant in India. This belief gave them that little bit of extra motivation to keep the chain alive.

In other words, say the researchers, the structure of the social network is not the whole story. Whether or not its small-world character will indeed be exploited to make connections depends on the actions and even on the perceptions of the people in it: a timely reminder that social physics cannot afford to do away entirely with individual psychology.

16

Weaving the Web

The shape of cyberspace

The Internet Age has been hailed as the end of geography. In fact, the Internet has a geography of its own, a geography made of networks and nodes that process information flows generated and managed from places. The unit is the network, so the architecture and dynamics of multiple networks are the sources of meaning and function for each place.

Manuel Castells (2001)

We are in the presence of a new notion of space, where physical and virtual influence each other, laying the ground for the emergence of new forms of socialization, new life styles, and new forms of social organization.

Gustavo Cardoso (1998)

'Tis true; there's magic in the web of it.
William Shakespeare (1602–4)

*

In May 2000 I received an email announcing its subject with the enticing words I LOVE YOU. Fortunately, by the time I saw it I was aware that a large proportion of the online community had found the same message in their inbox. What lay in store, if I chose to open the message, was not an amorous declaration but an insidious computer virus that would lay waste to my files before posting itself elsewhere. Computer viruses are not (yet) fatal, but they are a costly bore. They have wreaked havoc in commercial computer networks, and have caused individual users untold anguish. The I LOVE YOU virus immobilized the internal email systems of the US Senate and the British House of Commons, and afflicted several other governments and financial institutions. Another virus dubbed the Love Bug caused $6.7 billion worth of damage in 2000.

The Slammer computer worm more or less disabled all of South Korea's Internet network in early 2003, and the increasing ingenuity and elusiveness of these virtual parasites must make us fear that one day such an epidemic will assume global proportions, like a modern-day Black Death. The threat highlights the role of the Internet and its companion electronic network the World Wide Web (WWW) in maintaining today's social order. Increasingly the economy is a wired system, depending on computer-based communications for even the least of its machinations. Many governments share this dependency, as do facets of our social framework ranging from goods distribution systems to schools to policing. 'There's a new infrastructure out there,' says Larry Irving, head of the US Commerce Department's National Telecommunications and Information Administration, 'and we're much more dependent on this infrastructure than we've ever been or ever expected to be.'[1]

It is not fanciful to see in the intertwined branches of the Internet and the WWW a manifestation of Hobbes's Leviathan – a ruler whose body is composed not of the multitudes of its subjects but of silicon chips, glowing terminal screens and mile upon mile of copper wire and optical fibre. In his short story 'For a Good Purpose', Italian writer Primo Levi envisaged a similar beast – the automated telephone routing system – acquiring consciousness from the complexity of its connections, and engaging in a bout of seemingly random rewiring that, regardless of whether it made for a smaller world, caused chaos among the human population.

How robust is this electronic Leviathan, this Hobbesian Cyber-Commonwealth? 'Though nothing can be immortall, which mortals make', conceded Hobbes,

yet, if men had the use of reason they pretend to, their Common-wealths might be secured, at least, from perishing by internall diseases . . . Amongst the *Infirmities* therefore of a Common-wealth, I will reckon in the first place, those that arise from an Imperfect Institution, and resemble the diseases of a naturall body.[2]

Yes indeed: electronic viruses are just as capable of striking down the Internet as their biological namesakes are of leaving us bed-ridden. So too is the Internet potential prey to the many other afflictions that flesh is heir to: overload, transmission errors, local breakdowns causing bottlenecks, and attack from outsiders with malicious intent. Hobbes's message for us is that we would do well to build our Leviathans along the lines of a healthy 'naturall body'. In this chapter we shall discover something quite astonishing about the Internet: it seems to have this character already, although no one planned it that way. And while this does indeed lend the network a certain robustness, it also exposes an Achilles' heel, a weakness to a particular kind of attack. Identifying such weaknesses may be crucial to making the Internet safe from cyber-terrorism, a danger that grows daily as our reliance on computerized systems increases.

The Internet is just one example of how our social structures and institutions are laced with the kind of networks we encountered in the previous chapter, in which countless individual units or agents are linked by interactions into a surprisingly small world. The big picture is just beginning to emerge of what these networks really look like and how they arise unbidden and unplanned among multitudes of agents going about their local business without a thought for the global view.

There is an urgent need to understand this 'network society'. The global communication and information networks of the Internet and the WWW are shaping novel cultural and institutional structures. Sociologist Manuel Castells believes that the electronic web now being spun over the world – what he calls the Internet Galaxy – provides an entirely new way of conducting business and indulging pleasures. He claims that the decentralized, non-hierarchical 'geography' of this network will change the very nature of these activities:

As new technologies of energy generation and distribution made possible the factory and the large corporation as the organizational foundations of industrial society, the Internet is the technological basis for the organizational form of the Information Age: the network.[3]

The growth of this net has been more rapid than anyone could have anticipated, and certainly too fast to allow considered prognostications about its effects or future potential. Widespread use of the WWW did not begin until 1995; by the end of that year there were about 16 million users. By early 2001 the figure had grown to over 400 million; some estimates predict that it will top a billion by 2005, and 2 billion by 2010.

Thanks to the new physics of networks, we can now see that the structure of the Internet reflects the ethos that produced it – as Castells says, 'The culture of the producers shaped the medium.'[4] That is to say, only a network that grows 'organically' according to no master plan, observing principles of free access and meritocratic choice (the best or most useful WWW sites get the most connections), will develop the kind of architecture that the Internet and the WWW display – with its strengths and its pitfalls.

But this culture of 'freedom of information' does not guarantee that the network society will be utopian. The Internet is, you might say, free to those who can afford it, well over half of whom live in North America and Europe (and most of the rest in the Asia Pacific Rim). As the power of the Internet grows, those who lack the technological infrastructure, money or education to access it are increasingly dis-enfranchised, creating a 'digital divide' not just between developed and developing countries but between different socio-economic sectors of a single culture. Meanwhile, the far-reaching potential of the Internet is being exploited as avidly by criminals as by business entrepreneurs. Pranksters and malcontents have new opportunities for sabotage; advertisers have a new way to send you their junk.

These considerations make it essential that we understand the virtual anatomy of this information machine. For it and its influence will soon touch us all, in one way or another, as Castells foretells:

I imagine one could say: 'Why don't you leave me alone?! I want no part of your Internet, of your technological civilization. Of your network society! I just want to live my life!' Well, if this is your position, I have bad news for you. If you do not care about the networks, the networks will care about you, anyway. For as long as you want to live in society, at this time and in this place, you will have to deal with the network society.[5]

THE NET IS CAST

'It is no exaggeration', said US Judge Stewart Dalzell in 1996,

> to conclude that the Internet has achieved, and continues to achieve, the most participatory marketplace of mass speech that this country – and indeed the world – has yet seen . . . Federalists and Anti-Federalists may debate the structure of their government nightly, but these debates occur in newsgroups or chat rooms rather than in pamphlets. Modern-day Luthers still post their theses, but to electronic bulletin boards rather than the door of the Wittenberg Schlosskirche. More mundane . . . dialogue occurs between aspiring artists, or French cooks, or dog lovers, or fly fishermen.[6]

'Participatory' is the apt adjective. Although the rhetoric of global inclusiveness often overlooks the limited resources of developing countries, there is otherwise nothing elitist about the Internet. Anyone with access to a computer terminal can get an email address, free of charge, and send messages to others on the far side of the world. Judge Dalzell's comments came in a summing-up speech in which he ruled attempts to circumscribe the boundaries of messages transmitted over the Internet as unconstitutional. 'The Internet may fairly be regarded as a never-ending worldwide conversation', he said. 'The Government may not . . . interrupt that conversation.'[7]

It is a well-known irony that this open conversation grew out of the network devised in secrecy by the US military in the 1960s, called ARPANET. (ARPA, the Advanced Research Projects Agency, was the euphemism for scientific research funded by the US military.) The Net was originally intended for transfer of documents; its architects were taken by surprise when its facility for sending brief messages, initially instigated for their own convenience during the project's development, became its most attractive feature. Soon enough the non-military academics and industrial scientists who'd watched ARPANET grow wanted their own version, and Usenet was born, sending its packages of coded data down national and international telephone lines. On Usenet there emerged the anarchic, anything-goes philosophy that now characterizes the Internet, the mega-network that grew from the proliferation of smaller networks with their dedicated servers.

The Internet and the World Wide Web – the Net and the Web – are sometimes perceived as synonymous, but they are not. The Net is a communications network; the Web is an information bank. The man usually credited with launching the WWW is Tim Berners-Lee, an English computer scientist who was employed by the European particle-physics laboratory CERN in Geneva to help its physicists store and retrieve information. Berners-Lee's concept is summed up in his description of a program he devised in 1980 as a kind of personal 'memory substitute': 'It allowed one to store snippets of information, and to link related pieces together in any way. To find information, one progressed via the links from one sheet to another.'[8]

In 1989 Berners-Lee devised a means of implementing this scheme in an electronic information system of any size: hypertext. In a hypertext document, the words that appear on screen are interspersed with markers – push-buttons or text that is underlined or differently coloured – which allow the user to leap directly to a different document containing further relevant information. These *hyperlinks* are inserted by whoever creates the document, but they can link to a page posted by someone else. If the content of the hyperlinked page is later changed or updated, the hyperlink takes the user to the new version.

Berners-Lee's proposal for such a system at CERN was soon seized upon by others at the laboratory as the inspiration for a grander vision.

They had no wish to limit themselves to a mere institutional web; by 1990 there was talk of a World Wide Web, which would use the global infrastructure established by the Internet to access electronic documents stored on computer systems the world over, as well as articles generated within Internet news groups. To navigate around such a complex network, one needed a guide, which came in the form of a browser, a program that provided a user-friendly gateway to the universe of resources on the Web. Netscape, now owned by AOL (America Online), and Microsoft's Internet Explorer are the two most commonly used browsers.

It was estimated at the end of 2002 that there were around 3 billion documents then available on the WWW. Several million more web pages are being added every day. With a library that size, one needs an efficient means of getting quickly to the information one is seeking. This is the role of search engines – programs which can be regarded as robots trawling the Web's archives for documents containing key-words or text that match words fed to them as search criteria. Such is the scale and complexity of the Web that today's search engines are able to survey only a small part of it.

The issue of network topology was one with which the Internet's inventors had to grapple from the outset. In large measure this was not so much a mathematical as a technological question. Older readers will remember a time when the orthodox image of a computer system was a central mainframe linked to a series of terminals; the terminals had no autonomy but were mere portals to the central oracle. The concept of a free-standing personal microcomputer did not exist when the Internet's ancestor ARPANET was being planned. Thus the 'obvious' topology for a computer network in the 1960s was a highly centralized one in which all messages were sent to a central distributing mainframe which passed each of them to the appropriate node (Figure 16.1a).

But Paul Baran, the communications engineer who was set the task

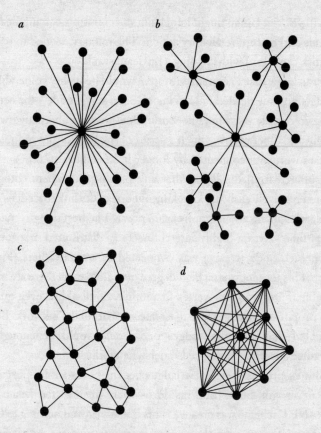

Figure 16.1 Possible network topologies for the Internet: (*a*) central-ized, (*b*) decentralized with local servers, (*c*) distributed and (*d*) max-imally connected.

of developing a communications network for the US military, realized that this would be a crazy way to build such a system. Baran was employed by the RAND Corporation, a US think-tank created in 1946 to supply strategic analysis and recommendations to the US Government for its nuclear defence policy. At the height of the Cold War, many of the US military's top brass feared that a nuclear strike from the Soviet Union might knock out the country's communication systems and disrupt retaliatory action. The military needed to have a network capable of withstanding a nuclear assault.

Baran knew that a centralized system would be highly vulnerable to attack, since one needed only to knock out a single node – the central hub – to disable the whole network. One possible alternative was a decentralized network in which a series of smaller, locally centralized clusters were interconnected by longer links (Figure 16.1b). Yet this still suffered from the defect that any part of the system could be isolated relatively easily by knocking out one of the links or 'edges' (we are clearly talking about graphs again now). The best design – radical for its time – was a highly interconnected, 'distributed' network in which each node (vertex) was connected to several others (Figure 16.1c). This topology has a high degree of redundancy: there are many possible routes from one vertex to another. So disabling even quite a high proportion of edges fails to isolate any part of the network. Baran estimated that just three edges per vertex would be enough to guarantee adequate resilience in such a network.

Baran's proposed network languished in bureaucratic purgatory and never got built. But his ideas were resurrected when the ARPANET scheme was created. This initiative owed nothing to Cold War fears of nuclear attack; it aimed merely to increase the efficiency of the computer resources used at ARPA-funded university sites by interlinking them. The initial plan was to connect the mainframe computers at each site directly to all the others. But in 1967 the computer scientist Wesley Clark realized that this 'maximally connected' network would rapidly get unworkably complex. With just 10 vertices there are already 45 edges (Figure 16.1d), and the number of edges increases faster than the number of vertices. This places huge demands on the mainframe computers at each vertex, and would be likely to cause 'freezing' of the information flow. Clark proposed that instead the mainframes should be linked to a sub-network of smaller, interconnected computers, all speaking the same language, that were dedicated simply to routing the information. Baran's work showed Clark and others how such a network could be wired up to function efficiently and robustly.

That was about as much topological planning as the Internet ever got. It has expanded far beyond anything these early pioneers could have envisaged, and has been subject to no regulatory authority or architectural design. The growth, like that of cities (Chapter 6), has been organic and uncoordinated. Indeed, some suggest that the Internet is now best regarded as an ecosystem every bit as complex as those in the natural world. It is no longer possible to draw an exhaustive map of its connectivity pattern – the structure is too huge and deeply embedded for it to be traced.

Nevertheless it seems certain that the Internet does have the highly redundant, distributed topology proposed by Baran. The packets of data into which messages and web pages are chopped are sent along many different paths between the source and the destination, perhaps even traversing the globe in opposite directions before being pasted together at their target point. This is why the Internet is able to cope with so much traffic at once: if one pathway is congested or blocked, another usually presents itself. On the other hand, we know that a map of the Internet would not look like Figure 16.1c. That is more like a regular grid (page 457), albeit one in which the number of links per vertex is not uniform but varies between about four and six. As we saw in the previous chapter, there are no short cuts in such a network: you get from A to B via a series of many short jumps.

So what, exactly, does the Internet look like? This is a critical question, relating to problems such as how to route mail most efficiently, how robust the system is in the face of local breakdowns, how easily users in one part of the world can reach those elsewhere, and whether the Internet can grow without limit or whether it will reach some saturation point beyond which the network cannot cope. As the economic impact of the network increases – already e-commerce in the United States turns over $100–200 billion a year, and similar figures are quoted for the economy-wide savings potentially available from electronic communications – these questions are becoming pressing ones.

The Internet network is a physical entity, like the London Underground. The nodes or vertices are computers, and the links or edges are the transmission lines or satellite channels that run between them. The World Wide Web is a more nebulous realm. Its vertices consist of document pages stored electronically in machines throughout the world, and its edges are the hyperlinks or Uniform Resource Locator (URL) addresses that connect a page directly to another in some other web site on some other machine. When we follow one of these links, the communication is conducted along the same physical transmission lines that the Internet uses. But the Web itself is not delineated by these lines. One transmission line can in principle connect a million web pages to a million others in a maze of URL links.

It is not obvious how the Internet or the WWW are structured. Could they be random graphs? Are they small-world networks like those we encountered in the previous chapter? Or maybe they are branching hierarchical trees, like river networks or the passageways of our lungs? The answer is that they appear to be something else entirely.

NO SENSE OF SCALE

Mapping the WWW is like mapping a maze. If you could rise above the maze in a balloon, the task would be easy: you would simply draw what you saw laid out below you. But there is no balloon that will give us a bird's-eye view of the WWW, because it exists only in cyber-space – it corresponds to no physical structure we can look at. So we have to do just what we would to map out the maze from the ground: we must enter it and keep track of where it takes us. What a strange contrivance it is that we have made for ourselves: we have built it, but we cannot easily tell exactly what it is we have built. Rather, we must investigate its shape as though we are blind, feeling the contours little by little. In 1999 Réka Albert, Hawoong Jeong and Albert-László Barabási of the University of Notre Dame in Indiana confronted this cartographic challenge by sending a robot into the maze of the WWW, charged with the task of mapping out its convoluted pathways.

The robot was a computer program which was instructed to enter a web site and follow all the hyperlinks. These took the robot to an assortment of other web sites, at each of which it would repeat the same process. On each foray the robot kept a record of the number of outgoing hyperlinks from each page it encountered.* To conduct this search for all 1 billion or so of the documents then on the WWW would have been far too much for the robot to accomplish. Instead, the researchers told their pet program simply to stay within the boundaries of the domain specific to the University of Notre Dame (www.nd.edu). This alone encompassed 325,729 HTML documents (HTML, or Hypertext Mark-up Language, is the standard language for writing web pages, devised by Tim Berners-Lee), interconnected by almost 1.5 million links. This, the researchers hoped, was a data set sufficiently large and representative to stand proxy for the Web as a whole.

Albert and her colleagues found that the probability distributions of both incoming and outgoing links in this graph were power laws (Figure 16.2). That is to say, the probability of a random page on the

---

* To construct the network, one also needs to know about the incoming links to each site. The robot could not identify these directly, but they are implicit in where each outgoing link *goes*.

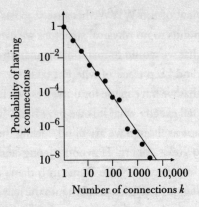

Figure 16.2 The probability distribution of incoming hyperlinks to HTML documents on the World Wide Web follows a power law, generating a straight line on this logarithmic plot. The outgoing links have a similar distribution. This implies that the WWW is a scale-free network.

university web site having a certain number of links to or from it depended on the number of links according to a power-law relationship. Most pages had few links; a few had many; and each time the number of links was doubled, the number of pages with that many links decreased by a constant factor. Although we'd intuitively expect fewer pages with many links than with just a few, a power law is by no means the only relationship consistent with this expectation. The result is, in other words, not obvious. Indeed we might even have supposed that the relationship would fit a bell curve, with most pages possessing an average of perhaps three or four links. But the power-law relationship says that there is no such preference – no scale to the connectivity of the network. A power law, as we saw earlier, is an indication that the system is scale-free.

There is nothing pre-ordained in this power law. Everyone who sets up a web page at the University of Notre Dame, just like anyone who does so on any other WWW domain, is completely at liberty to decide how many hyperlinks to make elsewhere. (Of course, they cannot determine how many such incoming links their page will receive.) Yet out of this multiplicity of individual choices comes a mathematical law which holds true, in the case of the Notre Dame data, for pages with from one to over a thousand links. The network seems to organize itself into this power-law state, just as the sand-pile model of self-organized criticality generates a critical state with a power-law distribution of avalanches (page 297).

The power law tells us at once that the WWW is not like the small-world networks conjured up by Steven Strogatz and Duncan Watts. Their 'rewiring' graphs do have a preferred connectivity: the probability distribution function of the number of edges per vertex rises to a maximum at some particular value and then declines. Neither is the WWW network like a random graph, for that displays a different probability distribution too. The probability of finding a vertex with a very high number of connections is far greater for a power-law distribution than it is for either a random graph or a Strogatz–Watts small-world graph – just as a power law in economic market fluctuations shows an appreciable chance of large events.

Does this mean that the WWW is not a small world? Not necessarily – it simply means that it is not the kind of small world Strogatz and Watts devised. Albert and her colleagues tested whether their data set showed the small-world signature: a small average (characteristic) path length between two randomly selected vertices, coupled to a high degree of vertex clustering. It turns out that these two properties combine to give a single criterion for a small world: as the number of vertices in the network grows, the average path length increases only very slowly.* The researchers found that a graph constructed to have

---

* Mathematically speaking, the characteristic path length is proportional to the logarithm of the number of vertices.

the same power-law distribution of connectivities as they had observed for their section of the WWW does indeed show this behaviour.

So the World Wide Web *is* a small world – but one with a very specific topology characterized by a power law: by scale-free connectivity. Albert and her colleagues estimate that if the entire WWW has the same structure as the Notre Dame domain, then any two of its web pages are separated by an average of just 19 links. Because of the Web's small-world structure, this average distance should increase by only two links even if, as predicted, the number of web pages increases by 1,000 per cent in the next few years. All the information we want will remain just a few clicks away.

If that's so, why can it be so damned hard to find? One of the commonest complaints from Web users is that a huge amount of useless information stands between you and your quarry. Unfortunately, search engines struggle to carve a path through the junk because their search criteria are necessarily rather blunt tools. Yes, that vital document might be just 19 clicks away – but because of the highly interconnected nature of the WWW, so are a vast number of other documents, all of which the search engine must trawl through. Moreover, the rapidly changing size, scope and content of the Web makes it hard for search engines to keep pace by indexing new pages and updating or discarding old ones which no longer exist or cease to be relevant. It has been estimated that the best search engines reach only about 30 per cent of the entire indexable Web, and that some of them trawl through only 3 per cent.

Lada Adamic, working at the Xerox Research Center in Palo Alto, California, has shown how the small-world character of the WWW can be exploited to design better search engines. These take advantage of the high degree of clustering between web pages devoted to related topics, which distinguishes the topology from that of a random network. An intelligent search engine could use this clustering to limit its sphere of enquiry, after finding the best 'hub' sites from which to launch the search. This smart scheme can be more efficient than a random walk through the maze of the Web.

Power laws seem to be a recurring leitmotif of the WWW. Adamic and her colleague Bernardo Huberman have uncovered this kind of probability distribution in the number of pages per web site. And they and their collaborators discovered in 1998 that users who surf the Web also obey power-law statistics. Surfing is the common alternative (or often the complement) to using search engines. You find a web site that looks as though it might contain the information you want, and then you follow the hyperlinks to other pages on the site until either you find what you seek or you conclude that it is not out there after all.

Most users will happily surf not just from page to page but from site to site. But Huberman, Adamic and their colleagues considered only the surfing pathways that surfers take within single sites. They were interested in how 'deep' people go – how many clicks they follow, on average, before quitting the site. By looking at various data sets – the behaviour of over 23,000 users registered with the service provider AOL, and visitors to the Xerox web site, for example – they found that the probability distribution function of the number of clicks on a site obeyed a power law, or something very close to it.* This information can assist web-site designers by helping to predict the likely number of hits per page.

* The Xerox team proposed a model of surfing behaviour in which visitors to the site execute something like a random walk through its pages until they come across a page whose information value to them lies below some threshold. At this point they quit the site. The model predicts a probability distribution function that has a so-called inverse gaussian shape. This is not the same as a gaussian (page 75), but is close to a power law, deviating from it gradually as the path length gets longer. The data aren't really good enough to distinguish between this relationship and a true power law.

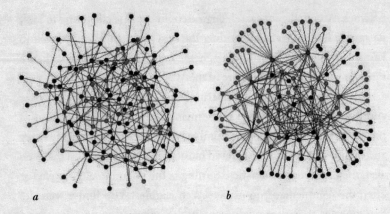

Figure 16.3 Random graphs (*a*) are rather homogeneous, whereas scale-free networks (*b*) seem to be 'pinched' at a few highly connected vertices.

What does a scale-free network actually look like? In a random graph most vertices have roughly the same number of edges, and the network looks rather uniform throughout (Figure 16.3*a*). In a scale-free network most vertices have only one or two links, yet a small but significant proportion are highly connected. Thus the structure is very uneven, seemingly dense or pinched in some places but sparse in others (Figure 16.3*b*). These highly linked nodes provide the short cuts that make the network a small world.

The Internet, like the WWW, also has this scale-free topology with a power-law distribution of connectivity between nodes (Figure 16.4). Herein lies its strength. Albert, Jeong and Barabási have shown that scale-free networks are much more resilient against random flaws, such as breakdown of a node, than are random networks or the small-world but scale-dominated networks of Strogatz and Watts. (The latter two types of network may both be classed as 'exponential', meaning that the probability of highly connected nodes decreases rapidly – exponentially – as the number of links increases.) Albert and

her colleagues found that a scale-free network barely 'notices' if up to
1 in 20 of its nodes is disabled – the characteristic path length is hardly
changed at all. In exponential networks, in contrast, a few mal-
functioning nodes can increase the path length substantially, making it
harder for all the other to nodes communicate. Moreover, exponential
networks tend to break into many isolated clusters when the fraction
of 'dead' nodes exceeds about 28 per cent. When this happens the
network is totally unable to convey information over any significant
distance. A scale-free network, on the other hand, does not 'shatter'
but 'deflates' slowly as nodes are knocked out. A relatively large

Figure 16.4 The structure of part of the Internet, mapped out by tracing the
shortest routes for messages sent from one central computer to thousands
others. Such maps can be seen at http://www.cybergeography.org/
atlas/topology.html

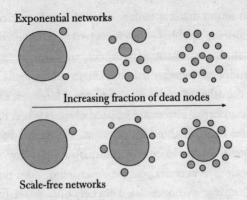

Figure 16.5 Collapse of networks with dead nodes. Exponential networks such as the random networks of Erdös and Rényi break up rapidly as nodes are disabled. In contrast, scale-free networks deflate slowly, retaining a central, highly connected web.

connected cluster still persists even when half the nodes are dead. Because most of the nodes in a scale-free network have only one or two links, breaking links at random mostly just isolates the nodes one at a time (Figure 16.5).

Thus the Internet has precisely the topology it needs to provide a reliable communications network even if some of its nodes are out of action. Such local failures do not arise only from computer mal-function – a node might become temporarily useless because it is jammed by a high volume of information passing through. An effective alternative route can be found quite easily in a scale-free network even if rather a lot of the nodes are simultaneously jammed. (In fact, around 3 per cent of data routers in the Internet network are disabled at any time.)

The remarkable thing is that the Internet has grown *unplanned* into this seemingly most robust of conceivable networks. No one designed it this way. Indeed, if anyone had possessed the authority to dictate the topology of the Internet, the chances are that they'd have chosen a far less robust architecture (like Paul Baran's: Figure 16.1*c*). The message is clear: sometimes it is best to let technology organize itself. The question remains, however, of *why* the Internet has this structure. I shall turn to this at the end of the chapter. First we shall confront a less happy prospect. For despite its robustness, the Internet has a potentially fatal flaw. And that flaw is keeping some people awake at night.

CYBER-ATTACK

It is ironic that the Internet, born out of the desire to provide a communications system robust enough to survive enemy aggression, is now becoming seen as a vulnerable structure on which the survival of the United States depends. In 1996 American President Bill Clinton set up the Commission of Critical Infrastructure Protection to assess ways of protecting the electricity, communications and computer networks from attack. The form of such attacks is quite different from those feared in the days of the Net's origins. No missiles need be launched, no armies or air force mobilized. The aggressors need not even leave their homes. They need not be foreign superpowers with immense military might; a single individual armed with conventional computer technology could, with the right know-how, bring a country to its knees. By disrupting the computerized communications of banks and financial institutions, a cyber-terrorist could push the American economy into free fall. Misinformation fed into the computer systems of the power and fuel distribution infrastructure could shut down entire states or blow up gas lines.

Terrorism has been defined by the FBI as 'the unlawful use of force or violence against persons or property to intimidate or coerce a government, the civilian population, or any segment thereof, in furtherance of political or social objectives.'[9] Cyber-terrorism is an attempt to do that using electronic information networks. It is a real threat, and some politicians are beginning to fear it more than nuclear war. Protection against cyber-terrorism has become a major pre-occupation in the USA: the CIA's Information Warfare Center has a staff of over a thousand, and the FBI, the Secret Service and the US Air Force all have their own anti-cyber-terrorism squads.

In an age when terrorism has become the spectre haunting civilization, this fear is not sheer paranoia. Cyber-terrorism is cheap, it puts the perpetrator in no immediate danger and it could be devastating. The Pentagon's computer system has already been attacked. In one military exercise, called Eligible Receiver, staff of the National Security Agency used freely available software to gain access to computer systems that would have allowed them to shut down the entire American power grid, and to gain partial control of the US military's Pacific Command. Most of the 'crackers' eluded attempts to trace them.* But this kind of network takeover is only one part of the problem, and possibly not the most significant. It would be far simpler, and perhaps just as effective, to use a blunter weapon: merely to stop the network from functioning. The scale-free small-world topology of the Internet guarantees that attempts to do this by knocking out nodes at random will probably be futile, because the network can withstand a considerable amount of such punishment without losing much of its connectivity. But a cyber-terrorist is unlikely to try such a misguided approach.

In an exponential network, the existence of a characteristic average

---

* A cracker, who aims to use the Internet for disruptive or destructive purposes, is not the same as a hacker, who is primarily interested in developing and improving software according to the unspoken customs of the freewheeling and meritocratic 'net culture'.

scale of connectivity means that none of the nodes is special. In a scale-free network, on the other hand, some nodes are definitely more equal than others. The most highly connected nodes are the crucial short cuts that keep the average path length low. Remove these nodes and you cut out many links at once, affecting a wide region of the network. If, instead of eliminating nodes at random, we begin with the most highly connected, a comparison of robustness for exponential and scale-free networks reveals a significant difference. In the former case there is barely any difference in the effects of random and targeted strategies for disabling nodes. In the latter case, the scale-free network breaks up rapidly as the most highly connected nodes are destroyed. Removing 1 in 20 of the nodes on this basis doubles the characteristic path length – suddenly big detours are necessary to get from place to place. So a cyber-terrorist who wants to trash the Internet need only identify a relatively small number of the most highly connected nodes and make those the targets of sabotage. In other words, when an attack is guided by intelligence it can do a disproportionate amount of harm to a scale-free network like the Internet. It is around the linchpin nodes that the most secure safety walls should be raised.

This topological structure makes the Internet more like a 'naturall body' than anyone could have guessed. These webs spun by information technology are most often compared with the brain, and perhaps with some justification. Strogatz and Watts have shown that the pattern of connections in the neural network of the parasite nematode worm *Caenorhabditis elegans* displays the characteristic features of a small-world net: large clustering and small characteristic path length between individual neurons. But Barabási and his colleagues have uncovered a still more vital network in the cells of a wide range of organisms that supports a striking analogy with the scale-free Internet.

Perhaps the deepest principle of life is metabolism: the conversion of raw materials from the environment into the energy and the

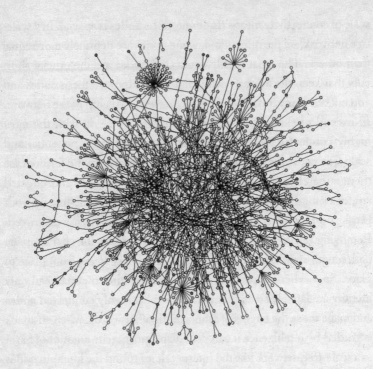

Figure 16.6 A part of the yeast metabolic network. The vertices are molecules either supplied to or manufactured by the network, and the links (edges) are defined by the enzymatic chemical reactions that convert one molecule to another.

molecules that cells need every moment of their existence. Cells need access to a variety of building blocks: our own cells, for example, are supplied with amino acids, sugars and lipids from food, as well as vitamins and mineral nutrients, water, oxygen and other essential substances. They use enzymes to rearrange the atoms in these molecules to form new enzymes, nucleic acids, hormones, energy-rich molecules, and so forth. The sequence by which a raw material is converted into a useful molecular form is called a metabolic pathway.

Almost without exception, these pathways are not linear but branched and interlinked. A single raw material such as glucose is reconfigured or fragmented in many different ways. The energy-rich molecules made during the breakdown of this sugar are used to power many other metabolic processes. So metabolism defines a large network of chemical reactions in which particular molecular substances can be regarded as nodes, and reactions (usually catalysed by enzymes) can be considered as the edges that link one node to another (Figure 16.6).

Barabási and his colleagues looked at the metabolic networks of forty-three different organisms ranging from bacteria to plants to 'higher' life forms like the nematode worm. They found that the connectivity distribution function in every case was scale-free: the probability of a node having a certain number of links followed a power law. This implies that there are a few highly connected hubs in the network which play a crucial role in holding the web together. Many of the molecules representing these hubs, and their relative importance in the network, are the same for all organisms: a reflection of life's common evolutionary origins.

The scale-free structure of metabolic networks makes sound evolutionary sense, because it makes metabolism relatively insensitive to a small incidence of random failures. If one or two enzymes are defective, perhaps due to a genetic defect, this weakens or even severs the corresponding edges in the graph. While this can have harmful consequences for certain biological functions, it need not, in a scale-free network, break up the whole web and make life impossible. Thus we can regard the scale-free network as an example of good 'engineering', wrought by the trial-and-error exploration of options that is natural selection.

On the other hand, the weakness of scale-free networks is their sensitivity to planned attack. If the hub nodes are removed, the network rapidly falls apart. This suggests opportunities for the intelligent design of drugs to combat bacterial infection. Drugs that interfere with the most highly connected molecules will have a disproportionate effect on the viability of the attacked cells. Understanding the network structure would be a first step towards selecting appropriate targets. In this scenario we are the 'cyto-terrorists', and our intentions are, from a human perspective, purely benign.

Could the massive breakdowns of power grids in the USA and Italy in 2003 have had anything to do with the topological vulnerabilities of the networks? Barabási thinks so, even though it is not clear that all power grids are scale-free. '[The August] blackout has little to do with faulty equipment, negligence or bad design', he says. 'The magnitude of the blackout is rooted in an often ignored aspect of our globalized world: vulnerability due to interconnectivity.'[10] Barabási points out how systems such as power grids are prone to cascades of failure, where a malfunction at one point shifts power to other lines, leading to an escalation of overloading. 'Cascading failures are common in most complex networks', he says. 'While celebrating that everybody on earth is only six handshakes from us, we need to accept that so are their problems and vulnerabilities.'[11]

SPREADING IT AROUND

Wilful destruction and accidental failure of nodes are not the only dangers facing our social networks. Computer viruses represent a more insidious threat to electronic communications than does cyber-terrorism: rather than destroying the network, viruses take advantage of its extreme connectivity to colonize it and wreak havoc at the nodes. Here again, in other words, a favourable property of the Internet's topology is turned against it. Computer viruses typically spread by email contact – by the transmission of parcels of information between one computer and another, just as biological viruses are transmitted between individuals as particles in breath or body fluids. Once a link is established between two nodes of the network, they can potentially infect each other.

The study of how diseases spread in a population – the science of epidemiology – has a long history. Computer scientists have borrowed the mathematical models developed by epidemiologists to try to understand the dissemination of computer viruses. One of the standard approaches in epidemiology is to assume that at any given moment an individual exists in one of two states: healthy and infected. Healthy agents are susceptible to infection from infected agents. An encounter between the two is assumed to lead to infection with a particular probability. At the same time, infected individuals have a chance of becoming healthy (which is to say, susceptible) again. The disease then spreads at a rate equal to the ratio of the probabilities of transmission and of recovery. This is known as the susceptible–infected–susceptible (SIS) model.

Epidemiologists using this model have found that it predicts an epidemic threshold. The disease spreads throughout the population, persistently infecting a constant proportion of people, if the spreading rate is greater than some threshold value; otherwise, it dies out quickly.* Some real epidemics seem to behave this way. The idea of a threshold is central to vaccination programmes: provided a certain proportion of the population is vaccinated, the spreading rate remains below the threshold value, and the disease cannot reach epidemic levels.

Duncan Watts, working with Mark Newman of the Santa Fe Institute in New Mexico, has shown that the small-world networks he

---

* The threshold is technically equivalent to a critical point in a non-equilibrium phase transition – like the liquid–gas critical point, except that the states it separates are non-equilibrium states.

devised with Steven Strogatz also possess thresholds for spreading of a 'disease' in an SIS-type model. In other words, if social networks are something like these webs, the standard ideas of epidemiology hold true. But the Internet is not a Strogatz–Watts small world: it is a scale-free small world. Stefan Bornholdt and his colleagues at the University of Kiel have demonstrated that email connections through their university server trace out a scale-free network, suggesting that not only is the Internet physically scale-free (that is, in terms of user nodes and links between them), but the 'acquaintance network' defined on this electronic web also has this same topology.

Physicists Romualdo Pastor-Satorras in Barcelona and Alessandro Vespignani in Trieste have found that this difference completely changes the way in which computer viruses spread themselves through the system. When they used computer simulations to investigate the SIS model on a scale-free network, they discovered that there is no threshold at all. No matter how slow their spreading rate, all viruses can pervade the system, infecting a certain proportion of the nodes. Because infected nodes can be 'cured' with antivirus software, the virus does eventually die out. But this process can be very slow. Software to combat a particular virus typically becomes available within days or weeks of the first infection, whereas viruses can survive in the Internet's dense web of connections for many months or even years. The researchers studied the prevalence data of viruses reported between 1996 and 2000 and found that, although the survival probability drops sharply in the first few months of a virus's life history, a low level of infection remains for long afterwards. (The immensely damaging Love Bug virus, for instance, is still the seventh most prevalent computer virus, even though it is supposed to have been eradicated.)

This behaviour is puzzling for conventional epidemiology. The standard models predict that a viral infection should either become an epidemic or die out quickly. Yet the slow, lingering decay, without an epidemic threshold, seems to be a peculiar characteristic of transmission on scale-free networks. The discovery has implications both good and bad. On the downside, even a slow-spreading virus can pervade the system and last for a long time. But it does so only at a very low level – the fraction of infected nodes rapidly becomes very small.

This result highlights the fact that network topology matters. For those seeking to combat computer viruses, standard epidemiological wisdom may not be sufficient. What if real social contact networks are scale-free too? If they are, we may need to rethink our notions about how real diseases spread. As we have seen, social contacts are notoriously hard to map. But a team of sociologists at Stockholm University in Sweden has joined forces with physicists Gene Stanley and Luis Nuñes Amaral in Boston to uncover the web of sexual contacts in a random sample of almost 3,000 Swedes aged between eighteen and seventy-four. They found that the distribution in the number of partners over the twelve months before the survey was conducted obeys a power law. In other words, this contact network is scale-free.

If this sample is representative of populations elsewhere, there are serious implications for strategies to combat sexually transmitted diseases such as Aids. It is extremely difficult to eradicate diseases completely in a scale-free network, since they can permeate the population regardless of the spreading rate. Where a vaccine is available, random immunization against the disease cannot contain it even if a large proportion of the population is treated (perhaps at considerable expense).

But the picture is not necessarily so gloomy. As we have seen, the Achilles' heel of scale-free networks is the disproportionate influence of just a few highly connected nodes: the 'hubs' which hold it all together. Sever the connections to these hubs and the web quickly falls apart. Pastor-Satorras and Vespignani have shown that in such circumstances, targeting immunization programmes at the most promiscuous individuals sharply lowers the network's vulnerability to epidemics of sexually transmitted diseases. (By the same token, an analysis of email networks shows that 'immunizing' just 10 per cent of the vertices selected on the grounds of their large number of connections can effectively suppress a computer virus outbreak entirely.)

In networks of sexual contacts it is often far from easy to identify and treat these key individuals. However, Barabási and his student Zoltan Dezso have found that even a rather inefficient targeting campaign has a crucial effect: it raises the epidemic threshold of a scale-free network above zero. In other words, even if only a few of the hub nodes are isolated there is a chance of eradicating the disease. The more effectively a policy reaches the most highly connected members of the sexual network (in the case of Aids), the higher the threshold – the easier it is to stamp out the disease. Faced with limited resources, it is therefore better to combat an infectious disease by making at least some effort to focus control strategies on the 'hub' individuals rather than to promote random, blanket immunization and hope for the best.

WORLD OF WEBS

Scale-free networks may be a much more widespread form of small world than the rewiring networks of Strogatz and Watts. (In fact it is not clear that *any* of the real small-world networks studied so far is truly like a rewiring network.) Barabási and Albert revisited the case studies that Strogatz and Watts first considered: the movie-actor network and the power grid of the western USA. For Strogatz and Watts the only indicator of small-world behaviour was the coexistence of large clustering of nodes and a small characteristic path length. These criteria do not uniquely specify the topology of connections, however – they apply both to rewiring networks and to scale-free networks. Barabási and Albert found scale-free distributions of connectivity both for movie actors and for power lines. Thus, for example, there is no meaningful 'average' connectivity between movie stars: the number of links decreases smoothly (as described by a power law) from Rod Steiger to the most obscure B-movie character actor.

The movie-actor network supplies probably the best-documented evidence we have so far for suspecting that real social networks are scale-free – although studies of scientific collaboration networks (like the Erdös network, page 445), musical collaborations in the jazz world and sexual-contact networks support that contention. On the assumption that this is so, Ricardo Alberich and colleagues at the University of the Balearic Isles on Mallorca have shown that social networks can provide a kind of watermark to distinguish fact from fiction. Spider-Man's friendship web, they say, just doesn't look real.

The Marvel Comics, which began in 1939 under the name of Timely Comics, describe a bizarre universe of superheroes: Captain America, the Fantastic Four, the X-Men, and thousands more. These characters move in the same world as one another – their stories overlap, so that many may appear together in the same comic strip. When this happens, say the Spanish researchers, it constitutes a 'social connection' – a link between two characters in the Marvel universe network. This network contains about 6,500 characters (nodes) in all, which are featured in around 13,000 comic books. Does it have the same topological features as a real-life web like the movie-actor network? Well, sort of. That's to say, the probability that any given comic book has a certain number of characters in it follows a power law.* To this extent, the Marvel universe network looks realistic. But it cannot completely hide its artificial origins, the researchers

---

* Actually the power law obtains only when the number of characters appearing together in a story exceeds ten. For smaller groups, the statistics deviate from the power-law distribution.

conclude. They find that the clustering of the network is much lower than that of scale-free or small-world nets, and indeed scarcely larger than that of a random graph. Clustering describes the extent to which characters form into 'friendship circles': in a highly clustered net, remember, two people who have a friend in common are more likely to know each other than are two people chosen purely at random.

The small degree of clustering in the Marvel universe reveals its ad hoc, fictional nature, showing that inventing a universe is harder than it seems. In a sense, this is perhaps not surprising. After all, the Marvel story-writers had no notion of what social networks look like, and were not trying consciously to create one that seemed 'realistic'. They simply put two or more characters together if they imagined the combination would be interesting. Of course, real social networks are also unplanned. It's just that Spider-Man's social network and yours or mine seem to have grown according to different principles. It's not yet clear how those differences arise, but if we can uncover the reasons, we might learn more about how social connections are forged in the real world.

Scale-free networks are now starting to look like such a fundamental aspect of human culture that eyebrows are raised and questions asked when they do *not* appear. The global trade network, the Indian railway system, the trading activity in online auctions such as eBay, the Chinese airline system – all display the familiar power-law distribution (though often with an important proviso, described below). When the emergence of leadership structures is studied using the minority-game model of Challet and Zhang (see page 418), it generates a scale-free hierarchy in which a few key individuals exercise inordinate influence. It seems that, once people begin to interact and establish connections, the ubiquitous gaussian distribution which so dazzled early social statisticians vanishes, and the scale-free distribution emerges in its place.

DO THE RICH ALWAYS GET RICHER?

When a pattern recurs in many different systems which bear no obvious relationship to one another, we must suspect a common causative principle, one which can be couched in the most general terms without reference to the specifics of this or that case. Abrupt phase transitions tend to arise, for example, when competing forces are at play in systems of many 'particles': a propensity towards order and a disrupting influence of noise. What, then, are the generic principles that create scale-free networks?

Both the random graphs of Erdös and Rényi and the rewired small worlds of Strogatz and Watts are constructed by purely random processes. One chooses vertices in the graph at random and links them together via an edge. But Barabási and Albert point out that most networks do not grow in this way. Instead there is a bias towards linking a new node to one which is already well connected. New actors are more likely to play supporting roles in films with established stars than they are to be cast in a film full of unknowns. Thus the greater your fame, the greater the likelihood that you will attract new links.

Similarly, a kind of 'magnetism of fame' operates in the evolution of the World Wide Web or the network created by scientific citations (another scale-free system – see page 302). Web pages that already have a high number of incoming hyperlinks become well known and are the natural choice for web designers who want a pointer from their new page to a source of further information. People cite famous papers because they are the ones everyone else cites, or because they are more likely to have read them. Fame breeds fame, in other words.

One would like to imagine that there is some meritocratic principle underlying all of this: that web pages receive lots of links, papers get many citations and actors are frequently cast because they are *good*. It would be surprising if that were not so. Nonetheless, fame brings increasing returns to good and bad alike. (You can no doubt supply your own examples.)

If every new vertex in a growing graph were invariably linked to the most highly connected vertex, the network would become not scale-free but centralized around a single hub. This is not usually what happens when networks grow unplanned. If the graph is large, it will be extremely difficult for a new vertex to 'find' the most highly connected vertex and the chances are that some less-than-maximal vertex will be selected instead. In the movie world this variety of choice is inevitable: even the most famous and well-connected star can't appear in every new movie. So the formation of new links to the most highly connected vertices can be only a bias, not a certainty.

Barabási and Albert have shown that this bias is the only ingredient needed to grow a scale-free network. They imagine a graph that is growing by the addition of new vertices, each one linked to an existing vertex chosen at random *but* with a bias that gives highly connected vertices a better chance of acquiring the new links. The resulting network is scale-free. Many human organizational networks develop on this 'rich get richer' principle: bigger businesses are more likely (but not guaranteed) to get new clients, for example, partly because they can afford more and better advertising.

If the connectivity of a vertex can be equated in some sense with its 'wealth' (for example, if we are thinking about the trading connections of a company), this suggests that in a society governed by freedom of choice, and where one's ability to gather a 'market share' is determined by how much of it one already has, a power-law distribution of inequality is the probable outcome. We would, of course, expect there to be differences in access to or command of resources in any free market, but the scale-free growth process produces greater disparities than a random distribution of wealth. It results in a significant number of extreme 'events': very wealthy individuals or huge companies. As the social scientist George Kingsley Zipf showed in the 1930s, this power-law behaviour is precisely what we see for many size distributions in society, from the size of firms (page 330) to the size of cities and the distribution of incomes (see page 308). This is not to say that power-law disparities in a free market are inevitable. But it does suggest that, if we decide they are undesirable, we shall probably need to restrict some of the freedom with which the market operates.

This type of network growth does not always guarantee quite such a high degree of inequality, however. Gene Stanley and colleagues at Boston University have looked more closely at the supposedly scale-free networks identified by Barabási and Albert and have discovered that they do after all appear to have a ceiling. For example, the power law in the movie-actor network, which seems inexorably to reward a few privileged individuals with ever better connections, begins to falter at the high extreme: the best-connected actors have rather fewer links than the power law would predict (Figure 16.7).*

What truncates the power law? Stanley and his colleagues say that in the real world some factor invariably imposes an upper limit on the number of links a node can acquire. Actors have limited time and finite careers – even the most hard-working of them cannot appear in a thousand films. Old scientific papers, even the most seminal, fall victim to neglect – researchers no longer read the old literature, but instead cite a more recent review article or textbook on the subject. Airports cannot handle an infinite amount of traffic, and cost or local

---

* The statistics of this network are affected, however, by the ambiguities of TV shows. Stanley and his colleagues included them; Barabási and his co-workers did not. A TV show is listed in the actors' database as a single title, even though it might constitute a long-running series with many guest stars who do not actually appear together in a single episode. This can generate some spurious links, and it may affect the extent to which power-law behaviour is displayed.

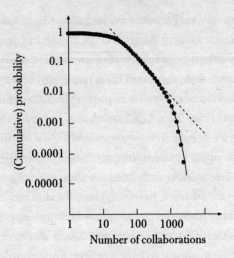

Figure 16.7 The most highly connected movie actors do not quite follow the power law that characterizes the rest of the network – those with more than 300 or so collaborations have rather fewer connections than the power law predicts (they plot below the straight line). There is, in other words, an upper limit to the number of collaborations an actor can attain.

demographics sets limits on how much they can expand. If vertices either have a capacity limit or get less able to form new connections as time goes by (a kind of ageing), the scale-free structure of the network disappears for the most highly connected nodes.

For some other social networks a power law does not even get established in the first place. Instead the connectivity declines rapidly for the most highly connected links, so that the probability of 'super-hubs' is vanishingly small. This is true, for example, of the Southern Californian power grid, and also of the network that links the world's airports by flight paths. Moreover, Stanley and his colleagues found it to be the case for some real social networks: the friendship networks of forty-three Mormons in Utah (where such ties are unusually close and well defined) and of several hundred students at Madison Junior High School in Wisconsin. All these networks seem to have a gaussian probability distribution: there is an average 'scale' to the connectivity. Nevertheless, they are also small-world networks, as revealed by the very slow increase in characteristic path length as the number of vertices increases.

Thus there are several types of small world. The rewiring networks defined by Strogatz and Watts are one example of a 'single-scale' small world in which there is a preferred average connectivity and a sharp decline in the numbers of highly connected vertices. At the other extreme are the scale-free networks found by Barabási and colleagues, where greed or promiscuity knows no bounds, and even very highly connected vertices are far from rare. In between these two are a 'whole zoo of networks',[12] in Barabási's words.

Whatever its precise topological properties, a small-world network is seldom if ever planned; it simply emerges from the laws that govern the connection of new vertices. Yet Nisha Mathias and Venkatesh Gopal in Bangalore, India, have argued that such a network never-theless has great attractions from an engineering point of view, because it typically provides an ideal compromise between two conflicting urges: to maximize the connectivity of vertices and to minimize the length of the 'wiring'. In general, connecting two nodes in a network has a financial cost proportional to the distance between them. You need twice as much cable to link two power plants 100 kilometres apart as you do if they are 50 kilometres apart. A network communicates most efficiently if every node is connected directly to every other; but that requires an awful lot of wiring. So network builders want fewer and shorter connections, and network users want more and longer ones (and hence many short cuts). Mathias and Gopal have shown that the first of these options corresponds to a regular grid, and the second to a random network. A small-world network is a way of getting good long-range connection at the cost of little more wiring than is needed for a grid that makes only 'local' links.

The physics of networks is a very young science, but already it has revealed some apparently profound rules which apply to our social patterns and institutions. The consequences of network topologies for processes of change such as the spread of epidemics or of cultural norms are at present only barely sketched. In economics, we have seen that the structure of trading networks has been identified as one of the key questions in understanding how the complex dynamics of the market arise. According to the organizers of a workshop on econophysics at the Santa Fe Institute in 1996, 'Markets actually function by means of networks of traders, and what happens in markets may reflect the structure of these networks, which in turn may depend on how the networks emerge.'[13]

In a sense we can see network theory as a natural extension of statistical physics – a kind of 'sticky' version of many-particle systems in which encounters between individuals lead not simply to collisions but to the formation of permanent, elastic links. The network then becomes a kind of map of the history of the system, just as the expansion of the Internet over time is recorded in the growing web that it weaves. In networks the past matters: a memory of it is frozen in place and shapes the present. These networks provide a graphic image of the extent to which our lives become entwined in ways as innumerable as they are barely fathomable.

17

Order in Eden

Learning to cooperate

Nor do I know, if men are like sheep, why they need any government: or if they are like wolves, how they can suffer it.

Sir William Temple (1751)

In transgressing the law of Nature, the offender declares himself to live by another rule than that of reason and common equity . . . and so he becomes dangerous to mankind; the tie which is to secure them from injury and violence being slighted and broken by him . . . every man upon this score, by the right he hath to preserve mankind in general, may restrain, or where it is necessary, destroy things noxious to them, and so may bring such evil on any one who hath transgressed that law, as may make him repent the doing of it, and thereby deter him, and, by his example, others from doing the like mischief.

John Locke (1690)

It would be child's play to shell the road behind the enemy's trenches, crowded as it must be with ration wagons and water carts, into a bloodstained wilderness . . . but on the whole there is silence. After all, if you prevent your enemy from drawing his rations, his remedy is simple: he will prevent you from drawing yours.

Ian Hay (1916)

*

What makes us civilized? The many answers to this question tell us much about those who supply them, but scarcely help us reach a consensus. Most commentators since the Enlightenment have focused on the tension between personal and collective freedoms in a civilized society – a tension which demands compromise. In Sigmund Freud's words,

The liberty of the individual is no gift of civilization. It was greatest before there was any civilization, though then, it is true, it had for the most part no value, since the individual was scarcely in a position to defend it. The development of civilization imposes restrictions on it, and justice demands that no one shall escape those restrictions . . . This replacement of the power of the individual by the power of a community constitutes the decisive step of civilization.[1]

How, though, to circumscribe those powers and those restrictions? Where should the compromise lie? For Hobbes, civilization can emerge from the barbaric State of Nature only if we voluntarily take up the yoke of dictatorship, relinquishing all rights except the right to exist. The social contract of John Locke was altogether more collaborative: citizens relinquish certain freedoms to the State only so long as the State serves their interests. In other words, the State does not have the almost unlimited power that Hobbes would grant it. A populace, said Locke in *Two Treatises of Government* (1690), has the right to rise up and displace a government that breaks the social contract – not capriciously, for that way lies anarchy, but after 'a long train of Abuses,

Prevarications, and Artifices'.[2] Moreover, Locke recommends safe-guarding against such abuses of power, to which Hobbes (perhaps strangely, given his pessimistic view of human nature) seems largely indifferent.

At whatever point the balance of power between the state and the people is set, there is a recognition that civil peace and order come at the expense of individual restraint – either self-imposed or state-enforced. All liberal philosophers since the seventeenth century have concurred that this must include, at the very least, the renunciation of attempts to harm others. Karl Popper had this to say:

Absolute freedom is nonsense . . . We need a society in which the freedom of each person is compatible with the freedom of other persons. The compatibility of my freedom with yours depends on our both renouncing violence towards each other. I won't knock you down, and you won't knock me down.[3]

Freud agreed, but acknowledged that, as Hobbes had asserted, this runs contrary to our aggressive and destructive impulses, our 'death instinct'. As a result, he claimed, mankind's aggression 'is introjected, internalized; it is, in point of fact, sent back to where it came from – that is, it is directed towards his own ego.'[4] The consequence is a pervasive sense of guilt, which manifests itself as proscriptive religion and the concept of original sin. Maybe so. Yet the common assumption in such arguments is that, by agreeing not to harm one's neighbour, one is thwarting one's best interests – acting contrary to reason, if 'reason' is to be equated with a Darwinian survival of the fittest rather than with Locke's divine law or Kant's categorical imperative. Does the cheetah, after all, display altruism towards the gazelle?

In the second half of the twentieth century, all these discussions about liberty, government and power have been shown to be short-sighted in a profound and astonishing way. Political philosophy has either seen humankind as basically bad, and therefore in need of some authority to enforce good behaviour, or as basically good, thus for the most part guaranteeing that our relations with our neighbours are civilized. But a theory which has its origins in the frivolity of parlour games has demonstrated how good behaviour can arise and persist even among enemies – and even in the total absence of moral considerations.

Game theory, as it is called, is not generally deemed to be a part of physics at all, but of mathematics (if of a decidedly empirical stamp). Yet it is very decidedly in the spirit of the kind of social physics we have considered so far, in which the behaviour of a collection of individuals arises in a non-intuitive but robust manner from the inter-actions between them. In exploring this theory in this chapter and the next, we shall come across some by now familiar phenomena: indifference to details, abrupt changes in patterns of behaviour or in statistical populations, sensitivity to fluctuations, generalizable 'laws'.

Some of the conclusions of game theory go to the very root of the issue of how we should live and govern our lives. They may challenge our fundamental systems of belief. And since they already *are* taken seriously in the corridors of power, it is important to try to understand as clearly as possible what we may and may not deduce from them, and to what extent they pertain to real human interactions. At one point in its development, game theory seemed to advocate a Cold War policy of belligerent international relations that would have been disastrous had it been rigorously adopted; only later did it emerge why such a conclusion was flawed. Both liberals and conservatives can find support for their beliefs within this discipline, particularly if they select carefully from its findings. More than in any other area of mathematical social science, game theory demands that we confront objective results with moral considerations and exercise great care when we are tempted to couch technical conclusions in anthropomorphic terms. Potentially, it is political dynamite.

IS GOVERNMENT NECESSARY?

In those countries where it has been put into practice, the democratic foundation of John Locke's political philosophy is now so taken for granted as the 'right' way to run a society that it is hard to imagine how anyone could have felt otherwise, except out of sheer self-interest. But Locke's *Two Treatises of Government* were written as a counter-argument to an alternative that was widely advocated at the time: the notion that kings ruled by divine right. This was asserted in Robert Filmer's *Patriarcha; or The Natural Power of Kings*, published in 1680 but written before the English Civil War, when the political climate was very different. Filmer's sovereign is, if anything, even more fearsome than Hobbes's. At least Hobbes would have him elected from (and by) the masses, whereas for Filmer the king can do as he will with impunity because he is a descendant of Adam and his authority comes from God. This was, naturally enough, music to the ears of Charles I, who knighted Filmer before discovering for himself the fate of kings who become a law unto themselves.

In the seventeenth century Filmer had the weight of history on his side. Most countries and states since antiquity had been monarchies, and the acme of civilization was considered to be embodied in the autocratic ancient Greek and Roman states, some of whose leaders proclaimed their own divinity. Plato acknowledged that monarchy could develop into tyranny under an immoral ruler, but his preferred alternative was an aristocracy – the rule of a few good men. Plato rejected and deplored democracy – literally, 'the rule of the people' – for the masses harboured many wicked and immoral people.

Not that it hadn't been tried. Following a succession of tyrannies in the Greek states of the seventh and sixth centuries BC, Athens established a democratic government in 507 BC. In many respects it was an appealing advertisement for this system of rule: Athenian democracy fostered Sophocles, Euripides, Aristophanes and Thucydides. This was the society that constructed the Parthenon and produced the paintings of Apelles, reputedly the greatest of the ancient artists. But the Athenians finally suffered devastating defeat at the hands of Sparta in the Second Peloponnesian War, and their democracy collapsed in 411 BC.

It's difficult to decide on the best system of government, however, until you have decided why government is necessary in the first place. For Plato and his contemporaries, the need for leadership was taken for granted and the central question was how to acquire good leaders rather than bad. For Hobbes, that was a secondary matter: almost any leader short of a monster was better than none, for how else could man be prevented from waging war on his neighbour? Locke was more inclined to see the good in humankind. He too spoke of a State of Nature that existed before nations and governments, but it was a benign Eden: 'Men living together according to reason, without a common superior on earth, with authority to judge between them.'[5] Here, 'reason' means adherence to a kind of divine principle that respects such 'self-evident' truths as 'thou shalt not kill': 'Reason . . . teaches all mankind . . . that being all equal and independent, no one ought to harm another in his life, health, liberty, or possessions.'[6]

Thus Locke's State of Nature is quite different from Hobbes's vision of every man for himself, and Locke probably had Hobbes in mind when he said:

And here we have the plain 'difference between the state of nature and the state of war', which, however some men have confounded [them], are as far distant, as a state of peace, goodwill, mutual assistance and preservation, and a state of enmity, malice, violence and mutual destruction are from one another.[7]

In his complaint against Hobbes's brutal view of human behaviour, Locke was by no means alone. His patron Lord Ashley Cooper, the third Earl of Shaftesbury, echoed him in 1711 by charging that Hobbes 'forgot to mention Kindness, Friendship, Sociableness, Love of Company and Converse, Natural affection, or anything of this kind.'[8] It is understandable that Locke imagined people to be guided by these things in a State of Nature, since his faith in human nature stems from God's power over men. Hobbes the atheist (by all appearances) took less rose-tinted a view.

Locke's optimistic axioms, then, allow him the luxury of a less oppressive system of rule, so that the social contract between people and government has authority only so long as the leaders serve the common good. And to ensure that the government does not gravitate towards tyrannical oligarchy, Locke insists on checks and balances to power. Leadership is to be divided into legislative, executive and judicial branches, each with the authority to restrain abuses by the others. In seventeenth-century England these roles were filled by Parliament, the King and the courts, respectively. In the United States today, their equivalents are Congress, the President and the Supreme Court.

Despite his generous view of human nature, Locke acknowledges that criminals and miscreants do exist. And his remedy, while men live in a State of Nature, is a Biblically ruthless 'natural law' which permits that 'whoso sheddeth man's blood, by man shall his blood be shed'.[9] Thus all is not necessarily so tranquil in Eden, after all. But when a State exists, such sanctions no longer fall to the individual: natural law gives way to 'positive law' decreed and enforced by the legislative

State. Indeed, this is the main purpose of the State: to maintain law, order and justice – particularly, as far as the materialist Locke is concerned, with regard to one's possessions: 'The great and chief end of men uniting into commonwealths, and putting themselves under government, is the preservation of their property.'[10]

So long as this is so, all is well. Thus from Locke's faith in human goodness springs the notion that a government should interfere as little as possible in civil life. This is the touchstone of most liberal political philosophy. Immanuel Kant argued similarly against the paternalistic state, even when (perhaps especially when) the state tells us that its dictates are for our own good. Kant's political beliefs were echoed by the Prussian Wilhelm von Humboldt in *Ideas towards a Definition of the Limits of State Action* (1851), and by John Stuart Mill who, in his essay *On Liberty* (1859), asserted that 'The only purpose for which power can be rightfully exercised over any member of a civilized community, against his will, is to prevent harm to others.'[11]

Mill even maintained the right of an individual to harm *himself* so long as this does not injure others. At the root of this libertarian philosophy is the benign social mathematics of utilitarianism, defined by Mill as:

The creed which accepts as the foundation of morals, Utility, or the Greatest Happiness Principle, holds that actions are right in proportion as they tend to promote happiness, wrong as they tend to produce the reverse of happiness. By happiness is intended pleasure, and the absence of pain; by unhappiness, pain, and the privation of pleasure.[12]

To some political philosophers even a minimally interventionist state was too much. Rousseau expounded a romantic anarchism based on the belief that civilization is degenerate and only the primitive 'savage' can be noble and good. Man is naturally good, claimed

Rousseau in 1754, and it is only institutions that make him bad. Thus the activities and pursuits of civilized society – the arts and sciences, organized agriculture and industry – are corrupting. Rousseau's own version of the 'Social Contract', in his book of that name from 1762, does not reach far beyond Locke's in promoting democracy and debunking the divine right of sovereigns (though even that much was dangerous in pre-Revolutionary France, and Rousseau was forced to flee to Germany). But his real agenda is more radical:

Europe is the unhappiest Continent, because it has the most grain and the most iron. To undo the evil, it is only necessary to abandon civilization, for man is naturally good, and savage man, when he has dined, is at peace with all nature and the friend of all his fellow-creatures.[13]

It seems here almost as though Rousseau has gone out of his way to show us the flaw in his dreams, while himself remaining oblivious to it. Lions too will gaze benevolently on their prey, *when they have dined*. But it is the finding of dinner that stokes the fires of conflict and war. When man is hungry, whether he be 'savage' or 'civilized', he is apt to do whatever he needs in order to get food. And we are many in a world of limited resources. Even in the eighteenth century that was clear, and Malthus's prediction that things could only get worse (page 68) led Darwin to present a very different picture of the State of Nature: red in tooth and claw, as Tennyson put it.

Darwin's theory of natural selection seemed to cast us back into a Hobbesian world of brutish nastiness, where nature advises us to grab what we can and let the devil take our neighbour. Restraint can be bred into people, but Darwinism makes selfishness the law of the wild. Darwin himself believed that humankind is exempt from the savage law of nature by virtue of our evolved tendency towards socialization. Indeed, he regarded this as an essential feature of our evolutionary fitness: '[T]he small strength and speed of man, and his want of natural weapons, &c., are more than counterbalanced by his . . . social qualities, which lead him to give and receive aid from his fellow-men.'[14] But perhaps predictably, natural selection was soon enough seen by others as a blueprint for justifying aggression rather than cooperation. Kill or be killed – that was the law.

Some sought to escape this deeply unpalatable prospect. In 1902 the Russian prince Peter Kropotkin published *Mutual Aid*, a book intended to show that cooperation, not competition, is innate to humans. He cited examples ranging from the customs of Polynesian islanders to the establishment of medieval guilds. Kropotkin was motivated less by a wish to argue for the best in humankind than by the need to establish human goodness as a foundation for his anarchist political opinions: if people have a natural inclination to cooperate, why should government be necessary? But Kropotkin had only anecdotes to support his case; the Darwinists apparently had a law of nature.

Besides, not even Kropotkin's examples of altruism can evade the unfortunate truth that civilization has its discontents – its freeloaders and its criminals. Not everyone plays the game. Locke's State was designed to apprehend and punish thieves, vagabonds and murderers. Of course, many people break the law because they are penniless and desperate, or in the throes of violent passion, or mentally disturbed. But others do so simply because they stand to gain, and deem the risk of punishment worth taking.

Refusal to cooperate with society need not involve obvious theft or the harming of other individuals. Another person's tax evasion has a negligible effect on me (or anyone else) personally; but for once that purse-lipped catchphrase of the nanny state – 'What if everyone did it?' – makes a valid point. The problem is as old as communal living. In the Middle Ages, villages had common ground where everyone could graze their livestock. The system depended on restraint. If someone grazed more animals than others did, they gained an advantage at the expense of a slightly faster depletion of the common grazing land. And if one person did it, others would follow their lead. The result was the 'tragedy of the commons', which were rapidly grazed bare. Are things any different now, with our seas over-fished to near depletion, our rivers polluted and our skies laden with greenhouse gases?

This is the basic problem for societies based on the principle of unenforced cooperation: they are ripe for exploitation by those who put self before community. So was Kropotkin's dream simply that? Is draconian legislation essential to prevent freeloading?

WHO SHOULD RUN THE WORLD?

There is no field in more urgent need of an answer to the problem of society's 'discontents' than that of international relations. For both Hobbes and Locke, a commonwealth of people becomes a single entity, a 'body politic' – a Leviathan. And Leviathan was a notoriously belligerent beast. Nations were like men in a State of Nature, since there was no overarching government to contain their actions. Will they then coexist, as Locke put it, in a state of peace, goodwill, mutual assistance and preservation, or a state of enmity, malice, violence and mutual destruction?

Hobbes had no doubts about the answer. For all his attempts at rigorous logic in analysing the interactions between individuals, when it came to the relations between states he threw up his hands in despair:

In all times, kings and persons of sovereign authority, because of the interdependency are in continual jealousies, and in the state and posture of gladiators, having their weapons pointing and their eyes fixed on one another;

that is, their forts, garrisons, and guns upon the frontiers of their kingdoms; and continual spyes upon their neighbours; which is the posture of war.[15]

The late seventeenth century, the dawn of the supposed Age of Reason, was marked by some of the most violent, petty, brutal and avaricious decisions ever made by the leaders of European states, and it seems that Hobbes regarded them as beyond hope of rational analysis. Yet how different is his description from the state of the world three hundred years later?

Viewed in this light, the case for world government – favoured by early twentieth-century scientific thinkers such as H. G. Wells and Leo Szilard – looks strong. Even if the majority of nations choose peace unilaterally, what can be done about rogue states in the absence of an enforceable legal framework that binds all? This is, of course, the purpose of the United Nations; but as we have recently witnessed, even Western democracies do not always observe the UN's dictates. The Declaration of Universal Human Rights is not enforced; nations disregard the International Court of Justice whenever it suits them. We cannot pretend there is a world government in anything like the same sense as there are national governments.

Could world government ever be established other than by war? And even if it were possible, would we want it? The idea of a World State goes back at least to Kant and Rousseau; but Kant rejected it, even though it would create a 'state of universal peace', because he feared this peace would soon become of a kind enforced by 'the most horrible despotism'.[16] Rousseau argued for a federal government to free Europe of its perpetual strife, but it was a far cry from today's European Union – capable of compelling every state to join and to do its bidding, an authoritarian vision which even the most fervent modern Europhobe cannot plausibly scare us with.

Robert Kagan of the Carnegie Endowment for International Peace in Washington, DC, argues that Europe now has its Kantian perpetual peace, and in a benign form – but that this has been made possible only because the USA has remained in a Hobbesian world where power matters above all else. While Europe advocates a world kept in check by international law and negotiation, argues Kagan, the reality is that at present only US military might maintains order. Europe's 'post-historical paradise' relies on the US remaining 'mired in history'.[17] Kagan's analysis is simplistic – he treats nations as though they were just so many Hobbesian brutes ready to wage war purely in pursuit of power, largely ignoring considerations of trade and economics, culture and history. (Some Islamic states are antagonistic towards the USA not because they want to conquer it, for example, but because they dislike its policies in the Middle East.) All the same, Kagan reminds us that the questions of international power politics debated in the Enlightenment are still as pressing, and as unanswered, today – and indeed that the terms of the debate have not shifted very much.

Can nations ever cooperate without a central authority? Why should they do so when powerful nations stand to gain by exploiting small ones? Is it better to act as an aggressive 'hawk' or a pacifistic 'dove'? Or should nations at least strive to *look* like hawks by accumulating massive military forces and nuclear arms? And what is the 'right' (Locke would have said 'natural') response to aggression? Can there be a just war? Game theory will not alone enable us to resolve these questions, but, by stripping them down to their bare bones, it can at least help us to formulate them more clearly.

WAR IN THE TRENCHES

We can start to explore such issues amidst the horrors of a war that few, if any, would now regard as just, or even sane. '1914 opens the age of massacre', says historian Eric Hobsbawm. 'Millions of men faced each other across the sandbagged parapets of the trenches under which they lived like, and with, rats and lice.' In this bloody stalemate, France lost one in five men of military age, and half a million Britons under thirty were killed. The Western Front, says Hobsbawm, 'naturally helped to brutalize both warfare and politics: if one could be conducted without counting the human or any other costs, why not the other?'[18] The total war of 1914–18 was arguably the precondition for Hiroshima and the Holocaust, for technologically assisted slaughter of civilians on an immense scale. Where, in all of this, is there scope to explore questions of cooperation and tolerance?

Actually, the answer is already the stuff of legend. Who does not know of the Christmas Day truces on the Western Front, when enemies greeted one another and played games of football between the mud and the barbed wire? And then, according to the legend, they returned to the war. But in fact many did not. On visiting the trenches in the middle of the fighting, one British officer was outraged at what he found. Not the squalor, the misery, the madness – but the casual attitude the men had to the business of winning the war. The officer confessed to being

astonished to observe German soldiers walking about within rifle range behind their own line. Our men appeared to take no notice. I privately made up my mind to do away with that sort of thing when we took over; such things should not be allowed. These people evidently did not know there was a war on. Both sides apparently believed in the policy of 'live and let live'.[19]

Of course, the soldiers knew all too well that there was a war on. They also knew that, as far as the Western Front was concerned, it was unwinnable. They were not acting out of cowardice, nor sloth or despair, nor even out of sporting regard for the enemy. They were simply being rational. They knew what was best for them.

This apparent laxity towards killing the enemy was, at first sight, dangerous as well as distinctly unmilitary. From time to time the Allied generals would order a push, compelling the troops to go 'over the top' and make a head-on assault on the German forces. If they had in the meantime been spending their days chatting, smoking and ignoring their 'duty' to eliminate enemy soldiers, they had to face greater numbers in these battles and casualties would be heavier. Surely it paid to kill the Germans at every opportunity?

But the point is that the Germans were in exactly the same position. So the choice was simple. Either you fought each other constantly, bombarding the enemy trenches with artillery fire and deploying snipers to pick off anyone foolish enough to show their head above the parapet – while enduring the same treatment from the other side. Or you held fire on the tacit understanding that the enemy would do the same – and meanwhile you hoped to be relieved from the front before the next push came. In the one case you endured fear of sudden death at every moment; in the other you had a quiet life and some hope of going home at the end of it.

Armies have always relied on propaganda to demonize the enemy and instil in the troops a loathing and hatred of their opponents, until the instruction to kill needs no enforcing. Many British soldiers no doubt went to the front despising the 'Boche', and vice versa. But this illusion becomes hard to sustain if you find that the enemy will leave you alone if you leave him alone – all the more so once you start to see how futile the fighting is. Indeed, some troops facing each other on the Western Front made formal truces. Naturally the military commands regarded such behaviour as treasonable, and court-martialled the culprits. But the live-and-let-live policy that flourished in the trenches did not have to depend on open compacts. It arose by stealth through the complicity of the opposing troops, to the fury and despair of the generals on both sides.

It was not – or it did not begin as – a humanitarian agreement not to kill. Quite the contrary: it was enforced by killing. Both sides came to realize that if they flouted an unwritten agreement by launching a bombardment in the hope of gaining some strategic advantage, they would get as good as they gave. 'If the British shelled the Germans,' says G. Belton Cobb in *Stand to Arms* (1916), 'the Germans replied, and the damage was equal: if the Germans bombed an advanced piece of trench and killed five Englishmen, an answering fusillade killed five Germans.'[20] The fighting was, in other words, conducted on a tit-for-tat basis. Such exchanges are a lethal form of communication: they say, 'We will do as we are done by.' This is at the same time both a threat and an olive branch, for it also implies that non-aggression will be greeted with the same.

Such live-and-let-live arrangements may have been triggered by basic necessities. Men must eat, and cannot fight and eat at the same time. Mealtimes thus provide a natural ceasefire. And one does not bombard the enemy supply wagons behind the front line (even though, ballistically speaking, it would not be hard) if one wants to avoid an equal response. In *Goodbye To All That*, Robert Graves recounts how each side would also sometimes hold its fire to allow soldiers to retrieve their dead and injured from no man's land.

Why doesn't all war end up this way? The very conditions that made the Western Front so terrible and futile – entrenched forces at deadlock, unable to advance – also promoted the limitation of hostilities. Normally enemy troops encounter each other as part of mobile armies or in sudden attacks by or between guerrilla groups. The same combatants don't enter into sustained confrontation with one another. In these circumstances it pays to kill as many of the enemy as you can. But if you know you are going to be facing the same enemy for a long time, you have a motive to cooperate instead of engaging in conflict: you can expect the same treatment yourself in the future. By the same token, businesses which deal with one another regularly are unlikely to cheat or default on payment because they know that the other business will have a chance to do the same to them in the future. If the deal is a one-off, an unscrupulous company can welch on it without fear of reprisal. In addition it takes time, in the absence of direct contact, to establish the terms of a live-and-let-live policy and to develop sufficient trust that the other side will adhere to it. Deadlocks grant that period of indirect negotiation.

Overt unsanctioned ceasefires in the trench war ran the risk of reprimand or of encouraging the High Command to force a unit onto the offensive with a direct order. So the soldiers at the front began to develop ways of appearing to be fighting while in fact doing no damage. Each side would recognize this for what it was: a way of sustaining the ceasefire while convincing their officers that conflict continued. These dummy attacks became not just a decoy to deceive the generals but a show of good will towards the 'enemy'. According to Tony Ashworth, a historian who has made a careful study of this live-and-let-live system in the First World War,

In trench war, a structure of ritualized aggression was a ceremony where antagonists participated in regular, reciprocal discharges of missiles, that is, bombs, bullets and so forth, which symbolized and strengthened, at one and the same time, both sentiments of fellow feelings, and beliefs that the enemy was a fellow sufferer.[21]

This ritual element of combat is in fact widespread in warfare and conflict. Among animals it can take the form of leadership contests where opponents engage in displays which obviate the need for life-threatening combat. Stags brandish their horns, and may proceed to the stage of locking them in push-and-shove; but only rarely does the confrontation progress to a lethal fight, even though both stags are armed with deadly weapons. These rituals are, however, designed to produce a winner. In the trenches they were designed to evade the senseless bloodshed that the generals believed to be necessary.

The live-and-let-live strategy of the trenches reputedly engendered not only tolerance and restraint but mutual goodwill between troops sent out to kill one another. But if this happened, it was a consequence, not a cause, of the cooperative behaviour. The soldier shooting into the air did not necessarily have moral compunctions about killing his enemy – he merely knew that this action increased his own chances of survival. Cooperation arose out of self-interest. We can recognize in a qualitative way how this could have come about. But to make a science of it, we need a tool developed, ironically enough, at a US military think-tank.

THE GRAND TOURNAMENT

You are on a train, and as you take your seat you find a wallet. It is fat with money. What do you do? At face value this is a simple case of binary choice, like those we looked at in Chapter 13. You can either try to return the wallet to its owner, for example by looking inside for an address or handing it to a member of staff, or you can place it quietly in your pocket.

The behavioural models we have looked at in earlier chapters have tended to assume that agents in a multi-agent scenario respond to the actions of their neighbours in what we might regard as a knee-jerk or at least a somewhat mechanical way. That is to say, stimulus A induces response B, either invariably or with a certain probability. But in a situation like this, choices aren't made so simply. True, some people are invariably honest and some invariably dishonest. But the terrain in between is not negotiated by the random throw of a die. What flashes through our mind, perhaps involuntarily, is the thought, 'Who would know?' And then, maybe, 'What if *I*'d lost my wallet, how would *I* feel?' In such cases, we weigh up our options according to some moral code – but that code is bedevilled by *temptation*.

Temptation is arguably the fundamental problem for human societies. It sometimes pays not to be the good, kind, considerate citizen but to rebel, to cheat, to fight, to do the dirty. If my neighbours are all meek and law-abiding, what is to stop me from appropriating some of their land, or goods, or cattle? A Hobbesian individual in a Hobbesian world is as miserable as everyone else. But a Hobbesian in Eden can run riot, amass a fortune, gorge himself, and fear no reprisal (unless he believes in God). Temptation is a part of the human condition, and that is the problem for all utopias: not everyone is nice, because sometimes crime pays.

It is not obvious how to devise a 'particle' that can be led into temptation. But in the 1950s Merrill Flood and Melvin Dresher at the RAND Corporation in California did more or less just that. They developed a simple mathematical model which incorporated the element of temptation into an interaction between two agents. The model was presented as a kind of game. Flood and Dresher were exploring the theory of games devised by the mathematical physicist John von Neumann in the 1920s. One of the most formidable mathematicians of the twentieth century, von Neumann helped to establish the theoretical basis of the computer and made crucial contributions to the Manhattan Project during the Second World War. He cultivated something of a reputation as a playboy genius, which his passion for gambling and poker did much to enhance. But von Neumann didn't just want to play these games – he wanted to understand them.

For sheer complexity, a mathematician can do no better than to study the game of chess. There is a sense, however, in which poker is much more challenging, for it incorporates the psychological element of bluffing. The question is not, as in chess, what the next best move is, but which move will anticipate, mislead or disconcert your opponent. The elements of risk and uncertainty in games such as poker led von Neumann to see a connection with economics, and in 1944 he set out his ideas in a book co-authored with economist Oskar Morgenstern, called *Theory of Games and Economic Behavior*.

The game devised by Flood and Dresher involved a gamble. It has become known as the Prisoner's Dilemma, and it introduced game theory into sociology, biology and political science. The game is played by two agents, who are depicted in the explanatory metaphor as prisoners suspected of committing a crime. Each is offered the inducement that if he testifies against the other, thereby securing the other's conviction, he will be set free. If neither agrees to testify, both will receive a sentence – but only a light one, because of the paucity of evidence. If both testify against the other, the sentences will be heavier – but not as heavy as that of the convicted party if only one of them testifies, since the evidence is equivocal.

The temptation is, of course, to testify against the other prisoner. But if both do so, they both get heavier sentences than if neither testifies. Should each prisoner refuse to testify, hoping that the other will do the same? If they both act 'rationally', then both should in fact testify against the other. This choice gives the best outcome for each agent no matter what choice the other prisoner makes. If prisoner 1 testifies and prisoner 2 does not, prisoner 1 gets off the charge – the best of all outcomes for him. If prisoner 2 testifies, on the other hand, then it is still better for prisoner 1 to testify than not to: he gets a lighter sentence in the former case than the latter. So the dilemma doesn't sound like a dilemma at all – it is always better to grass. The trouble is that, since this is equally true for both prisoners, it compels them both to testify and thus to be worse off than if they had cooperated with one another by not testifying.

The essence of the Prisoner's Dilemma can be expressed in terms of a choice either to 'cooperate' or to 'defect'. The best outcome – the maximal payoff – for one agent comes if he defects (testifies) while the other cooperates (that is, refuses to testify – the cooperation here is with the other prisoner, not the authorities). In that case, the other player is the sucker and gets the worst outcome. But if the agents play rationally, they get neither this optimal payoff nor the next best thing, which is the payoff from mutual cooperation. Instead, they get the meagre rewards of mutual defection, which are only a little better than the sucker's payoff.

To recast this dilemma in terms of individuals living in a society, we can regard cooperation as being law-abiding and defection as breaking the law for one's own gain at another's expense. The basic dilemma – that cooperation is good but defection can be even better – was recognized by Rousseau and Spinoza. In his *Discourse on the Origin of Inequality*, Rousseau imagined five men from pre-civilized times agreeing to cooperate on a stag hunt, on the understanding that each will get a fifth of the spoils. When a hare comes within reach of one of the men, he grabs it – but without his help, the stag escapes. The 'defector' has the immediate gratification of stewed hare, rather than sharing the difficulties and dangers of catching a stag – but his fellows have nothing.

At face value, the Prisoner's Dilemma seems to confirm Hobbes's pessimism: egoistic individuals guided by logic will always seek to exploit one another. What did it seem to say to the Cold Warriors who drew on the advice of the RAND Corporation? Get the first fist in, for your enemy will try to do the same. Build up your nuclear arsenal with all the resources you can muster, for your enemy is planning to defect, and so you must be ready to do so too. Indeed, you should even consider defecting first – making the first strike. If the other side 'cooperates' even to the extent of not launching a strike immediately, you can make them the sucker by doing so at once. You win; they lose.

This does not seem much of a basis for understanding how to construct a world in which people and nations live in harmony. Thankfully, there is more to say about the Prisoner's Dilemma.

The frustrating thing about this game is that the players – the prisoners, if you like – cannot communicate. It is obviously in their interests to agree to cooperate rather than to both defect. But since they cannot convey to each other a readiness to do this, they are better off assuming the worst of the other player, which implies that they must defect. If you play the game more than once, however, there is scope for communication of a kind. For even if the players cannot correspond directly, they can signal their intentions by the way they play. If one player reveals a willingness to cooperate by doing so in one round, the other player might decide to reciprocate in the next. The players, who, having both begun with ruthless defection, later begin to cooperate, find that they achieve better outcomes as a result.* They do not need to experience any sense of guilt or moral obligation in order to switch to cooperation. Pure self-interest is enough to make that the best choice.

This means that the impasse that compels defection from both players in a single round of the Prisoner's Dilemma can be broken simply by playing the game repeatedly. And that is how we commonly encounter comparable situations in real life. If I cheat on my neighbour, he has plenty of opportunity to retaliate. Most businesses deal again and again with the same clients. If two countries share a border, they cannot avoid ongoing political, economic and social interactions.

---

* In these repeated games a 'score' is kept of how well each player did on each round. A player scores highest for defecting when the other cooperates; moderately for mutual cooperation; poorly for mutual defection; and worst of all for cooperating when the other defects (the 'sucker's payoff'). One might imagine each player accumulating points or money, for example, rather than prisoners accumulating years of incarceration.

So in the repeated or *iterated* Prisoner's Dilemma game, the players have the chance to learn from their mistakes and to build up a relationship of mutual trust. Cooperation can evolve.

Is this how real people play the game? Psychologists have studied that question extensively in controlled tests, and found that co-operation does develop – but to a degree that varies widely, depending on the character of the players, the nature of the payoffs and the circumstances of the interaction. One can imagine, for example, that it is easier to defect anonymously than face to face. And let's not forget the element of temptation. If you think you are facing a nice player who will do their best to cooperate, you might be tempted to throw in an occasional defection, thereby boosting your own score at your oppo-nent's expense. If they are forgiving, you might get away with it if you don't try it too often. Sadly, in a cooperative world, defection pays. This then raises the question of what is the best way to play the iterated Prisoner's Dilemma. If you know nothing about your opponent, which strategy should you adopt?

In the late 1970s Robert Axelrod devised an experiment to try to answer this question. He asked professional game theorists to submit strategies for playing the iterated Prisoner's Dilemma, and then put each strategy to the test in a round-robin tournament conducted on a computer. Each strategy was played one-to-one against each of the others for many rounds; the winner was the strategy with the highest aggregate score. The fourteen entries came from psychologists, mathematicians, economists, sociologists and political scientists. Each strategy consisted of a set of rules for determining the choice of cooperation or defection. For example, one might simply choose always to cooperate. (This is obviously a bad choice, since it always comes off worst unless everyone else is also an unconditional co-operator. So no one chose this strategy.) Or one might cooperate on the whole but defect every fourth round. Many of the submitted strategies were more complex than this. But the tournament was won by the simplest of them all. It was submitted by Anatol Rapaport, who called it Tit For Tat. Its sole rule was to begin by cooperating, and thenceforth to do whatever its opponent did in the previous round.

Playing against an unconditional cooperator, Tit For Tat (TFT) cooperates throughout the entire encounter. So both players do equally well. Against an unconditional defector, TFT gets the sucker's payoff in the first round (where it cooperates), but then it defects consistently, as though determined not to be taken advantage of again. Because of the first round, TFT comes away slightly worse off than the inveterate defector – but only just. Both of them, in any event, do much worse than if they'd cooperated. By mirroring its opponent, TFT can adapt to whatever the situation calls for. Against cooperators it is nice; against defectors it is tough. When faced with a mixture of cooperation and defection, it gives as good as it gets. So TFT reaps the benefits of cooperation where possible, but cannot be exploited. Neither does it exploit: it never achieves a higher total payoff than its opponent. Some of the other strategies did well against those with a tendency to cooperate; others could hold their own against defectors. But by making the best of both situations, TFT came out on top in a diverse mixture of strategies. It was a modest, simple-minded victory.

Following the success of his first computer tournament, Axelrod decided to hold a second, with essentially the same rules as before. News had travelled, and this time there were sixty-two entries, from six countries. Some came again from professional scientists and academics, but others were from computer hobbyists, including a ten-year-old boy. All knew the outcome of the first round and so had a chance to consider the reasons for Tit For Tat's success. Anyone could submit any strategy, but only one person chose TFT: Rapaport. All the others decided that they could outdo TFT with something more sophisticated. They couldn't. TFT was again the winner.

Does this mean that Tit For Tat is the best way to play the iterated Prisoner's Dilemma? Not exactly. There is in fact no best way to play, for it depends who you're playing. It is very easy to illustrate that this is so. If you are playing against a colony of unconditional cooperators, you will do best to be an unconditional defector – that strategy will fare better than TFT, which would behave like one of the cooperators (except when it plays you). But the message of Axelrod's tournaments seemed to be that if you don't know who you're up against, TFT is the best default strategy.

What, then, makes Tit For Tat so special? For one thing, it is flexible: able to cooperate but not open to exploitation. Cooperation from the other player will immediately elicit cooperation from TFT in the next round. But defection is met unhesitatingly with defection. This sends out a clear message: TFT will do as it is done by. It is a strategy from the Old Testament, not the New: an eye for an eye, not turning the other cheek. This clarity of response is in itself a factor in TFT's favour. In the second tournament, one entry was a strategy designed to try to figure out the rules the other player is using, in order to find some way of exploiting them.* This often happens in real life: one person will check out the other, weighing up how much he or she can get away with. If you know beyond doubt that you can never defect without being treated the same way, you have a good incentive to cooperate. If you have reasons to doubt that the retaliation will be relentless, you might be tempted to try your luck. TFT, in contrast, guilelessly encourages cooperation and discourages defection.

But there is another telling aspect of Tit For Tat that contributed to its success: it is never the first to defect. All strategies can be broadly divided into two camps by this criterion: will they defect first or not?

* TFT takes the other player's actions into account too, but only to the extent of mirroring them – it does not seek any deeper understanding of the opponent's strategy.

Those that will not are generally called 'nice' strategies. (There is no consensus on what to call the others; but 'nasty' will do.) Axelrod found that nice strategies do consistently better than nasty ones. Indeed, in the first tournament the ranking produced a clear distinction: the eight top-scoring strategies were all nice, and the others, separated from the nice ones by a substantial gap in points scored, were all nasty.

THE SECRET OF COOPERATION

So the Prisoner's Dilemma starts to look less grim when it is iterated: niceness and cooperation fare better than nastiness and exploitation. Even individual selfishness need be no barrier to fair play. But being cooperative does not in itself guarantee success; Tit For Tat plays a much tougher game than that. Axelrod has identified four characteristics of a successful strategy:

> Don't be the first to defect (be nice).
> Always reciprocate.
> Don't be too clever.
> Don't be envious.

What is envy, in this context? It means not trying to do better than the other players, but simply doing as well as you can for yourself. The Prisoner's Dilemma is not what is called a zero-sum game: someone else's gain does not have to come at the expense of your loss. If you both cooperate, you can both do well (even if not as well as you would if there is an option to exploit the other players). Axelrod confesses that real players seem to find it hard to relinquish competitiveness and envy. In his tests of the iterated Prisoner's Dilemma with student volunteers, he finds that they tend to measure their performance against that of others, in which case any advantage gained can tempt the others into defecting to try to redress the balance. This can trigger recriminatory outbursts of defection.

The live-and-let-live behaviour in the trenches of the First World War can be seen as an example of cooperation arising from a Tit For Tat strategy. Belton Cobb's remark makes it clear that there was no compunction about retaliating in a lethal manner to hostilities from either side – mutual ceasefire did not depend on good feelings between enemies. (Yet, significantly, it seemed that self-interested cooperation allowed these feelings gradually to develop.) And on the whole both sides followed a 'nice' strategy, declining to fire first.

In case we should still wonder whether human sentiments, rather than the mathematical exigencies of game theory, led to this reciprocal cooperation, we might bear in mind that Tit For Tat strategies are also found in the natural world. There is evidence that vampire bats, stickleback fish, monkeys and even viruses behave according to the rules of TFT. No one can reasonably attribute altruism to viruses: their behaviour is purely the result of genetic selection. That is to say, those organisms with a genetic predisposition to show TFT-like behaviour gain an evolutionary advantage, and so natural selection works to their benefit – ensuring that this genetic trait becomes more widespread.

This implies that we too may be genetically hard-wired to co-operate, perhaps in a TFT-like manner. Indeed, it would be astonishing and puzzling if we were not. Edward O. Wilson argues that as civilization evolved, such modes of human behaviour will have become converted from instinctive impulses to social norms, then to legal imperatives, and ultimately to moral principles.

One could even argue that the case for a genetic embodiment of the lessons of the Prisoner's Dilemma is supported by the readiness with which we greet their optimistic aspects. We would be sorely dismayed if game theory were not capable of producing cooperative behaviour; in fact we might then be tempted to dismiss it as nonsense or as mendacious. We are, it seems, predisposed to look favourably upon altruism and to frown upon apparently selfish behaviour. That this might be a learnt response does not evade the issue; we learn it because those are the cultural norms of our society – and where did *they* come from?

Here, then, is a possible resolution to the divergent views of human nature evinced by Hobbes and Locke, which led them to such differing conclusions about systems of government. People do not, in the absence of a higher authority, necessarily seek to exploit one another in the way Hobbes envisaged. But neither do they desist from it because of a 'reason' instilled in them by God. The 'reason' can come from nature alone: from the inexorable mathematics of inter-action coupled to the winnowing effect of natural selection. The Prisoner's Dilemma is actually implicit in Hobbes's analysis, since he acknowledges the miseries of mutual defection and argues that men are better off cooperating if this can somehow be arranged:

That a man be willing, when others are so too, as farre-forth, as for Peace, and defence of himselfe he shall think it necessary, to lay down this right to all things; and be contented with so much liberty against other men, as he would allow other men against himselfe.[22]

This connection between *Leviathan*'s State of Nature and the Prisoner's Dilemma was pointed out in 1969 by political scientist David Gauthier. Without a contract to cooperate, says Hobbes, a man would 'expose himselfe to Prey'.[23] But such a contract is liable to dissolve unless there is some authority that can enforce it, since men's appetites will make them liable to defect the moment they see advantage in doing so. Thus, says Gauthier, Hobbes's omnipotent sovereign provides an escape from the Prisoner's Dilemma that men face in the State of Nature, since in a sovereignty defection no longer brings potential rewards but only certain punishment. Even if, as has been argued, it is somewhat misleading to cast Hobbes's scheme in game-theoretic terms when he had no interest in deducing the psychology of people faced with such behavioural dilemmas, it seems clear that Hobbes recognized the underlying problem that arises when antisocial actions offer potential rewards.

But game theory suggests that Hobbes's rather extreme solution – a capitulation of all individual powers and rights beyond self-preservation – may not be necessary. His error, if we may call it that, was to treat people as blind animals who cannot learn from experience – an 'experience' that can be handed down from previous generations as a genetic predisposition towards cooperation.

By the same token, we might expect to find other implications of game theory hard-wired into human experience. The tendency to form tribal groups increases the likelihood of repeated interactions with other group members and so enables cooperation to develop. Robert Axelrod endorses the notion of prolonged interaction – he calls it 'enlarging the shadow of the future'[24] – as a way of promoting and nurturing cooperative behaviour. The flipside of this principle is distrust of strangers, since it takes time to establish the mutual trust on which cooperation depends. But this apparent biological predisposition for xenophobia should be moderated by the realization that 'nice' strategies do best: even on the first encounter it is preferable to cooperate.

So by arranging to make future exchanges more probable, we can guide two parties towards the benefits of mutual trust. This could entail making a relationship more durable – it was the long-term confrontation of forces at the Western Front that made the tacit ceasefires possible. Or we might increase the rate of interactions: in small communities, the same people deal with one another day after day both socially and economically, and so trust is easier to establish than in large cities where interactions are more occasional and impersonal. Increased demographic mobility reduces the durability of interactions and so reduces the incentive for cooperation: transient neighbourhoods are rarely cohesive and 'neighbourly'.

It is clearly not news to businesses that their interests are served by developing good long-term relationships with clients. But the way in which such relations can break down gives us some reason to suspect that the reciprocity does indeed stem from a Prisoner's Dilemma style of exchange. A study in 1963 indicated that one of the commonest reasons for court action between businesses is the complaint of wrongful termination of a franchise. It is only when relations between businesses are about to cease – when the iterations of the 'game' are about to end – that one or both players decides it is worth their while to start a legal battle and risk bitter recrimination rather than to find a 'peaceful' way to resolve differences. In psychological tests, people who play the Prisoner's Dilemma will often sacrifice mutually established cooperation for a few rounds of defection when they know the game is about to finish. In the same way, companies about to go bust are at greatest risk from non-paying clients, and are themselves more likely to default on debts.

The durability of interactions has implications for modes of government. Karl Popper considers that the most important attribute of a true democracy is not what it does, but that it 'should keep open the possibility of getting rid of the government without bloodshed, if it should fail to respect its rights and duties, but also if we consider its policy to be bad or wrong.'[25] For as Pericles put it in democratic Athens, 'Even if only a few of us are capable of devising a policy or putting it into practice, all of us are capable of judging it.'[26] In a democracy, unpopular governments can be removed by elections – which seems incontestably proper. But the termination of any government carries some risks, for a departing government no longer has anything to lose from acting with blatant self-interest. Bill Clinton's outgoing US administration of 2000 demonstrated this with a display of political backhanding that the president would never have risked in mid-term.

To some extent this situation can be remedied by the existence of political parties, which carry long-term accountability for the short-termism of its members. There is no doubt that in 2002 the British Conservative Party was still paying the price for its deeply unpopular policies, its arrogance and its corruption while in office five years before, even though the perpetrators had largely disappeared from the political scene (several of them ignominiously). The US Republican Party paid the same long-term price for the Watergate affair in the 1970s. Thus a political system with a durable party structure might be expected to be less susceptible to corruption than one with more ephemeral kinds of political organization. Karl Popper called the party system 'horrible', since it makes parliamentarians tend primarily to serve their party rather than their constituents. 'I think', he wrote, 'that we should, if possible, go back to a state where MPs say: I am your representative, I belong to no party.'[27] But this could be a recipe for eliminating the accountability that political systems need if they are not to be plagued by abuses of power.

FOR YOUR OWN GOOD

It would be a foolish evolutionary psychologist who tried to argue that Tit For Tat is all there is to altruism. For one thing, the self-sacrifice we are capable of showing towards our own kin has biological roots which owe nothing to game theory – it seems to be an aspect of the 'selfish gene' idea, benefiting individuals who share a close genetic similarity to ourselves.

Moreover, behavioural economist Ernst Fehr and his colleague Simon Gächter in Switzerland have conducted experiments with human subjects which suggest that cooperation can arise in groups even when the individuals do *not* encounter one another repeatedly. Fehr and Gächter divided 240 students into groups of four, gave each of them an equal sum of money, and invited them to invest it (or not) in a group project. The project produced returns in proportion to the degree of investment. If all four members invested all their money, they all got a return that exceeded the investment. So it was in the *group's* interest for everyone to invest everything. But because each member received less than one monetary unit for investing each unit of his or her own, it was in each *individual's* interest not to invest but to freeload, relying on the contributions of the others.

This is analogous to a Prisoner's Dilemma insofar as it presents players with benefits for mutual cooperation but temptations towards individual defection. But Fehr and Gächter mixed up the groups after every round of investment and return, giving them no opportunity to establish mutual trust. They found that cooperation could nevertheless flourish if the rules included some provision for punishing defectors (those who invested little). Players would mete out such punishments even when they were charged a fee for doing so. Without the threat of punishment, cooperation was low; when this threat was introduced, cooperation increased sharply. The researchers call this 'altruistic punishment', since it is likely to be of no immediate and direct benefit to the punisher – even if punishment reforms the defector, the punisher is unlikely to encounter him or her again. It is nevertheless altruistic because it may benefit those others who will be grouped in later rounds with the player who defected.

This sort of behaviour suggests that the possibility of Tit-For-Tat-style retaliation or punishment may have a part to play in enforcing cooperation in society, even when encounters are not repeated. Fehr and Gächter reported that players seemed moved by a sense of injustice: they were simply angry at defectors, and acted on that anger irrespective of whether they stood to gain from it. Moreover, players reported that punishment was a deterrent to their own inclinations to defect. It is worth noting, however, that this is precisely the kind of behaviour one might expect from players predisposed towards a TFT strategy. By allowing punishment to be meted out after the game has been played, the researchers were in effect enacting a kind of two-round game, in which first-round defectors are themselves treated to defection from other players in the second round. The results could imply that we are all imbued with a desire for justice which we will exercise, if necessary, at our own expense.

In all these games, mutual cooperation pays best in the long term. If indeed cooperators are more 'successful', we would expect a pre-disposition towards cooperation to become an irreducible element of our neural circuitry. Sadly, however, this does not mean that Kropotkin was right: that no government is needed because people can be trusted to organize themselves. History shows us what people are capable of, and it does not look much like Eden. Human nature is diverse; it is also mutable, for better or worse. And it is influenced not just by one-to-one interactions, but by the multitudinous society in which each of us is embedded. To deal with that, game theory needs to get more sophisticated.

18

Pavlov's Victory

Is reciprocity good for us?

The fact is that a man who wants to act virtuously in every way necessarily comes to grief among so many who are not virtuous.

Niccolò Machiavelli (1513)

Can one wait with calm confidence for the day when the despotic states that have made wars in the past have been turned, by the social and economic forces of history, into peace-loving democracies? Are the forces of evolution moving fast enough? Are they even moving in the right direction?

Kenneth Waltz (1954)

In the interests of peace I am opposed to the so-called peace movement.

Karl Popper (1988)

*

If the Middle East today bears witness to harsh words and harsh deeds, it has done so before:

You must purge the evil from among you. The rest of the people will hear of this and be afraid, and never again will such an evil thing be done among you. Show no pity: life for life, eye for eye, tooth for tooth, hand for hand, foot for foot.[1]

The uncompromising reciprocity of a Tit For Tat rule may have worked for the children of Moses (although even for them it was not the lawful response to all insults). But can this really form the baseline of a civilized society? Game theory seems best suited to investigating the State of Nature that Hobbes feared as barbaric and Locke idealized as beneficent; it delivers the hopeful message that goodness can arise out of barbarism. But only at the cost, it seems, of uncompromising retaliation to all aggression. The whole point of government, says Locke, is to eliminate the necessity for every man to be his own judge and enforcer. But does that oblige a government to enforce co-operation among its subjects and its neighbours alike with the same policy of swift and unquestioning retribution? Where in this social calculus might we find room for negotiation, conciliation, mediation, even forgiveness?

To understand the implications of game theory at more than a superficial level, we need to subject Tit For Tat to a rigorous examination. That is my objective in this chapter. We shall circumscribe this strategy's advantages and probe its weaknesses. We shall release it into a community and watch the consequences. We shall ask – as we always must in social physics – not just what the models tell us but what we consider desirable, and whether the two can be reconciled. And so we shall come back to the enduring question: what choices do we have?

ACCIDENTS HAPPEN

After Axelrod's second computer tournament, Tit for Tat looked invulnerable. But it isn't. In the real world it has a fatal flaw: communications are imperfect. Mistakes are made; intentions are misunderstood. In 1983 the Soviet Union shot down a South Korean civilian aeroplane which had mistakenly strayed into Soviet air space, in the belief that it was a military craft. All 269 passengers, including several Americans, were killed. A Tit For Tat policy, rigidly applied, would dictate that this error could be avenged only with Russian blood. Fortunately it was not, although the incident did heighten Cold War tensions. The NATO bombing of the Chinese embassy in Belgrade during the attack on Serbian forces in 1998 was at face value another apparent 'defection' resulting from a mistake. (There is still debate about whether it was truly unintentional.)

The hair-trigger status of American and Soviet nuclear arsenals during the Cold War highlighted the awful risks of a retaliatory policy in the face of potential mistakes. Taken to its extreme, this creates the scenario gloriously and chillingly lampooned in Stanley Kubrick's film *Dr Strangelove*, in which a rogue US army general launches a pre-emptive strike on the Soviet Union. All but one of the B52s is recalled in time; but the one that cannot be contacted releases its warheads. This triggers global nuclear war even though the Soviets know the bombs were dropped 'by mistake', because they have automated their missile system with the Doomsday Machine, which retaliates to any nuclear attack without the option of human intervention. They believed that the absolute certainty of retaliation would enforce cooperation, but the system did not allow for errors.

The problem with mistakes, as far as Tit For Tat is concerned, is not simply that a lone, erroneous defection provokes the same in return. Tit For Tat's simplicity means that, if this happens between two players who are using this same strategy, they get locked into a cycle of mutual recrimination. One defects by mistake; in the next round it returns to cooperation (since that is what its opponent did in the last round), but the other player returns the defection. This causes the first player to defect in the round after that, and so on: the mistake echoes back and forth for the rest of the match, so that mutual cooperation is never restored. (In the *Strangelove* scenario, of course, a single round of defection from each side is enough to end the game once and for all.)

This sort of behaviour arises in many cultures and societies. Axelrod points to the example of family feuds in Albania and the Middle East, which can continue with mutual reprisals for many generations – even after the original incident has been long forgotten. Agonizing vicious cycles of mutual slaughter have plagued the Protestant and Catholic communities in Northern Ireland for decades, and currently seem to be destroying all hopes of a peaceful settlement between Israel and Palestine. Clearly, Tit For Tat does not guarantee a harmonious world.

Nor, amid the mess and confusion of reality, is it always the best strategy. This became apparent when Axelrod's tournament was repeated, this time allowing for the possibility that the players may make errors. The players would occasionally choose their response at random rather than according to the rules of their strategy. For an error rate (a 'noise' level) of 10 per cent – one in ten random choices – TFT is no longer the winner. In fact, TFT then fares even worse when playing against other TFT players than when playing a mixed bag of strategies, since occasional errors create ample scope for fruitless cycles of reprisal.

In such a situation, TFT needs modification if it is to score highly. One alternative, called Generous Tit For Tat (GTFT), lets a certain fraction of defections go unpunished. Another, Contrite Tit For Tat (CTFT), declines to retaliate to a defection that follows a defection of its own – it 'accepts' that it got what it deserved. GTFT outperforms all the other entrants in Axelrod's second tournament when there is 1 per cent 'noise'; CTFT comes sixth. For higher noise levels, CTFT outstrips GTFT. Tit For Two Tats (TFTT) is a strategy that retaliates only after suffering two consecutive defections: it waits to see whether a defection implies that the other player's intentions really are bad, rather than being simply a mistake (that is, noise). TFTT was devised by the evolutionary biologist John Maynard Smith, and it came twenty-fourth in Axelrod's second tournament. Maynard Smith did not enter it in the first tournament; if he had, it would have won, because that first mixture of strategies contained some which impaired TFT's performance by getting locked into mutual retaliation (even without errors). This reinforces the point that there is no best way to play the game.

Another strategy which copes well with noise is less benevolent. Pavlov is a strategy based on pure opportunism, and was named in 1988 (although invented earlier) by David Kraines of Duke University and Vivian Kraines of Meredith College, both in North Carolina. Its philosophy can be summarized as 'win-stay, lose-shift'. Like TFT, it bases its choice of action on what happened in the previous round. If it did well, it makes the same choice again; if it did poorly, it switches. 'Well' here means either the reward for mutual cooperation or the best payoff of all, that for unilateral defection. In short, Pavlov sustains behaviour that brings rewards but changes behaviour that brings punishment. This recalls the simple, conditioned responses of Russian physiologist Ivan Petrovich Pavlov's dogs.

With a tough customer like Tit For Tat, Pavlov is quite happy to cooperate. It doesn't fare well with an incorrigible defector – it will try to cooperate every other round. But Pavlov will mercilessly exploit a habitual cooperator once it realizes it can get away with it, whereas

TFT would nobly cooperate. Pavlov performs poorly against the contestants of Axelrod's original tournament – in 1965 Anatol Rapaport gave this strategy the dismissive label of 'simpleton'. And it doesn't do a great deal better even in the presence of noise. But it has the virtue of being able to recover quickly from an isolated error, and if the circumstances are right it can come into its own, as we shall see.

DARWIN'S ALGORITHMS

That history is a sound guide to policy is a cliché, although Friedrich Hegel doubted that nations and governments were so guided. Yet people, businesses, institutions and even countries surely do sometimes change their behaviour in the light of experience – just as the British and German troops whose job it was to eliminate one another on the Western Front ended up entering into unspoken truces of mutual self-preservation. Some lawbreakers *can* be reformed. It is this capacity for change that makes international relations both complex and worth arguing over. Some observers believed that Saddam Hussein's government in Iraq might not have been given inevitably to unconditional defection but could have been transformed into a more cooperative regime had it been engaged in dialogue rather than isolated with sanctions and then barraged with bombs.

One of the most interesting and important questions we can ask of the Prisoner's Dilemma is what kind of behaviour emerges when the players can evolve – when they are allowed to change their strategies. In reality, people apply all sorts of moral, ideological, habitual and whimsical criteria in deciding how to behave. But in the spirit of game theory, it is useful to begin by asking what players will do if they are purely pragmatic: that is, merely seeking to optimize their gains. It seems reasonable to assume that players will tend to adopt those strategies that are more successful. This can be simulated in Axelrod-style tournaments by including an evolutionary dimension. At the end of one full round-robin, for example, we might allow players to adopt a new strategy with a probability proportional to that strategy's overall score. In this way the more successful strategies will multiply, while ones that perform poorly will die out. It's not hard to see that this is a Darwinian 'survival of the fittest' scenario. It mimics the way in which genetic mutations spread in a population: those carrying a mutation that conveys a reproductive advantage generate more offspring, enhancing the prevalence of that 'adaptive' mutation.

Martin Nowak at the University of Oxford and Karl Sigmund at the University of Vienna conducted just such an experiment in game theory in 1992, with salutary results. They set up a diverse population of strategies, in all of which the choice of whether to cooperate or defect was determined by what the opponent did in the previous round. Some strategies were more inclined towards defection, others tended to cooperation. Nowak and Sigmund let them all compete against one another, and then altered the proportions of each of them in line with their relative successes. Naively, we might expect that this evolutionary model will be ruled by Tit For Tat, which appears generally to fare best in a mixed population. And this did at first seem to be the outcome. Early in the game the defectors had the upper hand; cooperative strategies died out, and the average payoff of the population fell towards the low payoff gained by mutual defection. But at some point a tiny band of TFT players began to grow rapidly until they dominated the population (Figure 18.1). This takeover was accompanied by an upsurge in cooperation and a rise in the average payoff.

The abruptness of this change is reminiscent of (although not strictly equivalent to) a phase transition. The rise of TFT is a collective effect, a result of many mutual interactions between players. Mutual defection eventually becomes so self-defeating that a small group of

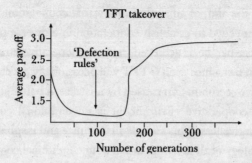

Figure 18.1 In a mixed population of Prisoner's Dilemma strategies subject to 'natural selection' of the most successful, Tit For Tat emerges from a population that has initially become mired in defection. This is accompanied by an abrupt rise in the average payoff for encounters between agents, since TFT allows a greater degree of cooperation.

TFT players gains more from their mutual cooperation than defectors do from exploiting the TFT players' initial attempts to cooperate. At this point the tables are turned, and it pays to cooperate with the doughty TFT group. Their presence helps seed the spread of cooperation throughout the population.* But Nowak and Sigmund found that the triumph of TFT is short-lived. Once it has established a culture of cooperation, TFT starts to suffer from its Achilles' heel: unforgivingness. These simulations contained an inherent amount of noise in the way the strategies worked, and this meant that TFT was gradually superseded by its more tolerant sibling, Generous Tit For Tat. In the end only GTFT remained.

Nowak and Sigmund concluded that 'Tit For Tat is the pivot,

---

* The crucial role of TFT-type strategies in bringing about this change is emphasized if the evolutionary game is replayed with no TFT players to start with. Then the prognosis is a gloomy one: cooperators die out and we are left with a colony of selfish defectors that plough their Hobbesian furrow to eternity.

rather than the aim, of an evolution towards cooperation.'[2] In other words, it is needed to establish cooperation in a diverse population, but once that has been achieved, 'softer' cooperative strategies will take over. In fact, since even GTFT will occasionally get caught up in unproductive recrimination caused by mistakes, a better strategy in a universally cooperative environment is unconditional cooperation: complete forgiveness. This sounds all very nice and inspirational. But the best strategy of all in a population of unconditional cooperators is unconditional defection: ruthless exploitation of the meek. Pitted like against like, cooperators do better than defectors, but cooperators are highly vulnerable to rogue defectors. A small band of defectors can wreak havoc in a cooperative culture. Tit For Tat can prevent this from happening, for it treats defectors severely while rewarding cooperation. It can be regarded as the police force of game theory, imposing cooperation with a firm hand. It models an ideal policing strategy in some respects, for (in the absence of noise) it only ever – and invariably – punishes defection, and never exploits cooperation. The implication seems to be that if we accept some level of defection as inevitable, we have to concede that a society needs at least some TFT players in order to maintain a general culture of cooperation.

Even that, however, may not guarantee a fair society. In 1993 Nowak and Sigmund discovered that TFT's implacable sense of justice does not always come out on top. In their earlier evolutionary games, players based their strategies for their next move on their opponent's previous move. But Pavlov, the opportunistic win-stay lose-shift strategy, does more than this: it takes into account the player's *own* last move too. When the two researchers pitched their earlier strategies against Pavlov, they found that Pavlov's opportunism triumphs. Pavlov does poorly against defectors and lacks TFT's ability to 'invade' a defecting population and spread cooperation. But in a (slightly noisy) community imbued with a spirit of cooperation,

Pavlov thrives. Nowak and Sigmund found that in such a circumstance Pavlov emerges victorious, even outstripping GTFT.

Both these strategies, Pavlov and GTFT, are somewhat tolerant to errors, unlike TFT. But Pavlov has another advantage. If in the model we allow strategies to randomly mutate into new forms, GTFT comes to share some of TFT's transience, becoming 'softened' by a gradual drift towards more unconditionally cooperative strategies. Pavlov, however, retains a hard edge. If it discovers by chance that it can get away with unilateral defection, it will continue to do so. So it is a wolf in sheep's clothing: it behaves well while cooperation becomes the norm under the firm authority of TFT, but it remains quite capable of exploiting a cooperative population once the TFT police have been transformed to unconditional cooperators. The motto of a Pavlovian society is no longer 'Do as you would be done by', but 'Never give a sucker an even break.'

The simulations from which Pavlov emerged as victor reveal a fascinating history. Because they involved the interplay of strategies which all based their next move in some way on the previous moves of *both* players, they took a more complex course than the earlier simulations. Most strikingly, there was far less of a sense of inevitability about the changes that took place over time (Figure 18.2). Each run of the simulation produced a different sequence of events. In the sample history shown here, there is an early attempt to establish cooperation: after a turbulent period this fails, and unconditional defectors reign for a long time. Then, after about 92,000 generations, the cooperators gain the upper hand.

This victory is short-lived and soon collapses into defection. Close inspection of the breakdown reveals that it is caused by the drift of TFT towards GTFT and thence to more forgiving strategies, creating a nation of 'softies' which is ultimately destroyed by rogue defectors. But this time the defectors are not quite unconditional: the dominant

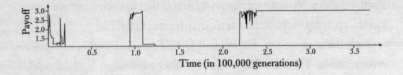

Figure 18.2 The changes in proportions of strategies over many generations of a Prisoner's Dilemma simulation are revealed by the changes in average payoff within the population: higher payoffs reflect a higher proportion of cooperative strategies. In these simulations the strategies evolve according to Darwinian selection, and they can take into account both their own last move and that of their opponent. There are several outbreaks of cooperation, the first two of which eventually collapse. The third is persistent.

strategy is instead one called Grim Trigger, which meets cooperation with cooperation – until it encounters a defection (as, in a noisy game, it inevitably must). Thereafter, Grim Trigger defects unconditionally. It is rather like *Dr Strangelove*'s Doomsday Machine. After about 220,000 generations, however, there is another resurgence of coopera-tion, which – after some initial adjustments – proves long-lasting. This switch is again triggered by TFT, but gradually it drifts towards a predominance of GTFT players before ultimately being taken over by mostly Pavlov agents, or close variants of it. This population is cooperative but potentially opportunistic, and is robust against invasion by defectors. It is not such a bad place to live – but the more virtuous of its citizens are not entirely safe from the threat of exploita-tion by superficially 'nice' Pavlovians.

These simulations display a mixture of chance and certainty. Cooperation always wins out if you wait long enough – there is always a happy(ish) ending. And Pavlov is not always the final dominant strategy, although it triumphs in about four cases out of five. Most remarkable, though, are the sudden revolutions that punctuate the course of events: we see 'good' and 'evil' empires rise and fall, and uprisings that falter and fail. Even in periods of apparent stability (for better or worse), the tally of strategies (and a detailed inspection of their characteristics) shows a certain amount of variation and a shifting of norms.

It is hard not to see in all of this an allegory of human history. Marx believed that the socialist revolution was inevitable. Game theory seems to be saying that nothing is so certain, since even if things *are* going to end up a particular way, we can't be sure just where along the evolutionary path we are at the moment. Did the players in the Second Cooperator Uprising (generation 92,000) think that the Age of Perpetual Cooperation, long forecast by the martyred philosophers of the failed First Uprising, had finally arrived? Were the commentators of the Third Cooperator Uprising (gen. 220,000) right to conclude that this was the 'end of history'?

MAGIC CARPETS

Empires rise and fall not only in time but in space. Rome once ruled from Portugal to the Black Sea, from the Scottish borders to North Africa. Charlemagne's Frankish realm reached deep into Germany, Italy and the Balkans. The Ottoman Empire overran lands from Transylvania to Egypt. The history of the world is a patchwork of borders, growing and shrinking. Imperialism seems mercifully to be a thing of the past, but the borders of NATO and of Europe are still liable to shift, and the maps of the Central and East European nations have changed more in the past ten years than in any other decade since the end of the Second World War. Does the Prisoner's Dilemma have anything to tell us about the way national and international boundaries move?

Introducing the element of space into the contests of game theory is no trivial matter, since it places constraints on the interactions that each player can have, and can therefore strongly influence the outcomes. In a tournament where everyone plays against everyone else, cooperators have the chance of deriving mutual benefit from their interactions; if they are on opposite sides of a map, they can no longer draw strength from the group and may be overwhelmed by defectors. So isolation can militate against cooperation. Israel's geographical location, surrounded by predominantly Islamic states, surely contributes to its self-perception as an embattled nation, and its supporters might argue that this situation prevents the country from being able to adopt the kind of conciliatory policies that European nations can afford to pursue. On the other hand, players in fixed locations may find a greater incentive to cooperate than do itinerant players, since they are compelled to interact repeatedly with their neighbours rather than moving on after one exchange. One of the difficulties faced by travelling people is that they have little opportunity to establish mutually trusting (and trustworthy) relations with those they encounter on their journeys – the 'shadow of the future' is not long enough.

Axelrod began to explore the notion of territoriality in the Prisoner's Dilemma in the 1980s. He considered a chessboard-like world in which each player occupies one grid space and interacts with the four neighbours with whom it shares an edge. Colonization of the board by successful strategies can take place by means of an 'evolutionary' mechanism. In each round of the game, all players play against all four of their neighbours. If one or more neighbours of a particular player gets a better score, the player converts to the strategy that was most successful.

Axelrod was primarily interested in how cooperation can spread in an exploitative society by means of Tit For Tat – or conversely, how a cooperative society can be undermined by defectors. He found that, for certain values of the payoffs, a single defector can spread its baleful

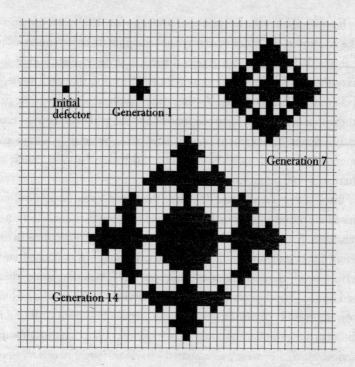

Figure 18.3 The changing shape of a colony of defectors (black squares) growing in a population of Tit for Tat players (white squares).

influence throughout a community of Tit For Tat players, providing the seed from which defecting strategies gradually expand. Curiously, however, the growing colony of defectors* is not so much like a spreading stain as like a snowflake: it sends out branches which split and rejoin to form a complex tapestry (Figure 18.3).

* Strictly speaking, the 'spreading of defectors' is really the spread of the use of defection strategies among players that stay immobile at each grid point. But one could equally choose to regard this as the replacement of cooperative players by defecting players, as though the 'defectors' kill off the 'cooperators' and colonize their grid spaces. The two perspectives are entirely equivalent.

If one considers the full gamut of strategies involved in a tournament like Axelrod's, the number of ways of arranging them into juxtaposed 'nations' is immense. But Axelrod tried it anyway, distributing the sixty-three strategies at random on a 14 × 18 grid so that each strategy was awarded four grid cells. As the contest evolved, all nasty strategies (those that defect first) were eliminated, and the grid ended up covered with a frozen arrangement of nice strategies. Since they always co-operated with their neighbours, none did better than any other, so there was no tendency for further change.

But Tit For Tat did not dominate the board, even though it dominated the round-robin tournament staged between the same selection of strategies. Each distinct initial arrangement of strategies produced a different final configuration, but all these configurations contained large territories which used strategies other than TFT, including some which had hitherto fared rather poorly in the tournament (Figure 18.4). Some 'less successful' strategies were able to survive and even to thrive in the spatial contest because they were lucky enough to have neighbours against which they were well matched – it was no longer necessary to do well against all comers, but only against one's neighbours. TFT is itself not an inherently expansionist strategy: it has no inclination to invade cooperative, nice territories, but only defective ones.

One might interpret this as implying that there is no 'absolute' best way to conduct one's affairs with other nations, except for the proviso that it pays to adopt a cooperative demeanour as one's default position. Moreover, the eventual distribution of strategies depends on where the map starts from: once again, history matters.

In 1992 Martin Nowak and Robert May at the University of Oxford devised a simpler spatial game to explore the effect of space on cooperation. In their version of the Prisoner's Dilemma there was no allowance even for strategies with the low level of complexity of Tit For Tat; each player was either an uncompromising defector or a

6	6	6	1	44	44	44	44	44	6	6	7	7	6	7	6	6	6
6	1	31	1	1	44	44	44	44	3	6	3	6	6	6	6	6	6
6	6	31	31	1	1	1	1	1	1	3	3	3	52	52	6	6	6
6	1	31	31	31	31	31	31	31	31	31	3	3	6	6	6	6	6
6	9	31	31	31	31	31	31	31	31	31	31	3	6	6	6	6	6
6	31	31	31	31	31	31	31	31	31	31	31	31	6	6	6	6	6
6	31	31	6	6	9	31	31	6	9	41	31	31	6	6	31	31	6
6	31	31	6	6	9	9	9	6	41	41	31	4	31	31	31	31	6
6	31	31	9	9	9	9	9	6	41	41	41	41	31	31	31	31	31
6	31	6	9	9	9	6	6	6	41	41	41	17	31	31	31	31	31
6	6	9	7	9	9	6	6	6	41	41	41	41	31	31	31	31	7
6	6	7	7	7	6	6	6	9	41	41	7	7	7	7	7	7	6
6	6	7	7	7	6	6	6	6	41	6	7	7	7	7	7	7	6
6	6	7	7	6	6	44	6	6	6	6	7	7	7	7	6	6	6

Figure 18.4 The territories claimed by different strategies in a Prisoner's Dilemma tournament enacted on a grid. The numbers indicate the ranking of each strategy in Axelrod's original round-robin tournament. All of these strategies are 'nice' – they are never the first to defect.

persistent cooperator. But again, players could switch from one strategy to the other if it was beneficial to do so. The players were distributed on a square grid, and each interacted with all eight of their neighbours – those with shared corners as well as edges. Each player's behaviour in a round copied that of the most successful of these nine players in the previous round: they cooperated or defected depending on which was locally the best option. This is a kind of two-state lattice model of interacting particles, reminiscent of the Ising model used in statistical physics (page 111).

How the game plays out depends on the size of the various payoffs, which may bias the interactions in favour of cooperation or defection. If the bonus for defecting in the face of cooperation is small, co-operation dominates; but this benevolent landscape is threaded with chains of defectors which shift a little from one round to the next

(Figure 18.5a). If defection is more lucrative, however, it becomes the commonest mode of behaviour. Yet small islands of cooperation constantly appear and vanish across the grid (Figure 18.5b). The fraction of cooperators now quickly evolves to a stable average level, irrespective of the starting configuration – there is a kind of 'irrepressible' tendency to cooperate even in a relatively selfish environment. Thus defection and cooperation need not inevitably annihilate the other, but can coexist indefinitely in patterns which are unpredictable in detail but entirely predictable in terms of averages.

The spread of defection and the spread of cooperation are not equivalent. Cooperators do best in dense clusters, where they can benefit from their mutual support. As Edmund Burke put it in 1770, 'When bad men combine, the good must associate; else they will fall one by one, an unpitied sacrifice in a contemptible struggle.'[3] But defectors in the midst of cooperators do best on their own, since they do far better to interact with cooperators than with other defectors. So although defection breeds more defection, the exploiters tend to repel one another, which leads to the formation of the thin threads of defection. Nowak and May found that a single defector spreads through a cooperative colony in much the same kind of snowflake formations as Axelrod had seen (Figure 18.5c). Again, these ramified 'magic carpet' patterns can be regarded as the result of the 'repulsion' between defectors, which militates against their forming a dense colony. You might say that each defector prefers to find its own patch, as isolated as possible from its rivals.

GOVERNED BY REASON?

The Tit For Tat policy and its more generous variants have defined most thinking about how cooperation evolves. Although TFT can

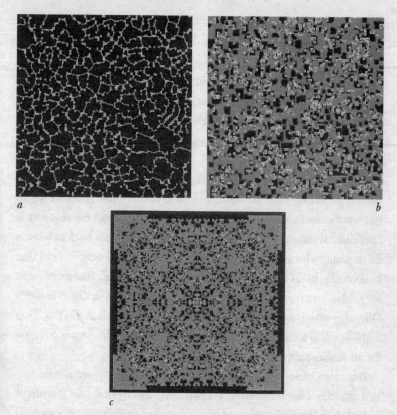

a

b

c

Figure 18.5 Spatial games between unconditional cooperators (black squares) and defectors (grey squares) show a complex behaviour which depends on how well it pays to defect against a cooperator. If the payoff is small, cooperation dominates (*a*). (Here, white squares show sites that have changed from cooperators to defectors in the previous round – that is, those sites where the boundaries are shifting.) For a larger payoff, defectors are more prevalent but communities of cooperators constantly arise and decline (*b*). A single defector can exploit a community of cooperators, seeding a community of defectors. Because of their mutual repulsion, the defectors grow not as a solid mass but in a ramified pattern (*c*).

yield to softer 'nice' strategies or to the opportunistic Pavlov once cooperation is established, there is still no better way to initiate cooperation in a Hobbesian world of exploitation than to meet tit with tat. On this basis, some argue that firm and immediate reprisals for bad behaviour are the only way to make 'rogue states' act responsibly: hence the recent bombings of Serbian Belgrade and the invasions of Afghanistan and Iraq. Popper put this conclusion into words which sound almost shockingly brutal, coming from one who had a reputation as a liberal: 'What is happening in Bosnia is proof of the failure, the cowardice, the blindness, of us in the West. It shows we do not want to learn what this century should have taught us: that war is prevented with war.'[4] The idea of 'preventive war' goes back at least as far as Kant, who advocated it in his essay 'Perpetual peace' – not that kings and princes in times past needed any philosophical endorsement for taking up arms. It is certainly true that strategies in the Prisoner's Dilemma which make softer or delayed reprisals, such as Tit For Two Tats, fare less well in a diverse population. By itself this seems to argue for air strikes rather than sanctions.

But as we've seen, TFT has its drawbacks too. It is painfully evident in both the Israel–Palestine conflict and the troubles in Northern Ireland how reprisals can hold back progress towards cooperation and peace. They can simply serve to undermine the establishment of trust. There is clearly a need for mechanisms that can 'damp out' the echoing cycle of retaliation, if good relations are ever to be resumed between TFT players who, for whatever reason, have broken rank. The Contrite or Generous Tit For Tat strategies offer some solutions; another is to operate a Partial Tit for Tat approach, where the reprisal is slightly less severe than the event that provoked it.

Human nature seems likely to complicate attempts to generate cooperation via TFT exchanges. It would be foolish indeed to ignore the strength of passions or the longevity of resentments when human lives are lost in episodes of 'defection', while TFT remembers nothing beyond the previous round and will 'forgive' at the first sign of cooperation. And consider the proposal in the United States that the custodial parent in separated couples who have children might be given the right to deny visits if the other parent does not keep up maintenance payments. This scope for retaliation could be regarded as an incentive for the maintenance-payer to co-operate. But quite aside from the fact that this scheme would logically have to allow the converse (payments legally withheld if access to the child was denied), it would be short-sighted not to take into account the irrational behaviour that can arise in partnership breakdown and arguments over childcare responsibilities, which may override the ability of the 'players' to decide dispassionately what course of action is most advantageous to them in the long run. More pertinent still is the matter of whether there can possibly be any justification for letting children become bargaining chips in such exchanges.

The threat of a Tit For Tat response has been advanced, both implicitly and explicitly, as a theoretical rationale for the policy of nuclear deterrence. The argument here is that, even if the awful prospect of an exchange between nuclear powers does not materialize, the evident capacity and stated readiness of a nation to retaliate to such an attack is an essential element of a peaceful status quo. Popper again supported the notion: 'We should have learnt by now that peace on earth needs to be backed up with weapons . . . You could never get peace inside a country by reaching a compromise with the criminals.'[5] This might be valid for war in general; but for nuclear war – and in particular the Cold War concept of mutually assured destruction – the iterative process that is essential for TFT's superiority (indeed, for its very definition) is not an option.

Strategist Hermann Kahn describes with compelling clarity the kind of confused and irrational thinking he often encountered at the RAND Corporation during the early years of the Cold War:

One *Gedanken* [thought] experiment that I have used many times and in many variations over the last twenty-five or thirty years begins with the statement: 'Let us assume that the President of the United States has just been informed that a multi-megaton bomb has been dropped on New York City. What do you think that he would do?' When this was first asked in the mid-1950s, the usual answer was 'Press every button for launching nuclear forces and go home.' The dialogue between the audience and myself continued more or less as follows:

KAHN: What happens next?

AUDIENCE: The Soviets do the same!

KAHN: And then what happens?

AUDIENCE: Nothing. Both sides have been destroyed.

KAHN: Why then did the American President do this?

A general rethinking of the issue would follow, and the audience would conclude that perhaps the President should not launch an immediate all-out retaliatory attack.[6]

In other words, according to political theorist Brian Skyrms, 'A strategy that includes a threat that would not be in the agent's interest to carry out were she called upon to do so, and which she would have the option of not carrying out, is a defective strategy.'[7]

In any event, it would be naive in the extreme to assume that real players in the international politics game will display the rigorous rationality of the idealized players in game theory. Supporters of an 'assured Tit For Tat' deterrent, for example, must grapple with the strong possibility (as Popper admitted) that Khrushchev had every intention of using his missiles in 1962, had he been able to get enough of them into Cuba secretly – deterrent or no deterrent. Looked at in this light, TFT simply creates the climate and the conditions for such a crisis. Moreover, the spectacle of many long-term adherents of the deterrence argument for American nuclear proliferation who are now advocating a missile defence system which undermines the very basis of this argument – what is mutually assured destruction if it is not mutual? – should remind us that the formal logic of game theory is no more than an expendable tool in the face of political ideology and expediency.

Another of the less attractive implications of Tit For Tat is that in effect it compels cooperation from defectors only if they 'know' how implacable it is. This implies that one needs to acquire a reputation for being ready to go on the offensive, which can become manifest as hypersensitivity – refusing to tolerate the slightest perceived threat or insult – or as a bullying propensity to adopt gunboat diplomacy. In the 1960s the United States was prepared to fight a bitter war on the other side of the world primarily to maintain, during the height of the Cold War, its reputation for toughness. This was admitted in a memo sent by John McNaughton, Assistant Secretary for International Security Affairs, to Secretary of Defense Robert McNamara, outlining US aims in Vietnam: they were '70 percent – To avoid a humiliating US defeat to our *reputation* as a guarantor)' (my italics), and only '10 percent – To permit the people of [South Vietnam] to enjoy a better, freer way of life.'[8]

In other words, the success of Tit For Tat could be regarded as an incentive to act belligerently. After all, the disastrous consequences of facing an opponent who does not appreciate one's ruthlessness were brought home in a gloriously sardonic manner in Kubrick's classic study of Cold War diplomacy, when Dr Strangelove explodes down the phone at his Soviet counterpart, 'You fools! A Doomsday Machine isn't any good if you don't tell anyone you have it!'

Anyone who considers using the Prisoner's Dilemma as a basis for deciding policy should feel duty-bound first to enumerate all the factors it neglects. Most obviously, as I have indicated, it takes a highly simplistic view of human nature: the assumption that people act rationally to seek the best gain for themselves neglects not only irrational passions, the fallibility of our reasoning powers and sheer foolishness, but also the positive influence of moral codes of conduct. Both experience and evolutionary biology lead us to expect that many people have an innate instinct to cooperate with their fellow beings, and do not have to learn that this serves their best interests before they will do it. On the other hand, a few people are probably pathologically inclined to defect within society, sometimes even when they see it does them no good in the long run. Moreover, the Prisoner's Dilemma provides no scope for negotiation: the 'prisoners', remember, are not allowed to collude, but must deduce each other's motives only from the way they play the game. Suspicion thrives in such circumstances, and in real life we generally seek to quell it by conducting our transactions more collaboratively.

All the same, there is no denying the strong implication from game theory that an unswervingly retaliatory strategy is the best way to bring about cooperation. As game theorist Karl Sigmund says:

It would of course be rather silly to attempt to reduce all human interactions to the Iterated Prisoner's Dilemma, or to negate the role of superior authority in civilized communities. But with all due restraint, it is worth pointing out that the brutally simple principle of paying back in kind leads to cooperation in a society of egoists, while the apparently higher summons to dispense with reprisals undermines such cooperation . . . The harsh law of retaliation seems to have been the foundation stone of many, possibly all, stable societies.[9]

At the risk of posing a question which evolutionary psychologists might regard as tautologous, one surely has the obligation then to ask: is this a moral way to behave?

560

The notion of Tit For Tat sits uncomfortably with liberal sensibilities. 'I came to this project', said Robert Axelrod in 1984, 'believing one should be slow to anger. The results of the Computer Tournament for the Prisoner's Dilemma demonstrate that it is actually better to respond quickly to a provocation.'[10] But pacifist thinkers from St Francis to Gandhi have asserted that to meet violence with violence is self-defeating. That is surely the message of the New Testament: love thine enemy, for it is the meek, not the vengeful, who shall inherit the earth. Many pacifists would argue that non-violence is a choice based not on cold logic but on higher moral imperatives, such as 'Thou shalt not kill.' When faced with behaviour that can only be considered as exploitative, even murderously so, this can induce agonies of doubt which, if honestly confronted, belie any suggestion that pacifism is a soft option. David Jones, a conscientious objector in the Second World War, explained the dilemma:

The pain of being a conscientious objector was the increasing knowledge of the enormity of what the Germans were doing. So that was the really challenging thing, not the war, so much as how one can justify not somehow trying to do something about it.[11]

Cecil Davies, a kindred spirit in the same conflict, found the question irresolvable:

Wilfred Owen said he was a conscientious objector with a bad conscience, and I think lots of COs often had a bad conscience, and while I still think I was right to do what I did when I did, I suppose if I had known about the Holocaust it might have been different . . . Life isn't simple.[12]

Indeed not – and the Prisoner's Dilemma should not let us pretend otherwise. But one cannot avoid the conclusion that, *on its own terms,* game theory implies that a retaliatory policy to defection might truly be more 'moral' insofar as it serves the greater good. A Tit For Tat strategy not only protects oneself from exploitation but also helps to safeguard the entire community. For it is only when TFT is eroded by softer cooperative strategies that the community becomes endangered by exploitative defectors or Pavlovian opportunists. Unconditional cooperation might seem more noble and kindly but it also places the burden of policing on the rest of the community. A utilitarian would be hard pressed to find an objection to TFT.

All of this seems like knowledge worth having. But we must beware of how easily it might be twisted or misconstrued. Naively, one might argue that TFT justifies the death penalty for murder – even if we allow that the state, not the individual, should implement the punishment. But the whole point of the iterated Prisoner's Dilemma is that it teaches cooperation through experience and adaptation: defectors convert to nice strategies because they are better off for it. Capital punishment simply eliminates the player, so that no further iterations take place. There is nothing in the Prisoner's Dilemma which suggests that one player learns from the mistakes of *another* – that the death of one defector warns others away from defecting. This could happen, of course, but game theory is silent about it – so the success of TFT is irrelevant to debates about the death penalty. The same surely applies to nuclear stockpiling during the height of the Cold War. Game theory and the Prisoner's Dilemma were very popular at the Pentagon, but one could hardly draw lessons from TFT when the game was, by its very nature, one that would be over in a single reciprocal exchange.

John Locke had the wisdom to see that, while he advocated the 'tooth for a tooth' form of Tit For Tat in his State of Nature, it would not do in a civilized society, and that relinquishing matters of policing, justice and punishment to the state both lightens the burden on the individual and lessens the need for savagery in enacting them. In an anarchic state you would not, in general, have the option of protecting yourself from assault by locking your would-be assailant away, and so you may be forced to do them harm. By implying that punishment is necessary to maintain a peaceable society, the success of TFT does not say much that will surprise anyone. It cannot tell us a great deal about the form that punishment should take, nor can it prescribe a course that will convert a sinner into a saint.

What the Prisoner's Dilemma does is help us to move beyond the pessimism of Hobbes without recourse to the rose-tinted assumptions of Locke. If we know that cooperation is possible, even in a world that lacks altruism, we have no reason to despair. Getting there is another matter, on which it is appropriate to let Popper have the last word: 'We should cautiously feel the ground ahead of us, as cockroaches do, and try to reach the truth in all modesty.'[13]

19

Towards Utopia?

Heaven, hell, and social planning

All successful men have agreed in one thing – they were *causationists*. They believed that things went not by luck, but by law; that there was not a weak or a cracked link in the chain that joins the first and the last of things . . . The most valiant men are the best believers in the tension of the laws. 'All the great captains', said Bonaparte, 'have performed vast achievements by conforming with the rules of the art – by adjusting efforts to obstacles.'

Ralph Waldo Emerson (1860)

[I]t may be that the next great developments in the social sciences will come not from professed social scientists but from people trained in other fields.

George Lundberg (1939)

We know that going to the Moon was a simple task indeed, compared with some others we have set for ourselves, such as creating a humane society or a peaceful world.

Herbert Simon (1996)

'We don't impose Heaven on people any more', she said. 'We listen to their needs. If they want it, they can have it; if not, not. And then of course they get the sort of Heaven they want.'

'And what sort do they want on the whole?'

'Well, they want a continuation of life, that's what we find. But . . . better, needless to say.'

'Sex, golf, shopping, dinner, meeting famous people and not feeling bad?' I asked, a bit defensively.

'It varies. But if I were being honest, I'd say that it doesn't vary all that much.'

Julian Barnes (1989)

*

Utopias come in many forms – as many, perhaps, as the human mind is capable of devising. Some have been socialist and egalitarian, some hierarchical and dictatorial. In some, people commit no crimes; in others all miscreants are exterminated. Some are ruled by women, or have no men at all, or alternatively no women. Sex plays a remarkably large role in many of them. Virtually all defy Voltaire's Dr Pangloss in deviating markedly from the world in which we live – which is perhaps unsurprising when we remember that one way of understanding the word 'utopia' is 'any place but here'.

Very few utopian visions have actually been realized, and those that have been tended to founder abruptly and often squalidly. The Civil War that prompted *Leviathan* also spawned the Diggers, a group of communist radicals led by Gerrard Winstanley, who in 1649 cultivated wasteland to escape the poverty endemic to rural England. Their colony was hounded by landlords and by Cromwell's army; within a year it collapsed. The noble experiment of Robert Owen's socialist commune at New Lanark in Scotland in the early nineteenth century enjoyed some success, but could not outlive its charismatic founder in the face of ruthless Victorian industrialism. (The Co-operative Societies formed to promote Owen's ideas have proved more robust.) The greater experiment in socialism that began with the October Revolution of 1917 has rather less to be proud of today. Marx and Engels condemned utopias as foolish fantasies because they ignored the historical inevitability of capitalism's demise and socialism's ascendancy. But the Marxist vision is, in retrospect, no less an expression of an arbitrary world considered desirable by its creators. And it was largely among the European socialists of the 1930s, such as the biologist Julian Huxley, that eugenics gained favour, until it became a part of the terrible society outlined in *Mein Kampf*.

That is the problem with utopias: one can never quite be sure how they will turn out. Human nature is the unpredictable element that sends the best laid plans awry, which is no doubt why, as the critic John Carey has put it, 'the aim of all utopias, to a greater or lesser extent, is to eliminate real people.'[1] Many of them are populated by mild-mannered, passively happy citizens among whom dissent and unrest have been savagely repressed, banished or somehow spirited away. Revealingly, the same characteristics appear in many dystopias, such as Aldous Huxley's *Brave New World*, where bovine passivity is maintained by biological engineering, the provision of free sex and drugs, and mindless leisure activities. One might argue that it is in this sense that traditional microeconomic theory is a utopian vision too, replacing real people with omniscient, rationally maximizing automata so that the market can seek its fictitious equilibrium.

The most striking aspect of utopias through history is that very few of them are at all appealing from today's perspective. Plato's *Republic* is a bellicose, class-ridden aristocracy in which censorship is the norm, art is discouraged and children are removed from their parents at birth. Thomas More's original *Utopia* might have been enticing to the sixteenth-century commoner, with its absence of poverty and equality of rights, but it comes at the cost of strict regimentation and conformity. The world of William Morris's *News from Nowhere* (that is, news from Utopia) looks pleasant enough but it is a naive romantic fantasy whose architect has made little attempt to hide the strings from which it hangs. The America of the year 2000 as idealized in Edward Bellamy's *Looking Backward* (1888) is chillingly plausible: a land of nationalistic conformity enforced by the Religion of Solidarity, run as a vast business syndicate in which all citizens are compelled to enrol in the militaristic 'industrial army'. Dissidents are sentenced to solitary confinement on a diet of bread and water. Bellamy evidently thought that this was all a splendid idea.

And so it goes on, underscoring the old truth that one man's utopia is another's dystopia – for there is too much variety in human nature for it all to fit one mould. 'What has always made the state a hell on earth', says the poet Friedrich Hölderlin, 'has been precisely that man has tried to make it his heaven.'[2] The notion that we could ever construct a scientific 'utopia theory' is, then, doomed to absurdity. Certainly, a 'physics of society' can provide nothing of the sort. One does not build an ideal world from scientifically based traffic planning, market analysis, criminology, network design, game theory and the gamut of other ideas discussed in this book. Concepts and models drawn from physics are almost certainly going to find their way into other areas of social science, but they are not going to provide a comprehensive theory of society, nor are they going to make traditional sociology, economics or political science redundant. The skill lies in deciding where a mechanistic, quantitative model is appropriate for describing human behaviour, and where it is likely to produce nothing but a grotesque caricature. This is a skill that is still being acquired, and it is likely that there will be embarrassments along the way.

But properly and judiciously applied, physical science can furnish some valuable tools in areas such as social, economic and civic planning, international negotiation and legislation. It may help us to avoid bad decisions; if we are lucky, it will give us some foresight. If there are emergent laws of traffic, of pedestrian motions, of network topologies, of urban growth, we need to know them in order to plan effectively. Once we acknowledge the universality displayed in the physical world, it should come as no surprise that the world of human social affairs is not necessarily a tabula rasa, open to all options.

Society is complex but that does not place it beyond our ken. As we have seen, complexity of form and organization can arise from simple underlying principles if they are followed simultaneously by a great many individuals. John Stuart Mill already recognized this in the nineteenth century:

The complexity does not arise from the number of the laws themselves, which is not remarkably great, but from the extraordinary number and variety of the data or elements – of the agents which, in obedience to that small number of laws, cooperate towards the effect.[3]

On the other hand, complexity is not the inevitable result of a multitude of interactions. Perhaps that is one reason to be wary of making complexity a buzzword for a new type of science. The real surprise to emerge from the physics of society is that patterns of social behaviour can sometimes be so simple, so observant of mathematical 'laws'. The steady averages and recurrent error laws of the pioneers in social statistics arouse no fears today that free will has been demolished; but the new discoveries of social physics have added further, more profound and less intuitive motifs to the universals of human activity: phase transitions, power laws, self-organizing patterns, collective motions, scale-free networks.

All the same, we should not forget that there is little truly new in a vision that recognizes society and culture as characteristics which emerge from the intercourse of many autonomous individuals, guided only by simple and local concerns. The German sociologist Georg Simmel summed up the philosophy of today's physics of society in 1908: 'Society is merely the name for a number of individuals, connected by interaction.'[4] Robert E. Park, an American social scientist, refined this notion in 1927 by pointing to the way in which individuals act together: 'Institutions and social structures of every sort may be regarded as products of collective action.'[5] In a sense, all we are now doing is broadening that idea, highlighting the collectivity that lies behind not just static institutions but also dynamic, changing features of society – traffic jams, economic fluctuations, the evolution of cultural traits and norms.

GOOD FROM BAD

One of the strongest themes to emerge from the physics of society is that individual behaviour is often no predictor of its social consequences. Society is not just a scaled-up person: Leviathan is more than the sum of its parts. Evil flowing unintentionally from acts of benevolence; good ends resulting from evil intentions – these are two of the universal themes of art and literature. *The Wealth of Nations* could serve as their prototype. Acting from pure self-interest, said Adam Smith, traders contrive to furnish goods for all, at an affordable price, leaving no gaps in the market:

It is not from the benevolence of the butcher, the brewer or the baker that we expect our dinner, but from their regard of their own interest. We address ourselves not to their humanity, but to their self-love, and never talk to them of our necessities, but of their advantage.[6]

Meanwhile (says Smith), governments that seek to regulate trade in the interests of society as a whole succeed in doing nothing but harm by impairing market efficiency.

Smith's beneficent invisible hand is prefigured in the remarks of Montesquieu, that actions undertaken for one motive can have quite different effects – for better or worse. Montesquieu was in turn influenced by a satirical poem, *Fable of the Bees: Or Private Vices, Public Benefits*, written by the Dutch physician Bernard Mandeville (1670–1733) and first published in 1705 as a pamphlet titled *The Grumbling Hive*. It was a mischievous work, but like most satire it aimed to make a serious point. Mandeville suggested that 'vices' such as vanity can work to the good of society, in this case by increasing the demand for luxury goods and thus providing employment in their manufacturing:

> To enjoy the World's Conveniences,
> Be famed in War, yet Live in Ease,
> Without great Vice, is a vain
> Utopia seated in the Brain.
> Bare Virtue can't make nations
> live in Splendour.[7]

In other words, society positively *needs* a little vice to keep its wheels lubricated. When vice is eliminated by an ethos of self-denial in the beehive that Mandeville described, the bees' society goes into decline, eroded by idleness and poverty. Mandeville was not promoting Hobbesian selfishness, but he despised the hypocrisy that presented virtue as something other than a necessary conceit that made the system work. Let's not kid ourselves, he insisted. He was, naturally enough, vilified as much as Hobbes had been, and his poem was denounced as a public nuisance. Yet Mandeville gradually expanded the work, which began as a divertissement, into a substantial two-volume treatise which provoked serious discussion about its implied moral philosophy. There is no better example of Mandeville's principle in action than the emergence of altruism from selfishness in game theory. Hard-line evolutionary psychologists would claim that there is no virtue that is not at some level motivated by self-interest – although they are a long way from proving it.

The physics of society is full of actions that generate effects other than what was intended or expected. The rush to escape a crowded room leads to a slower average rate of exit. In a corridor where people move in both directions, an increased randomness in the motion of individuals (greater 'noise' or 'heating') can cause a counter-intuitive blockage ('freezing') by making it harder to form counter-flowing lanes. Highway lane rules which try to segregate the fast from the slow can end up reducing the efficiency of traffic flow. There are few easier targets than governmental, regulatory or planning decisions that have had the opposite of their intended effects. In many such cases these unwanted outcomes can be put down to a failure to appreciate the interconnected and interactive nature of the system concerned. The increased congestion that can follow from the building of new roads is a classic example. Statistical physics may help to liberate planners and policy-makers from their propensity for linear thinking and to encourage a greater sophistication in their perception of cause and effect.

IS A SCIENTIFIC SOCIOLOGY GOOD FOR US?

When William Petty applied strict mathematical reasoning to social phenomena, particularly in making recommendations on taxation and the government of Ireland, some of his contemporaries found his approach ridiculously naive. And indeed it was. Petty's insistence on developing policy based on rationality and quantitative data rather than subjective intuitions and prejudices was pioneering; when things are not done this way, we are often the worse for it. But Petty tended to neglect any consideration of what people could, should or would tolerate, or of the need to adapt policy and legislation to existing custom and situation.

His claim that labouring people could afford to pay more taxes if they skipped their dinner on Friday evenings (thus saving on the cost of procuring it) was of course perfectly true. But only someone deeply lacking in an understanding of the circumstances and psychology of his fellows would venture to make it a serious proposal. In 1729 Jonathan Swift satirized this dry, detached style of the 'scientific' political philosophers in his proposal that the poor people of Ireland might make ends meet and at the same time serve the public good by selling some of their children to be cooked and eaten – for 'a young healthy Child, well nursed, is, at a Year old, a most delicious, nourishing, and wholesome Food; whether *Stewed*, *Roasted*, *Baked*, or *Boiled*; and, I make no doubt, that it will equally serve in a *Fricasie*, or a *Ragoust*.'[8]

By bringing mechanistic physics into social science, do we risk becoming like Petty? If people are smoothed down to billiard balls interacting through mathematically defined forces, where do we find room for compassion, charity, for the thousand and one parts of our daily lives that cannot be reduced to numbers but are what make our lives worth living? This cold idealization, the epitome of Hobbesian man, was in Lewis Mumford's view a disastrous outcome of the Enlightenment view of society:

The new order established in the physical sciences was far too limited to describe or interpret social facts . . . Real men and women, real corporations and cities, were treated in law and government as if they were imaginary bodies; whilst presumptuous fictions, like Divine Right, Absolute Rule, the State, Sovereignty, were treated as if they were realities. Freed from his sense of dependence upon corporation and neighbourhood, the 'emancipated individual' was dissociated and delocalized: an atom of power, ruthlessly seeking whatever power can command.[9]

And yes, the responsibility for this change must ultimately lie with Thomas Hobbes. Whether they agreed with him or not, most later moral philosophers inherited Hobbes's rational approach to some degree. The consequences are most apparent in the most explicitly 'scientific' and mathematical of the social sciences: economics.

Although Hobbes had rather little to say about how economic markets worked, his supposition that men make their decisions based on a rational assessment of the opportunities for private gain – that people act out of selfishness – underlies all of conventional economic thinking. Hobbes's ruthless world is the natural habitat of the rational maximizer, the so-called *Homo economicus*. One could argue that people surely *do* act in this way, and that is scarcely Hobbes's fault. Yet it is fair also to wonder whether the prevailing political and economic climate of our times does not actively cultivate, encourage and even compel this kind of behaviour – whether the basic assumptions of economics have not become self-defining and self-sustaining.

After all, it was not always this way. In the late Middle Ages, the idea of a 'just price' dominated economic thought. Goods were priced at a level that was congenial both to buyer and seller: the price gave the vendor enough to live on, and allowed the buyer to acquire necessities at inexorbitant rates. The just price was set by guilds, town councils or other regulators – the economy was extremely interventionist. It was no doubt a system riddled with inefficiencies and abuses, but the principle was based on an idea of social welfare. Merchants accepted at least to some degree that they had social obligations as well as profit motives. Hobbes, however, wrote at a time when the expansion of international trade had put the economy beyond the control of individual governments or states. At the international markets that flourished in European towns, there was no one to regulate prices; merchants could sell at whatever rate the buyers would accept. In this soil, self-interest could and did blossom. What it needed, and what Adam Smith provided, was a vindicating philosophy.

That prices nevertheless set their own level through the balance of supply and demand was first discussed by the Frenchman Richard Cantillon in his *Essai sur la nature du commerce en général* (1730–34), to which Smith's *Wealth of Nations* was much indebted. The appeal of the 'hidden hand' to merchants was obvious: they wanted to be left free to do whatever they could get away with, unhindered by regulations or price limits. But the attractions of such freedom were less evident to Smith himself. He had no interest in being an apologist to greedy businessmen; indeed, he was striving not for an economic theory but for a moral and social philosophy. He accepted only reluctantly the idea that men were motivated by self-interest, and deplored the fact that, in consequence, the labour and time of the poor, in civilized countries, were sacrificed to maintaining the rich in ease and luxury.[10]

Yet accept it he did. Why? Because, by simplifying economic theory and making it accessible to quantification, *Homo economicus* permits a 'scientific' analysis of the market. And that was part of the Enlightenment's programme: to bring science into human affairs.* This desire to simplify for the sake of analysis is, as we have seen, why

* Adam Smith had strong scientific connections himself: he studied Newton's *Principia* closely and was a friend of the chemist Joseph Black and the geologist James Hutton.

Homo economicus has survived for so long. Many economists today still have a Smithian objection to any form of market regulation, clinging determinedly to his notion that all intervention is necessarily harmful to the market and thus – because economic growth has been enthroned as the barometer of a nation's health – harmful to society as a whole. Yet there is not a single economic theory that can show that a totally free market sets the socially most beneficial price for goods, or leads to their optimal distribution.

Indeed, it may be in some ways quite the contrary. The demonstration that new 'interactive' models of trade and economics accentuate inequalities in wealth (page 438) is perfectly in accord with what Montesquieu pointed out in the eighteenth century: that the greatest selfishness is found in 'civilized' commercial societies. In contrast, nomadic cultures retain a strong sense of social obligation and make hospitality to strangers an almost sacred duty. Inequality is, in other words, to some degree the signature of commerce. This view echoes Rousseau in implying that the nastiness and brutishness of people, far from being most prevalent in the Hobbesian State of Nature, is partly a product of civic society itself. That too is unquestionably an oversimplification, as the most cursory familiarity with the ruthless social behaviour of other animals will indicate. But nevertheless, the idea that 'laws' of society follow from fundamentals of basic human nature must surely be modified to embrace the fact that this nature is itself moulded by the customs and norms of the society in which it operates. If we allow that much, it is fair to ask what biases might have been introduced into our cultural structures and institutions by a society shaped by the scientific rationalism of the eighteenth century.

Perhaps this seems a strange thing to say in a book about the value of physics-based models in the social sciences. But it is simply a note of caution. I believe that the current state of conventional economic theory, for example, which continues to accept questionable 'first principles' framed in the glow of the Enlightenment's understandable overconfidence, alerts us to the need for such caution.

CHOICE AND CERTAINTY

The issue is really whether we can trust ourselves to distinguish between moral and physical law. What a physics of society cannot do is tell us how we should live our lives, how we should define our individual and collective responsibilities, how to decide what is important. There are few things quite so misguided and dangerous as looking to science for moral guidance. There are no laws of nature that tell us how to behave or how to govern. John Stuart Mill recognized the folly of trying to build a utopia on the basis of 'laws' that supposedly determine how things ought to be, rather than how they are:

A large proportion of those who have laid claim to the character of philosophic politicians have attempted, not to ascertain universal sequences, but to frame universal precepts. They have imagined some form of government, or system of laws, to fit all cases; a pretension well meriting the ridicule with which it is treated by practitioners.[10]

The Scottish philosopher Adam Ferguson, an associate of Adam Smith, made this distinction clear in 1766. While a physical law tells us what *is*, said Ferguson, a moral law dictates what *ought to be*. The latter is a law 'in consequence of its rectitude, or the authority from which it proceeds . . . not in consequence of its being the fact'.[11]

The problem, of course, is that if there is no objective way of establishing moral laws, one person's law has no stronger claim than another's – and so primacy depends on the distribution of power. To Charles I it was a moral law that he should rule absolutely. The attempts of Bentham and Mill to avoid such relativism by using the principle of utilitarianism – the greatest good for the greatest number* – were never quite convincing: social policies cannot simply be determined by arithmetic.

Nonetheless, Ferguson identifies the central ethical axis of a physics of society: the balance between choice and determinism. Ferguson was no less a product of his times than Smith, Comte or Condorcet in believing that there *are* 'natural' laws that govern society. 'Nations', he wrote, 'stumble upon establishments, which are indeed the result of human action, but not the execution of any human designs.'[12] He did not believe that we must be ruled by these establishments, just that such laws as might exist are in the nature of inevitable consequences of certain actions. If we recognize these consequences, we can then decide whether they are to be desired, according to whatever moral precepts we elect to adopt. There is no reason to believe that unwelcome consequences cannot be alleviated or even avoided. The physics of society supports this idea. Economists who argue that large wealth inequalities are just one of the things we must live with if we want a society that has a better standard of living than the Stone Age are speaking not as scientists but as dogmatists. Better models of the economy can tell us whether, by making this or that change to trading patterns, we can tilt the income distribution towards greater equality. Similarly, we might hope to determine which changes to driving restrictions on a stretch of highway might lessen the tendency for jams to form (jams being undesirable, if not exactly immoral).

Ralph Waldo Emerson, another believer in the idea that natural

---

* George Kingsley Zipf has pointed out that 'the greatest good for the greatest number' is a meaningless notion, the 'fallacy of double optima'. Is great good for a few as desirable as moderate good for many?

laws direct some of the affairs of humankind, felt that these laws need not prescribe a particular kind of government. Insofar as our institutions are concerned, choice prevails. 'In dealing with the State,' he says, 'we ought to remember that its institutions are not aboriginal, though they existed before we were born . . . they all are imitable, all alterable; we may make as good; we may make better.'[13] In other words, things don't have to be this way. Karl Marx agreed: 'Philosophers have sought to interpret the world. The point, however, is to change it.'[14] Yet economist Paul Ormerod suggests that Marx had it the wrong way round: 'Politicians have sought to change the world. But the point is to interpret it correctly.'[15] In the end, we must do both: understanding and the power to effect change go hand in hand. Politicians are all too ready to seek or promise changes they do not know how to bring about, or might be impossible within the rule systems they have created. (Marx, in contrast, saw only a single, inevitable future – all he wanted was to hasten it.)

The idea of using physics to empower change and to predict its limits has no intrinsic politics. It is neither libertarian nor repressive, neither left-wing nor right. It might occasionally help us to cut through such ideologies. Since most political thinking starts off with a particular world-view and then looks for ways to make the real world conform with it, the suggestion that there may be unavoidable fluctuations and inevitable laws when many individuals get together and interact might be a deeply uncomfortable one for politicians. But if they do not confront it as a real possibility, they risk making choices that are both futile and wasteful. Historian Richard Olson summarizes the point perfectly in his account of Adam Ferguson's conclusions about how to deal with natural laws of society:

One way of expressing the relationship between physical and moral laws – i.e., between science and morality – in the formation of society as Ferguson understood the situation is to say that social systems are 'softly' deterministic. Left alone, they will inevitably develop along certain lines; but the possibility of changing those lines by conscious and intentional intervention does exist. The whole point of a 'social science,' then, is to explore the opportunities for and likely consequences of intentional moral action. Without the science, morality is blind; but without the morality, science is useless, pointless, and paralytic.[16]

PLANNING FOR FREEDOM

In antiquity, Ferguson's distinction between 'natural' and 'moral' was meaningless. Cicero (106–43 BC) had a touching faith in the natural morality of the universe. In *De legibus* ('On Laws'), he said, 'Now we must entirely take leave of our senses, ere we can suppose that law and justice have no foundation in nature, and rely merely on the transient opinions of men.'[17] History lends little support to Cicero's optimism. If we want law and justice, we must build a society that promotes them and protects them from abuses. That is where the State comes in.

To Aristotle it seemed obvious that humankind had a natural compulsion to form groups: 'man is by nature a political animal',[18] he asserted. Thomas Aquinas, in his little book *On the Governance of Rulers* (c.1259), could only agree: 'it is natural for man to be a social and political animal, to live in a group, even more so than all the other animals, as the very needs of his nature indicate.'[19] But he acknowledged the drawback: 'Where there are many men together, and each one is looking after his own interest, the group would be broken up and scattered unless there were also someone to take care of what appertains to the common weal.'[20] Thus, a State cannot simply be a group; it must be a group with a ruler. Government, in other words, is inevitable.

Yet this is still a long way from Hobbes's necessary dictatorship. The harshness of Hobbes's solution to government obviously reflects his much more jaundiced view of human nature. Unlike Cicero, he saw no redeeming socializing tendencies in man, but only a will to exploit and dominate: 'men have no pleasure, (but on the contrary a great deal of grief) in keeping company, where there is no power able to overawe them all.'[21] The twentieth century saw plenty of regimes with the capacity to overawe, and they were not a good advertisement for Hobbes. How do we ensure that society is not just stable but moral?

It is generally taken for granted in the West that the answer, or at least a good part of it, is to make society democratic. But this is a modern view; or at least, it has passed in and out of favour over time. The liberal democracies of Locke and Mill by no means represented the obvious political future of Europe in the 1930s, when it looked vulnerable both to fascism and to a warped and dictatorial socialism. Against such a backdrop we can understand why Austrian economist Friedrich von Hayek felt compelled to write his famous defence of capitalist freedom, *The Road to Serfdom* (1944). But Hayek's critique of socialism, which he regards as the beginning of an inevitable descent into totalitarianism, is not just a warning to beware of Hitler, Mussolini and Stalin. It also examines different notions of democracy, and addresses the crucial question that all democrats must ask: *how much* should they rule?

Hayek echoes Adam Smith in making the free market the ultimate arbiter of political and economic society. Since in Hayek's view money is the key to freedom (and the extreme naivety of that attitude must be seen in the context of the times), to impose constraints on the ways in which capital can be used is to compromise liberty. Hayek's approval for Smith is apparent as he quotes him thus:

The statesman who should attempt to direct private people in what manner they ought to employ their capitals, would not only load himself with a most unnecessary attention, but assume an authority which could safely be trusted, not only to no single person, but to no council or senate whatever, and which would nowhere be so dangerous as in the hands of a man who had folly and presumption enough to fancy himself fit to exercise it.[22]

Thus 'planning', to Hayek, means economic planning, and anything of that nature smacks of socialism. Far better to entrust the forces of competition to bring about a market which, if not necessarily equitable, is at least 'fair' in the sense that it equalizes opportunity within the boundaries of the cultural climate. Clearly, this debate about regulation versus market freedom is as relevant today as it was sixty years ago.

Underlying Hayek's advice is a firm belief in self-organizing principles which maintain a balance. It would be easy to read it as an advocacy of laissez-faire, if Hayek did not take such pains to explain the distinction: 'The liberal argument is in favour of making the best possible use of the forces of competition as a means of co-ordinating human efforts, not an argument for leaving things just as they are.'[23] Thus he feels that the economy needs *tending* so that these 'spontaneous force of society' can operate to greatest benefit:

The attitude of the liberal towards society is like that of the gardener who tends a plant and in order to create the conditions most favourable to its growth must know as much as possible about its structure and the way it functions.[24]

Here is the crux of the matter. One can take exception to Hayek's denial of an absolute system of values – his assertion that each person must be their own judge seems a recipe for disaster (or for a Hitler). One can question his belief that market forces inevitably operate for the 'best', and so should be given free rein. But it is surely incumbent on anyone who seeks to lead or govern a society that they 'know as much as possible about its structure and the way it functions'. Hayek is right at least to believe that there *are* spontaneous forces of society. In this book we have seen some of the modern efforts to establish what those forces are.

No doubt some of these forces are what we might call exogenous: they arise from outside, from technological change for example (such as mechanization, birth control, information technology), or from the environment (drought and famine, climate change). But many of the forces that act on society are the collective consequence of interactions between one person and another, whether that be in conducting trade, spreading new fashions, waging war, getting married or avoiding collisions. Thomas Schelling points out that these forces are typically a compromise of conflicting wishes – a kind of interplay of attraction and repulsion. What we wish for ourselves may not always be what we wish for others:

A good part of social organization – of what we call society – consists of institutional arrangements to overcome [the] divergences between perceived individual interest and some larger collective bargain . . . What we are dealing with is the frequent divergence between what people are individually motivated to do and what they might like to accomplish together.[25]

It is particularly (although not uniquely) when we are confronted with such divergences between individual and collective goals that our intuitions about the likely modes of aggregate behaviour may be a poor guide, and instead we need simple yet realistic models in order to predict the outcomes of our actions.

The wider challenge is then to decide whether or not we desire the consequences that these forces bring about. Hayek fails to make a case for why a free market is best for society. It might be the most efficient in some sense – although new models of the economy cast doubt even on that. But if it permits (indeed, promotes) huge discrepancies in earnings, out of all proportion to efforts or responsibilities, if it allows child pornography (or children) to be traded like bananas, if it places small nations at the mercy of big corporations, then surely we are entitled to ask whether this is what we want. As Schelling points out, 'How well each [person] does for himself in adapting to his social environment is not the same thing as how satisfactory a social environment they collectively create for themselves.'[26]

That is where the utility of social physics ends. If it were to become a justification for moral choices, it has exceeded its brief. This ought to be obvious, but today science is granted such authority that the small step into morality and ethics can sometimes pass unnoticed. Already there is a tendency to divide up science into 'good' and 'bad', for example by branding it 'holistic' or 'reductionist'. This is disheartening; as British writer Kenan Malik says, 'There are few things more dispiriting than turning science into faith.'[27] Hayek himself railed against the inappropriate application of scientific ideas to social problems, and in an age when liberals espoused pseudo-Darwinian arguments for eugenics he had fair cause for complaint. Knowledge of how things *are* can never supplant the obligation to justify our preferences for how they should *be*.

And so the physics of society is and can only be a tool, never a moral compass. John Stuart Mill identified with great elegance the way these tools should be employed:

The aim of practical politics is to surround any given society with the greatest possible number of circumstances of which the tendencies are beneficial, and to remove or counteract, as far as practicable, those of which the tendencies are injurious. A knowledge of the tendencies only, though without the power of accurately predicting their conjunct result, gives us to a considerable extent this power.[28]

It cannot, surely, be anything other than wise to search for these tendencies; indeed, one might regard this as a moral obligation. The American sociologist Franklin Giddings acknowledged in 1924 that this search is the prime objective of the social sciences, and that the goal might be all the more readily achieved if the discipline can indeed evolve into a true science:

[S]cience is nothing more nor less than getting at the facts, and trying to understand them, and . . . what science does for us is nothing more nor less than helping us to face facts . . . Facing the facts that the social sciences are making known to us, and will make better known, should enable us to diminish human misery and to live more wisely than the human race has lived hitherto.[29]

This, in the end, is all Thomas Hobbes wanted. That his 'facts' pointed to a political institution we would today regard as intolerable and more likely to contribute to than to diminish human misery is a warning of how hard it is to escape the tenor of one's times, however 'scientific' one strives to be. Nonetheless, many aspects of Hobbes's methodology and reasoning are remarkably modern, and Hobbes could justifiably claim to be at the head of a long succession of attempts to deduce collective human behaviour from individual tendencies.

As you can see, the implications of a physics of society can be formulated very nicely in the words of earlier ages. Indeed, it is often to such words that we must look for a wider perspective on what statistical physicists today are doing, for scientists now are encouraged, with good reason, to interpret their findings with the utmost conservatism and humility. Politicians and policy-makers, meanwhile, are tending to withdraw from the visionary political philosophy of their predecessors and to settle for quick solutions to short-term challenges. The most urgent question, to which the physics of society can contribute, is whether we can construct a society blessed with the wisdom and compassion that others, often in harsher or more difficult times, were able to glimpse and to demand.

Epilogue

Curtain call

The performer who provides his own applause is surely either deluded or desperate; but where better than in conclusion to describe this delightful manifestation of social physics?

In some countries and cultures, particularly in Eastern Europe, the applause with which a gratified audience expresses its appreciation of a performance tends to slew back and forth between random and synchronized. At one moment each person is clapping to their own rhythm, and the hundreds of overlapping pulses of sound create a continuous clattering roar like the sound of surf on shingle. But then something remarkable happens: this wash of noise resolves itself into a regular beat, as each pair of hands claps in unison with the others. The synchronization lasts for perhaps a minute or so, then dissolves again into chaos.

There is no one conducting *this* performance, no one to set the pace or to signal when synchrony should begin. It just happens – not once but several times during an ovation. Now, it is no great feat for two or three people to bring their handclaps into phase with one another. Indeed, it can be hard for them to avoid doing so, just as two people tend to synchronize their steps when they walk side by side. But the synchronized clapping of an audience of many hundreds is a challenge of another order. That it can crystallize so quickly is surprising enough; but why, once synchronized, does the clapping not stay that way, given that each member of the audience can consciously sustain it with little effort? Why this see-sawing alternation between order and chaos?

Tamás Vicsek, Albert-László Barabási and their colleagues have asked this same question. They made sound recordings of the post-performance applause in several theatres and opera halls in Hungary and Romania, and looked at how the sound volume changed as clapping veered between incoherent and synchronized. They found that, although synchronized clapping produces 'spikes' of noise that can exceed the sound level during incoherent clapping, nevertheless synchronization decreases the *average* noise intensity in the hall (Figure E.1).

This decrease happens not because the audience members are clapping any less vigorously when they are in phase, but because they clap *less often*. Each person spaces their claps roughly twice as far apart in the synchronized mode than when they are clapping freely. This is presumably because it is harder to clap in time with everyone else if the rhythm is too fast – synchronization of a slow handclap disintegrates if the pace quickens. No one in the audience thinks consciously about this; the measured rhythm of the synchronized clap essentially selects itself.

The researchers suggest that the difference in average sound volume explains why applause at concerts oscillates between in-coherent and synchronized. An appreciative audience wants to make a lot of noise. It also appears to enjoy the communal experience of syn-chronized clapping, which lends the crowd a single voice. But these two things are in conflict, because synchronization results in a drop in the average noise intensity. No one registers this drop consciously, but the audience, having switched into synchronized mode, nonetheless

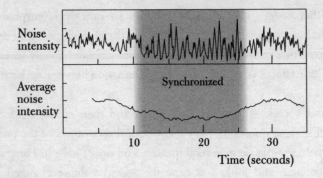

Figure E.1 Sound levels during audience applause in an Eastern European theatre. The switch from random to synchronized clapping is marked by the appearance of sharp, regularly spaced peaks in the noise intensity (top). When this happens, the average noise intensity diminishes (bottom), only to rise again as the synchronized clapping dissolves.

begins to speed up the rhythm so as to restore the sound volume to its earlier, unsynchronized level. In doing so, they lose the ability to keep their claps in step, and the synchronization dissolves into incoherence. Moments later, the attractive tug of synchrony reasserts itself and the cycle repeats.

A push and a pull; a tension between conflicting desires. This is all it takes to tip our social behaviour into complex and often unpredictable patterns, dictated by influences beyond our immediate experience or our ability to control. Regardless of what we believe about the motivations for individual behaviour, once we become part of a group we cannot be sure what to expect.

Notes

INTRODUCTION

1. W. Petty (1690). *Political Arithmetick*. Robert Clavel, London.
2. *Ibid.*, Preface.
3. K. Waltz (1954). *Man, the State, and War*, p. 75. Columbia University Press, New York.

CHAPTER 1

Epigraphs

B. Fontenelle. Quoted in A. G. R. Smith (1972). *Science and Society in the 16th and 17th Centuries*, p. 153. Science History Publications, New York.

B. Fontenelle (1686). *Conversations on the Plurality of Worlds*, transl. W. Gardiner (1715), pp. 11–12. London.

R. M. MacIver (1947). *The Web of Government*. Macmillan, New York. Quoted in R. Bierstedt (ed.), *The Making of Society*, p. 493. Random House, New York, 1959.

S. Cotgrove (1967). *The Science of Society*, p. 181. George Allen & Unwin, London.

1. Snorri Sturluson (13th century). *The Prose Edda*, transl. J. I. Young (with minor changes), p. 86. University of California Press, Berkeley, 1964.

2. J. Aubrey (1681). *Brief Lives*, ed. J. Buchanan-Brown, pp. 427–8. Penguin, London, 2000.

3. T. Hobbes (1651). *Leviathan*, ed. C. B. Macpherson, pp. 110–11. Penguin, London, 1985.

4. T. Hobbes (1651). *Philosophical Rudiments Concerning Government and Society*, Chapter 1, section 7, p. 26. Quoted in *Leviathan*, p. 28. [The *Rudiments* is a translation by Hobbes of his Latin text *De cive*, 1642.]

5. Hobbes, *Leviathan*, p. 82.

6. R. Descartes (1641). 'Meditations on first philosophy'. In *Meditations and Selections from the Principles of Philosophy*, p. 98. Open Court, La Salle, IL, 1952.

7. Hobbes, *Philosophical Rudiments*, Preface, pp. 10–11.

8. T. Hobbes (1642). *De cive*, ed. S. Lamprecht, p. 15. Appleton-Century-Crofts, New York, 1949.

9. F. Bacon (1627). *New Atlantis*. Quoted in J. Carey (ed.), *The Faber Book of Utopias*, p. 64. Faber & Faber, London, 1999.

10. R. M. MacIver (1947). *The Web of Government*. In Bierstedt (ed.), *The Making of Society*, p. 493. Random House, New York, 1959.

11. L. Mumford (1946). *The Culture of Cities*, p. 93. Secker & Warburg, London.

12. Hobbes, *Leviathan*, p. 151.

13. *Ibid.*, p. 161.

14. *Ibid.*, p. 186.

15. *Ibid.*, p. 190.

16. Job 41: 8–34.

17. Hobbes, *Leviathan*, p. 228.

18. F. L. Nussbaum (1965). *The Triumph of Science and Reason 1660–1685*, p. 61. Harper & Row, New York.

19. Hobbes, *Leviathan*, p. 261.

20. *Ibid.*, p. 379.

21. R. Olson (1990). *Science Deified and Science Defied*, Vol. 2, p. 41. University of California Press, Berkeley.
22. D. Hume. Quoted in R. Tuck (2002). *Hobbes: A Very Short Introduction*, p. 107. Oxford University Press.
23. T. Hobbes (1682). *Behemoth, or The Long Parliament*, p. 126. Ed. F. Tönnies, reprinted by University of Chicago Press, 1990.
24. Hobbes, *Leviathan*, p. 208.

Figure 1.2 caption: Hobbes, *Leviathan*, p. 227.

CHAPTER 2

Epigraphs
P. Hein (1966). *Grooks*, p. 24. MIT Press, Cambridge, MA.
A. Einstein (1900). Letter to Mileva Maric, 13 September. In A. Beck and P. Havas (eds), *The Collected Papers of Albert Einstein*, Vol. I, p. 154. Princeton University Press, 1987.
J. C. Maxwell (1856). Inaugural Lecture at Aberdeen, 2 November. *Notes and Records of the Royal Society of London* **28**, 69 (1973).

1. I. Kant. Quoted in F. A. Hayek (1962). *The Road to Serfdom*, p. xx. Routledge, London.
2. R. J. Boscovich (1763). *Theoria philosophiae naturalis*. Venice. Transl. as *A Theory of Natural Philosophy* (Open Court, Chicago, 1922), reprinted by MIT Press, Cambridge, MA (1966), p. 141.
3. P. S. Laplace (1814). *Essai philosophique sur les probabilités*, p. 3. Paris. Reprinted by Gauthiers-Villars, Paris, 1921.
4. S. Carnot (1824). *Réflexions sur la puissance motrice du feu et sur les machines propre à développer cette puissance*. Bachelier, Paris. English transl. by R. H. Thurston in S. Carnot, *Reflections on the Motive Power of Fire and Other Papers on The Second Law of Thermodynamics*, ed. E. Mendoza. Dover, New York, 1960.

5. C. P. Snow (1965). *The Two Cultures*, pp. 14–15. Cambridge University Press.

6. W. Thomson (1852). 'On a universal tendency in nature to the dissipation of mechanical energy', *Philosophical Magazine*, series 4, 4, 304–6.

7. E. A. Moelwyn-Hughes. Quoted in J. M. Thomas (1986). 'Royal Institution inaugural lecture', *Royal Institution Proceedings*, p. 3.

8. L. Boltzmann (1872). 'Weitere Studien über das Wärmegleichgewicht unter Gasmolekülen', *Sitzungsberichte der Akademie der Wissenschaften, Wien II*, 66, 275–370. In *Wissenschaftliche Abhandlungen*, Vol. 1 (ed. F. Hasenöhrl), p. 345. J. A. Barth, Leipzig, 1909.

CHAPTER 3

Epigraphs

W. Wundt (1862). *Beitrage zur Theorie der Sinneswahrnehmung*, p. xxv. Leipzig.

J. Herschel (1850). 'Quetelet on probabilities', *Edinburgh Review* 92, 42.

H. Spencer (1880). *The Study of Sociology*, 9th edn, p. 39. Otto Zeller, Osnabrück, 1966.

1. L. Boltzmann, in W. Hoeflechner (ed.) (1994). *Ludwig Boltzmann: Leben und Briefe*. Publikationen aus dem Archiv der Universität Graz, Akademische Druck- und Verlaganstalt, Graz. [English translation by F. Rohrlich (1992). 'A poem by Ludwig Boltzmann', *American Journal of Physics* 60, 972–3.]

2. D. L. Goodstein (1975). *States of Matter*, p. 1. Prentice Hall, New York.

3. R. Musil (1953–60) (written 1930–42). *The Man Without Qualities*. Transl. E Wilkins and E. Kaiser. Secker & Warburg, London.

4. W. Petty (1691). 'The political anatomy of Ireland'. In C. H. Hull (ed.), *The Economic Writings of Sir William Petty*, Vol. 1, p. 129. Cambridge University Press, 1899.

5. J. Graunt (1662). *Observations upon the Bills of Mortality*. In Hull, *The Economic Writings of Sir William Petty*, Vol. 2, p. 334.

6. Quoted in D. J. Boorstin (1985). *The Discoverers*, p. 668. Vintage, New York.

7. T. Short (1767). *A Complete History of the Increase and Decrease of Mankind*, p. i. London.

8. I. Kant (1784). *On History*, p. 11. L. Beck (ed.), Indianapolis, Bobbs-Merrill, 1963.

9. J. A. N. de Condorcet (1785). *Essai sur l'application de l'analyse à la probabilité des décisions rendues à la plurité des voix*. Quoted in J. Carey (ed.), *The Faber Book of Utopias*, pp. 161–2. Faber & Faber, London, 1999.

10. J. A. N. de Condorcet (1795). *Esquise d'un tableau historique des progrès de l'esprit humain*. Quoted in Carey, p. 165.

11. Condorcet, *Esquise*. Quoted in R. Bierstedt (ed.), *The Making of Society*, p. 68. Random House, New York, 1959.

12. J. T. Desaguliers (1728). *The Newtonian System of the World*, lines 17–18. Quoted in M. C. Jacob, *The Radical Enlightenment: Pantheists, Freemasons and Republicans*, p. 124. Allen & Unwin, London, 1981.

13. D. Hume (1741). 'That politics may be reduced to a science'. In S. Copley and A. Edgar (eds), *Selected Essays*, pp. 13–24. Oxford University Press, 1993.

14. J. Swift (1726). *Gulliver's Travels*, Book II, p. 176. Penguin, London, 1985.

15. E. Burke (1790). *Reflections on the Revolution in France*, para. 98. J. Dodsley, London. Republished (ed. C. C. O'Brien) by Penguin, London, 1982.

16. A. Comte (1830–42). *Cours de philosophie positive*. Quoted (in English) in Bierstedt, p. 192.

17. A. Quetelet (1830–31). 'Lettre à M. le Bourgmestre, 15 dec. 1831'. *Annuaire de l'observatoire de Bruxelles* 1, 285 (1834); 'letter to Bouvard', 5 November 1830, quoted in J. Lottin, *Quetelet: Statisticien et Sociologue*, p. 52. Institut Supérieur de Philosophie, Louvain, 1912.

18. A. Quetelet (1840). 'Notice scientifique' for his book *Sur l'homme*, *Annuaire de l'Observatoire de Bruxelles* **7**, 230.

19. A. Quetelet (1832). 'Recherches sur le penchant au crime aux différens ages', *Nouveaux mémoires de l'académie royale des sciences et belles-lettres de Bruxelles* **7**, 80.

20. A. Quetelet (1835). *Sur l'homme et le développement de ses facultés, ou essai de physique sociale*, p. 289. Bachelier, Paris.

21. Quetelet, 'Recherches sur le penchant au crime aux différens ages', p. 6.

22. J. Herschel (1850). 'Quetelet on probabilities', *Edinburgh Review* **92**, 14.

23. J. S. Mill (1862), *A System of Logic*. In *Collected Works*, Vols 7–8, ed. J. M. Robson, p. 932. University of Toronto Press, 1973.

24. H. T. Buckle (1857–61). *History of Civilization in England*, Vol. 2, p. 244. Hearst's International Library, New York, 1913.

25. I. Kant (1784). 'Idea of a universal history from a cosmopolitan point of view'. See I. Kant, *On History* (ed. L. Beck), p. 11. Bobbs-Merrill, Indianapolis, 1963.

26. Buckle, *History of Civilization in England*. See F. Stern (ed.), *The Varieties of History from Voltaire to the Present*, 2nd edn, pp. 121–32. Macmillan, London, 1970.

27. W. Newmarch (1860). 'Some observations on the present position of statistical inquiry with suggestions for improving the organization and efficiency of the International Statistical Congress', *Journal of the Statistical Society of London* **23**, 362.

28. N. W. Senior (1860). 'Opening address of Nassau W. Senior, Esq . . . ', *Journal of the Statistical Society of London* **23**, 359.

29. F. K. Hunt (1850). 'A few facts about matrimony', *Household Words* **1**, 374.

30. R. W. Emerson (1860). 'Fate', *The Conduct of Life & Other Essays*, p. 159. J. M. Dent & Sons, London, 1908.

31. M. Twain (1924). *Autobiography*, Vol. 1, Ch. 20. Harper & Brothers, New York.

32. F. Nietzsche (1874). *Untimely Meditation. Second Part: Of the Use and Disadvantage of History for Life*. E. W. Fritzsch, Leipzig.

33. Quoted in L. Campbell and W. Garnett (1882). *The Life of James Clerk Maxwell*, pp. 294–5. Macmillan, London.

34. J. C. Maxwell (1873). 'Molecules' (a lecture), *The Scientific Papers of James Clerk Maxwell*, ed. W. D. Niven, Vol. 2, p. 374. Cambridge University Press, 1890.

35. Quoted in Campbell and Garnett, pp. 438–9.

36. L. Boltzmann (1872). Quoted in 'Weitere Studien über das Wärmegleichegewicht unter Gasmolekülen', in *Wissenschaftliche Abhandlungen*, Vol. 1 (ed. F. Hasenöhrl), p. 317. J. A. Barth, Leipzig, 1909.

37. P. G. Tait (1886). 'On the foundations of the kinetic theory of gases', *Scientific Papers*, Vol. 2, p. 126. Cambridge University Press, 1898–1900.

38. C. S. Peirce (1877). 'The fixation of belief', *Popular Science Monthly* **12**, 1–15. Reprinted in *Collected Papers*, Vol. V, p. 226. Harvard University Press, Cambridge, MA, 1974. [I am grateful to Vincent Bauchau for bringing this citation to my attention.]

39. F. Galton (1875). *English Men of Science: Their Nature and Nurture*, p. 17. D. Appleton, New York.

40. A. A. Cournot (1843). *Exposition de la théorie des chances et des probabilités*, p. 181. Hachette, Paris.

41. J. J. Fox (1860). 'On the province of the statistician', *Journal of the Statistical Society of London* **23**, 331.

42. A. Taillandier (1828), Review of '*Compte général de l'administration de la justice criminelle en France*', *Revue encyclopédique* **40**, 612.

43. 'Introduction', *Journal of the Statistical Society of London* **1**, 3 (1838).

44. W. Farr (1861). Quoted in M. Diamond and M. Stone, 'Nightingale on Quetelet', *Journal of the Royal Statistical Society A* **144**, 70 (1981).

45. A. De Candolle (1830). 'Considérations sur la statistique des délits', *Bibliothèque universelle des sciences, belles-lettres et arts* **104**, 160.

46. Kant, *On History*, p. 11.

47. A. Quetelet (1847). 'De l'influence de libre arbitre de l'homme sur les faits sociaux', *Bulletin de la commission centrale de statistique* **3**, 142.

48. Quetelet, *Sur l'homme . . .* , p. 9.

49. W. Cyples (1864). 'Morality of the doctrine of averages', *Cornhill Magazine* **10**, 224.

50. Anon (1860). 'The address of the Prince Consort on opening as President the Fourth Session of the International Statistical Congress', *Journal of the Statistical Society of London* **23**, 280.

51. F. Dostoevsky (1864), *Letters from the Underworld*, transl. C. J. Hogarth, p. 32. J. M. Dent & Sons, London, 1913.

52. *Ibid.*, pp. 29–30, p. 32.

53. *Ibid.*, p. 31.

54. L. N. Tolstoy (1869). *War and Peace*, transl. R. Edmunds, Vol. 2, p. 1404. Penguin, London, 1969.

55. *Ibid.*, p. 1426.

56. *Ibid.*, p. 1440.

57. M. E. Hare (1905). Quoted in N. Ferguson, *Virtual History*, p. 446. Picador, London, 1997.

58. A. Tennyson (1868). 'Lucretius', p. 10. Printed for private circulation, Cambridge, MA.

59. Maxwell, 'Molecules', p. 373.

60. J. C. Maxwell (1867). Letter to P. G. Tait, in C. G. Knott, *Life and Scientific Work of Peter Guthrie Tait*, pp. 213–14. Cambridge University Press, 1911.

61. M. Smoluchowski (1918). 'Über den Begriff des Zufalls und den Ursprung der Wahrscheinlichkeitgesetz in der Physik'. In *Oeuvres*, Vol. 3, p. 87. Cracow, 1924–8.

CHAPTER 4

Epigraphs

R. W. Emerson (1860). *The Conduct of Life*, p. 210. J. M. Dent & Sons, London, 1908.

J. C. Maxwell (1878), in *The Scientific Papers of James Clerk Maxwell*, ed. W. D. Niven, Vol. 2, pp. 715–17. Cambridge University Press, 1890.

I. Newton (1704), 'Queries', in *Opticks*. Reprinted by Dover, New York, 1952.

1. K. Vonnegut (1965). *Cat's Cradle*, p. 163. Penguin, London.
2. M. Gladwell (2000). *The Tipping Point*. Little, Brown & Co., London.
3. C. Jencks (1995). *The Architecture of the Jumping Universe*. Academy Editions, London.
4. T. S. Kuhn (1962). *The Structure of Scientific Revolutions*. University of Chicago Press.
5. J. F. W. Herschel (1830). *Preliminary Discourse on the Study of Natural Philosophy*, p. 188. Longman, Rees, Orme, Brown & Green; and Taylor, London.
6. T. Andrews (1869). 'On the continuity of the gaseous and liquid states of matter', *Philosophical Transactions of the Royal Society* **159**, 575.
7. J. C. Maxwell (1874). 'Van der Waals on the continuity of the gaseous and liquid states', *Nature* **10**, 477–80, here p. 478.

CHAPTER 5

Epigraphs

S. Hales (1727). *Vegetable Staticks*. W. & J. Innys & T. Woodward, London.
H. Spencer (1876). *The Principles of Sociology*. Quoted in R. Bierstedt (ed.), *The Making of Society*, p. 262. Random House, New York, 1959.
G. C. Lichtenberg. Quoted in F. Cramer, *Chaos and Order*, p. 2. VCH, Weinheim, 1993.

1. H. D. Thoreau (1856). *The Journal of Henry David Thoreau*, ed. B. Torrey & F. Allen, Vol. 8, pp. 87–8. Houghton Mifflin, Boston, 1906.
2. J. W. Gibbs (1906). 'On the equilibrium of heterogeneous substances'. In H. A. Bumstead and R. G. Van Name (eds), *The Scientific Papers of J. Willard Gibbs*, Vol. I. Longmans, Green, New York.

3. D'A. W. Thompson (1917). *On Growth and Form*, p. 503. Revised edition published by Cambridge University Press, 1942, reprinted in 1992 by Dover, New York.

4. Thompson, p. 505.

5. I. Prigogine (1980). *From Being To Becoming*, p. 106. W. H. Freeman, New York.

6. J. L. Borges (1956). 'The Garden of Forking Paths'. In *Labyrinths*, pp. 44–54. Penguin, London, 1970.

7. Prigogine, p. 106.

CHAPTER 6

Epigraphs

G. C. Lichtenberg. Quoted in F. Cramer, *Chaos and Order*, p. 114. VCH, Weinheim, 1993.

E. A. Ross (1901). *Social Control*. In R. Bierstedt (ed.), *The Making of Society*, p. 336. Random House, New York, 1959.

H. von Kleist (1810). 'On the Marionette Theatre', *Berliner Adendblätter*, 12–15 December. Transl. I. Parry, *Times Literary Supplement*, 20 October 1978.

1. A. van Leeuwenhoek (1674). In C. Dobell (transl. and ed.), *Antony van Leeuwenhoek and His 'Little Animals'*. Russell & Russell, New York, 1958.

2. T. Hobbes (1651). *Leviathan*, ed. C. B. Macpherson, p. 228. Penguin, London, 1985.

3. J. K. Parrish and L. Edelstein-Keshet (1999). 'Complexity, pattern, and evolutionary trade-offs in animal aggregation', *Science* **284**, 99–101, here p. 100.

4. C. Reynolds. Quoted in S. Levy, *Artificial Life*, p. 74. Jonathan Cape, London, 1992.

5. E. O. Wilson. Quoted in R. Lewin, *Complexity*, p. 178. Macmillan, New York, 1992.

6. Hobbes, *Leviathan*, p. 225.

7. *Ibid.*, pp. 225-6.

8. L. F. Henderson (1971). 'The statistics of crowd fluids', *Nature* **229**, 381.

9. B. Hillier and J. Hanson (1984). *The Social Logic of Space*, p. 266. Cambridge University Press.

10. *Ibid.*

11. M. Batty, J. Desyllas and E. Duxbury (2003). 'Safety in numbers? Modelling crowds and designing control for the Notting Hill Carnival', *Urban Studies* **40**, 1573-90, here p. 1588.

12. Parrish and Edelstein-Keshet, p. 101.

13. P. Ackroyd (2000). *London: The Biography*, p. 2. Chatto & Windus, London.

14. *Ibid.*, p.103.

15. H. Kett (1787). Quoted in *Ibid.*, p. 517.

16. D. D. T. Chen (2000). 'The science of smart growth', *Scientific American*, December, 84-92, here p. 84.

17. *Ibid.*

18. L. Mumford (1938). *The Culture of Cities*, p. 233. Secker & Warburg, London.

19. *Ibid.*, p. 7.

20. H. Spencer (1876). *Principles of Sociology*. Quoted in Bierstedt, p. 262.

21. H. E. Stanley. Quoted in I. Petersen, 'The shapes of cities', *Science News*, 6 January, p. 9 (1996).

22. H. Simon (1996). *The Sciences of the Artificial*, 3rd edn, pp. 33-4. MIT Press, Cambridge, MA.

23. Ackroyd, p. 588.

CHAPTER 7

Epigraphs

T. C. Schelling (1978). *Micromotives and Macrobehavior*, p. 121. W. W. Norton, New York.

J. W. Gibbs. Quoted in A. T. Winfree, *The Geometry of Biological Time*. Springer-Verlag, New York, 1980.

R. Kipling (1890). L'Envoi to *The Story of the Gadsbys*. A. H. Wheeler & Co., Allahabad.

1. R. Moe (1999). Speech on urban sprawl, Red Hills Spring Event Dinner, Tall Timbers Research Station, Tallahassee, Florida, 24 March.

CHAPTER 8

Epigraphs

S. Johnson. Quoted in R. Heilbroner, *The Worldly Philosophers*, 7th edn, p. 41. Penguin, London, 2000.

G. B. Shaw (attributed).

J. W. von Goethe. Quoted in R. W. Emerson (1860). *The Conduct of Life*, p. 197. J. M Dent & Sons, London, 1908.

1. Heilbroner, p. 57.
2. D. Defoe (1706). *Review*, vol. ii, p. 26. Quoted in D. Donoghue, *England, Their England: Commentaries on English Language and Literature*, p. 65. Alfred A. Knopf, New York, 1988.
3. Quoted in J. G. A. Pocock (1985). 'Josiah Tucker on Burke, Locke, and Price'. In *Virtue, Commerce and History: Essays on Political Thought and History, Chiefly in the Eighteenth Century*, pp. 157–91. Cambridge University Press.
4. Emerson, p. 202.
5. *Ibid.*
6. J. Kay (1995). 'Cracks in the crystal ball', *Financial Times*, 29 September.
7. I. Fisher (1929). Quoted in Heilbroner, p. 251.
8. P. Krugman (1994). *Peddling Prosperity*, p. xi. W. W. Norton, New York.

9. *Ibid.*

10. A. Smith (1776). *An Inquiry into the Nature and Causes of the Wealth of Nations*. Abridged version, ed. L. Dickey, p. 33. Hackett Publishing Co., Indianapolis, 1993.

11. K. Marx and F. Engels (1848). *The Communist Manifesto*, transl. S. Moore, p. 25. Junius Publications, London, 1996.

12. *Ibid.*, p. 48.

13. Krugman, p. 26.

14. P. H. Cooter (ed.) (1964). *The Random Character of Stock Market Prices*. MIT Press, Cambridge, MA.

15. Quoted in G. Stix (1998). 'A calculus of risk', *Scientific American*, May, 70–75.

CHAPTER 9

Epigraphs

R. Heilbroner (1999). *The Worldly Philosophers*, 7th edn, pp. 316–17. Penguin, London, 2000.

R. W. Emerson (1860). *The Conduct of Life*, p. 199. J. M. Dent & Sons, London, 1908.

W. B. Arthur. Quoted in M. M. Waldrop, *Complexity*, p. 328. Penguin, London. 1994.

1. T. Carlyle (1849). 'The Nigger question'. In *Miscellaneous Essays*, Vol. 7, pp. 79–110. Chapman & Hall, London, 1888. [I recommend P. Groenewegen's 'Thomas Carlyle, "the dismal science", and the contemporary political economy of slavery', *History of Economics Review* **34**, 74–94 (2001) for a discussion of Carlyle's critique of 'economic science' and his views on the economic implications of the emancipation of slaves.]

2. F. Y. Edgeworth (1888). 'Tests of accurate measurement'. In *Papers Relating to Political Economy*, Vol. 1, p. 331. Macmillan, London, 1925.

3. F. Y. Edgeworth (1881). *Mathematical Psychics: An Essay on the Application of Mathematics to the Moral Sciences*, pp. 12–13. Kegan Paul & Co., London.

4. Edgeworth, *Mathematical Psychics*, p. 50.

5. Heilbroner, p. 176.

6. J. M. Keynes (1926). 'Francis Ysidro Edgeworth', *Essays in Biography*, p. 224. W. W. Norton, New York, 1963.

7. P. Krugman (1994). *Peddling Prosperity*, p. xi. W. W. Norton, New York.

8. Heilbroner, p. 317.

9. D. Howell (2000). *The Edge of Now*. Macmillan, London.

10. J. Kay (2000). 'Economic with the truth', *Prospect*, October, p. 70.

11. J. M. Keynes (1936). *The General Theory of Employment, Interest and Money*, p. 161. Macmillan, London, 1973.

12. G. Soros (1994). Quoted in the editorial introduction to W. B. Arthur, S. N. Durlauf and D. A. Lane (eds), *The Economy as a Complex Evolving System II*, p. 15. Addison-Wesley, Reading, MA, 1997.

13. Kay, p. 70.

14. A. Kirman (1996). 'Some observations on interactions in economics.' Paper presented at Workshop on Structural Change, Manchester Metropolitan University, UK, 20–21 May.

15. *Ibid.*

16. Arthur *et al.*, p. 9

17. C. Davenant (1696). 'An Essay on the East-India Trade', p. 25. London.

18. A. Smith (1759). *The Theory of Moral Sentiments*, p. 82. Eds D. D. Raphael and A. L. Macfie. Clarendon Press, Oxford, 1976.

19. E. Burke (1797). Letter to Arthur Young, 23 May. In *The Works and Correspondence of the Right Honourable Edmund Burke*, Vol. II, p. 398. Francis & John Rivington, London, 1852.

20. Y. Louzoun, S. Solomon, J. Goldenberg and D. Mazursky (2002). 'The risk at being unfair: World-size global markets lead to economic instability', Preprint.

21. J. M. Epstein and R. Axtell (1996). *Growing Artificial Societies*, p. 136. MIT Press, Cambridge, MA.

CHAPTER 10

Epigraphs

G. K. Zipf (1949). *Human Behavior and the Principle of Least Effort*, p. 27. Hafner, New York, 1965.

H. Simon (1996). *The Sciences of the Artificial*, 3rd edn, p. 2. MIT Press, Cambridge, MA.

P. Anderson (1997). In W. B. Arthur, S. N. Durlauf and D. A. Lane (eds), *The Economy as a Complex Evolving System II*, p. 566. Addison-Wesley, Reading, MA.

1. L. Laloux, M. Potters, R. Cont, J.-P. Aguilar and J.-P. Bouchard (1999). 'Are financial crashes predictable?', *Europhysics Letters* **45**, 1–5.

2. P. Bak (1997). *How Nature Works*, pp. 187–8. Oxford University Press.

3. *Ibid.*, p. 191.

4. A. Tennyson (1842). 'Locksley Hall', p. 20, line 137. In *Locksley Hall and Other Poems*. Ernest Nister, London.

5. B. Mandelbrot (1983). *The Fractal Geometry of Nature*. Freeman & Co., New York.

6. E. Wolf (1996). 'Trends in household wealth during 1989–1992'. Submitted to the Department of Labor, New York. See also J. Diaz-Gimenez, V. Quadrini and J. V. Rios-Rull (1997). 'Dimensions of inequality: Facts on the US distributions of earnings, income and wealth', *Quarterly Review of the Federal Reserve Bank of Minneapolis* **21**, 3–21.

7. C. Snyder (1940). *Capitalism the Creator*, p. 417. Macmillan, New York.

8. Z.-F. Huang and S. Solomon (2001). 'Finite market size as a source of extreme wealth inequality and market instability', *Physica A* **294**, 503–13.

603

9. Zipf, p. 543.
10. *Ibid.*, p. 544.

CHAPTER 11

Epigraphs

C. Handy (1993). *Understanding Organizations*, 4th edn, p. 313. Penguin, London.

J. B. S. Haldane (1949). 'Shapes and weights'. In *What Is Life?*, p. 186. Alcuin Press, Welwyn Garden City.

E. F. Schumacher (1974). *Small Is Beautiful*, p. 55. Sphere, London.

1. Quoted in G. Monbiot (2000). *Captive State*, pp. 166–7. Macmillan, London.

2. *A Review of Monopolies and Mergers Policy: A consultative document.* Cmnd 7198, pp. 136–7. HMSO, London, 1978.

3. A. Smith (1776). *An Inquiry into the Nature and Causes of the Wealth of Nations.* Abridged version, ed. L. Dickey, p. 5. Hackett Publishing Co., Indianapolis.

4. Monbiot, p. 179.

5. R. Axtell (2001). 'Zipf distribution of US firm sizes', *Science* **293**, 1818–20.

6. J. Sutton (1997). 'Gibrat's legacy', *Journal of Economic Literature* **35**, 42–3.

7. R. Axtell (1999). 'The emergence of firms in a population of agents: Local increasing returns, unstable Nash equilibria, and power law size distributions'. Center on Social and Economic Dynamics, Working Paper No. 3, p. 3. Brookings Institution, Washington, DC.

8. Sutton, p. 57.

9. J. Kay (2003). 'The real economy', *Prospect* May, p. 28.

10. Axtell (1999), p. 87.

CHAPTER 12

Epigraphs

J. S. Mill (1843), *A System of Logic*, pp. 572–3. Longman, Green & Co., London, 1884.

E. Hobsbawm (1994). *Age of Extremes*, p. 147. Abacus, London.

L. N. Tolstoy (1869). *War and Peace*, transl. R. Edmunds, Vol. 2, p. 1426. Penguin, London, 1969.

1. Thucydides, *History of the Peloponnesian War*, 2nd edn, transl. B. Jowett, Book III, Para. 11. Oxford University Press, London, 1900.

2. R. Axelrod and D. S. Bennett (1993). 'A landscape theory of aggregation', *British Journal of Political Science* **23**, 211–33, here p. 219.

3. K. Waltz (1979). *Theory of International Politics*, p. 167. Addison-Wesley, Reading, MA.

4. G. H. Snyder (1984). 'The security dilemma in alliance politics', *World Politics* **36**, 461–95.

5. Hobsbawm, p. 162.

6. M. Oakeshott (1933). *Experience and its Modes*, p. 128. Cambridge University Press.

7. Quoted in N. Ferguson (ed.) (1997). *Virtual History*, p. 1. Picador, London.

8. H. Trevor-Roper (1981). 'History and imagination'. In *History and Imagination: Essays in Honour of H. R. Trevor-Roper* (ed. V. Pearl, B. Worden and H. Lloyd-Jones), p. 364. Duckworth, London.

9. Ferguson, p. 85.

10. *Ibid.*

CHAPTER 13

Epigraphs

K. Mannheim (1936). *Ideology and Utopia*. Harcourt, Brace & Co., New York. Quoted in R. Bierstedt (ed.), *The Making of Society*, p. 505. Random House, New York, 1959.

J. Epstein (2001). 'Learning to be thoughtless: Social norms and individual computation', *Computational Economics* **18**(1), 9–24.

M. Gladwell (2001). *The Tipping Point*, p. 259. Abacus, London.

1. *The Onion* **36**(40), 9 November 2000. <www.theonion.com/onion3640/bush_or_gore>.

2. O. von Bismarck (1863). Quoted in M. J. Hinich and M. C. Munger, *Analytical Politics*, p. 3. Cambridge University Press, 1997.

3. Epstein, 'Learning to be thoughtless . . .'.

4. T. C. Schelling (1973). *Micromotives and Macrobehavior*, p. 20. W. W. Norton, New York.

5. *Ibid.*, p. 23.

6. P. Ormerod (2002). 'Sense on segregation', *Prospect*, February, 12–14.

7. M. Lind (2001). 'Are there global political values?', *Prospect*, December, p. 20.

8. F. Dostoevsky (1911). *Crime and Punishment*, p. 64. J. M. Dent & Sons, London.

9. Gladwell, p. 8.

10. G. B. Shaw (1908). *Getting Married*, p. 111. In *Selected Passages from the Works of Bernard Shaw*, p. 142. Constable & Co., London, 1912.

11. *Ibid.*, p. 130. In *Selected Passages*, p. 145.

12. G. S. Becker (1991). *A Treatise on the Family*, p. 325. Harvard University Press, Cambridge, MA.

13. *Ibid.*, p. 232.

14. *Ibid.*, p. 32.

15. *Ibid.*, p. 59.

16. *Ibid.*, p. 4.

17. *Ibid.*

18. *Ibid.*, p. 19.

19. Schelling, p. 36.

20. W. Newmarch (1860). 'Some observations on the present position of statistical inquiry with suggestions for improving the organization and efficiency of the International Statistical Congress', *Journal of the Statistical Society of London* **23**, 362–3.

21. W. B. Arthur (1994). 'Inductive reasoning and bounded rationality', *American Economic Review* **84**, 406–11.

22. *Ibid.*

CHAPTER 14

Epigraphs

J.-J. Rousseau (1772), 'Considérations sur le Gouvernement de Pologne'. In C. E. Vaughan (ed.), *The Political Writings of Jean Jacques Rousseau*, Vol. II, p. 432. Cambridge University Press, 1915.

C. Gilday. Quoted in J. Mander, 'Technologies of globalization'. In E. Goldsmith and J. Mander (eds.), *The Case Against the Global Economy*, p. 52. Earthscan, London, 2001.

S. George (2002), Quoted in 'Globally locally', *New Scientist* 27 April, p. 44.

1. O. Wilde (1877). 'The Canterville ghost'. In *The Works of Oscar Wilde*, p. 194. Collins, London, 1948.

2. J. M. Epstein and R. Axtell (1996). *Growing Artificial Societies*, p. 163. MIT Press, Cambridge, MA.

3. J. M. Epstein. Quoted in I. Peterson, 'The gods of Sugarscape', *Science News*, 23 November, p. 332 (1996).

4. Epstein and Axtell, p. 52.

CHAPTER 15

Epigraphs

E. Durkheim (1879). *Revue philosophique*, December.

M. E. J. Newman (2003). 'Ego-centered networks and the ripple effect', *Social Networks* **25**, 83–95.

1. S. Milgram. Quoted in M. Gladwell, *The Tipping Point*, p. 36. Abacus, London, 2001.

2. J. Guare (1990). *Six Degrees of Separation: A Play*. Vintage, New York.

Figure 15.2
Kevin Bacon – Eddie Albert: *The Big Picture*
Kevin Bacon – Meryl Streep: *The River Wild*
Kevin Bacon – Jack Nicholson: *All the President's Men*
Kevin Bacon – Maximilian Schell: *Telling Lies in America*
Kevin Bacon – Allison Caine: *Diner*
Kevin Bacon – Diane Lane: *My Dog Skip*
Rod Steiger – Humphrey Bogart: *The Harder They Fall*
Rod Steiger – Eddie Albert: *The Longest Day*
Rod Steiger – Jack Nicholson: *Mars Attacks!*
Rod Steiger – Diane Lane: *Cattle Annie and Little Britches*
Rod Steiger – Karl Malden: *On the Waterfront*
Rod Steiger – Marlon Brando: *On the Waterfront*
Rod Steiger – Maximilian Schell: *The Chosen*
Marlon Brando – Karl Malden: *On the Waterfront*
Marlon Brando – Maximilian Schell: *The Freshman*
Marlon Brando – Jack Nicholson: *The Missouri Breaks*
Jack Nicholson – Meryl Streep: *Ironweed*
Humphrey Bogart – Eddie Albert: *The Wagons Roll at Night*
Karl Malden – Allison Caine: *Nuts*
Maximilian Schell – Meryl Streep: *Julia*

CHAPTER 16

Epigraphs

M. Castells (2001). *The Internet Galaxy: Reflections on the Internet, Business, and Society*, p. 207. Oxford University Press.

G. Cardoso (1998). *Para uma Sociologia do Ciberespaço: Comunidades Virtuais em Português*, p. 116 (transl. M. Castells). Celta Editora, Oeiras, Portugal.

W. Shakespeare (1602–4). *Othello*, III. iv. 69.

1. L. Irving (1998) Quoted in E. Wasserman, 'Feds take steps against threat of cyber terrorism'. CNN, 25 September 1998. <http://www.cnn.com/TECH/computing/9809/25/cyberterrorism.idg/>.

2. T. Hobbes (1651). *Leviathan*, Chapter XXIX, p. 353. Ed. C. B. Macpherson. Penguin, London, 1985.

3. Castells, p. 1.

4. *Ibid.*, p. 36.

5. *Ibid.*, p. 282.

6. S. Dalzell. Quoted in J. Naughton, *A Brief History of the Future*, pp. 190–91. Weidenfeld & Nicolson, London, 1999. See also <www2.epic.org/cda/cda_dc_opinion.html>.

7. *Ibid.*, p. 192.

8. T. Berners-Lee (1990). 'Information management: A proposal', CERN Internal Report, May, p. 9. See also <www.w3.org/History/1989/proposal.html>.

9. FBI Definitions, 28 Code of Federal Regulations Section 0.85.

10. A.-L. Barabási (2003). 'We're all on the Grid together', *New York Times*, 16 August.

11. *Ibid.*

12. A.-L. Barabási (2002). Personal communication.

13. W. B. Arthur, S. N. Durlauf and D. A. Lane (1997). *The Economy as a Complex Evolving System*, Vol. II, p. 9. Addison-Wesley, Reading, MA.

CHAPTER 17

Epigraphs

W. Temple (1751). 'An essay upon the origin and nature of government', *Works*, Vol I, p. 99. Quoted in introduction to T. Hobbes (1651), *Leviathan*, ed. C. B. Macpherson, p. 61. Penguin, London, 1985.

J. Locke (1690). *Essay on Civil Government* (second *Treatise on Government*). In *Two Treatises of Government*, ed. M. Goldie, p. 118. J. M. Dent, London, 1993.

I. Hay (1916). *The First Hundred Thousand*, pp. 224–5. William Blackwood, London.

1. S. Freud (1930). *Civilization and its Discontents*, pp. 32–3. Hogarth Press, London, 1973.
2. J. Locke (1690). Second *Treatise on Government*. In *Two Treatises of Government*, 2nd edn (ed. P. Laslett), p. 433. Cambridge University Press, 1970.
3. K. Popper (1997). *The Lesson of This Century*, p. 35. Routledge, London.
4. Freud, p. 60.
5. Locke, p. 124.
6. *Ibid.*, p. 117.
7. *Ibid.*, p. 124.
8. A. A. Cooper (Third Earl of Shaftesbury) (1711). *Characteristicks of Men, Manners, Opinions, Times*, Vol. ii, p. 67. Wyat, London. Republished (ed. P. Aynes) by Clarendon Press, Oxford, 1999.
9. Genesis 9:6.
10. Locke, p. 178.
11. J. S. Mill (1859). *On Liberty*, p. 68. Ed. G. Himmelfarb. Penguin, London, 1985.
12. J. S. Mill (1861). 'Utilitarianism'. In J. S. Mill and J. Bentham, *Utilitarianism and Other Essays*, ed. A. Ryan, p. 278. Penguin, London, 1987.

13. J. J. Rousseau (1754). 'Discourse on inequality'. Quoted in B. Russell, *A History of Western Philosophy*, p. 663. Unwin Paperbacks, London, 1984.

14. C. Darwin (1871). *Descent of Man, and Selection in Relation to Sex*, p. 64. John Murray, London, 1874.

15. T. Hobbes (1651). *Leviathan*, ed. C. B. Macpherson, Chapter 51. Penguin, London, 1985.

16. I. Kant (1795). *Perpetual Peace*

17. R. Kagan (2002). 'Power and weakness', *Policy Review*, June/July, 113.

18. E. Hobsbawm (1994). *Age of Extremes*, pp. 24–6. Abacus, London.

19. G. Dugdale (1932). *Langemarck and Cambrai*, p. 94. Wilding & Son, Shrewsbury.

20. G. Belton Cobb (1916). *Stand to Arms*, p. 74. Wells Gardner, Darton & Co., London.

21. T. Ashworth (1980). *Trench Warfare 1914–1918: The Live and Let Live System*, p. 144. Holmes & Meier, New York.

22. Hobbes, p. 190.

23. *Ibid.*

24. R. Axelrod (1984). *The Evolution of Cooperation*, p.124. Basic Books, New York.

25. Popper, p. 70.

26. Pericles' funeral oration is recounted by Thucydides in *The History of the Peloponnesian War*, Book II. This paraphrase is in Popper, p. 72.

27. Popper, pp. 36–7.

CHAPTER 18

Epigraphs

N. Machiavelli (1513). *The Prince*, Chapter XV, p. 91, transl. L. G. Bull. Penguin, London, 1981.

K. N. Waltz (1954). *Man, the State, and War*, p. 108. Columbia University Press, New York.

K. Popper (1988). 'Reflections on the theory and practice of the democratic state', talk given in Munich, 9 June. Reproduced in *The Lesson of This Century*, p. 79. Routledge, London, 1997.

1. Deuteronomy 19:19–21.
2. M. A. Nowak and K. Sigmund (1992). 'Tit for tat in heterogeneous populations', *Nature* 355, 250–53.
3. E. Burke (1770). 'Thoughts on the causes of the present discontents'. Quoted in M. J. Hinich and M. C. Munger (1997), *Analytical Politics*, p. 136. Cambridge University Press.
4. Popper, p. 49.
5. *Ibid.*, p. 55.
6. H. Kahn (1984). *Thinking About the Unthinkable in the 1980s*, p. 59. Simon & Schuster, New York.
7. B. Skyrms (1996). *Evolution of the Social Contract*, p. 24. Cambridge University Press.
8. Quoted in N. Sheehan and E. W. Kenworthy (eds) (1971). *Pentagon Papers*, p. 432. Times Books, New York.
9. K. Sigmund (1995). *Games of Life*, p. 191. Penguin, London.
10. R. Axelrod (1984). *The Evolution of Cooperation*, p. 184. Basic Books, New York.
11. D. Jones. Quoted in F. Goodall (1997). *A Question of Conscience*, p. 198. Sutton Publishing Group, Stroud.
12. C. Davies. Quoted in Goodall, p. 199.
13. Popper, p. 91.

CHAPTER 19

Epigraphs

R. W. Emerson (1860). *The Conduct of Life*, p. 176. J. M. Dent & Sons, London, 1908.

G. Lundberg (1939). *Foundations of Sociology*. Macmillan, London. In R. Bierstedt (ed.), *The Making of Society*, p. 518. Random House, New York, 1959.

H. Simon (1996). *The Sciences of the Artificial*, 3rd edn, p. 139. MIT Press, Cambridge, MA.

J. Barnes (1989). *A History of the World in 10½ Chapters*. Jonathan Cape, London.

1. J. Carey (ed.) (1999). *The Faber Book of Utopias*, p.xii. Faber & Faber, London.
2. Quoted in F. A. Hayek (1962). *The Road to Serfdom*, p. 18. Routledge, London.
3. J. S. Mill (1843). *A System of Logic*, Book 6. In Bierstedt, p. 205.
4. G. Simmel (1908). In *The Sociology of Georg Simmel*, ed. and transl. K. H. Wolff. Free Press, Glencoe, IL, 1950.
5. R. E. Park (1927). 'Human nature and collective behavior', *American Journal of Sociology* 32, 733.
6. A. Smith (1776). *An Inquiry into the Nature and Causes of the Wealth of Nations*. Abridged version, ed. L. Dickey, p. 11. Hackett Publishing Co., Indianapolis, 1993.
7. B. Mandeville (1705). *The Fable of the Bees: Or, Private Vices, Publick Benefits*. London. Quoted in B. Willey (1940), *The Eighteenth Century Background*, p. 96. Columbia University Press, New York.
8. J. Swift (1729). 'A modest proposal for preventing the children of poor people in Ireland from being a burden to their parents or country and for making them beneficial to the publick'. In J. Hayward (ed.), *Selected Prose Works of Jonathan Swift*, p. 430. Cresset Press, London, 1949.
9. L. Mumford (1938). *The Culture of Cities*, p. 93. Secker & Warburg, London.
10. Mill, *Logic*, in Bierstedt, p. 203.
11. A. Ferguson (1766). *Institutes of Moral Philosophy*. Quoted in G. Bryson, *Man and Society*, p. 35. Princeton University Press, 1945.

12. A. Ferguson (1767). *An Essay on the History of Civil Society*, ed. D. Forbes, p. 122. Edinburgh University Press, 1966.

13. R. W. Emerson (1844). 'On politics'. In *The Essays of Ralph Waldo Emerson*, p. 325. Grant Richards, London, 1903.

14. K. Marx (1845). *Theses on Feuerbach*, No. 11. (Published, edited by F. Engels, as an appendix to F. Engels, *Ludwig Feuerbach und der Ausgang der klassischen deutschen Philosophie.* J. H. W. Dietz, Stuttgart, 1888.)

15. P. Ormerod (1998). *Butterfly Economics*, p. 182. Faber & Faber, London.

16. R. Olson (1990). *Science Deified and Science Defied*, Vol. 2, pp. 210–11. University of California Press, Berkeley, CA.

17. Cicero, *De regibus*. In Bierstedt, p. 41.

18. Aristotle, *Politics*, transl. W. D. Ross. Quoted in Bierstedt, p.24.

19. St Thomas Aquinas (*c.*1259). *On the Governance of Rulers*, transl. G. B. Phelan. In Bierstedt, p. 52.

20. *Ibid.*, p. 53.

21. T. Hobbes (1651). *Leviathan*, ed. C. B. Macpherson, p. 185. Penguin, London, 1985.

22. Smith, p. 130.

23. F. von Hayek (1962). *The Road to Serfdom*, p. 27.

24. *Ibid.*, p. 14.

25. T. C. Schelling (1978). *Micromotives and Macrobehavior*, pp. 127–8. W. W. Norton, New York.

26. *Ibid.*, p. 19.

27. K. Malik (2000). 'Natural science', *Prospect*, August/September, p. 37.

28. Mill, *Logic*, in Bierstedt, p. 209.

29. F. Giddings (1924). *The Scientific Study of Human Society*. In Bierstedt, p. 362.

Bibliography

A. Y. Abul-Magd (2002). 'Wealth distribution in an ancient Egyptian society', *Physical Review E* **66**, 057104.

L. Adamic (1999). 'The Small World Web', Preprint, Xerox Palo Alto Research Center, <http://www.hpl.hp.com/shl/papers/smallworld/smallworldpaper.html>.

R. Alberich, J. Miro-Julia and F. Rosselló (2002). 'Marvel Universe looks almost like a real social network', Preprint, <http://arxiv.org/abs/cond-mat/0202174>.

R. Albert and A.-L. Barabási (2000). 'Topology of evolving networks: Local events and universality', *Physical Review Letters* **85**, 5234–7.

R. Albert and A.-L. Barabási (2002). 'Statistical mechanics of complex networks', *Reviews of Modern Physics* **74**, 47–97.

R. Albert, H. Jeong and A.-L. Barabási (1999). 'Diameter of the World-Wide Web', *Nature* **401**, 130–31.

R. Albert, H. Jeong and A.-L. Barabási (2000). 'Error and attack tolerance of complex networks', *Nature* **406**, 378–82.

M. Allingham (2002). *Choice Theory: A Very Short Introduction*. Oxford University Press.

V. Allsop (1995). *Understanding Economics*. Routledge, London.

L. A. N. Amaral, A. Scala, M. Barthélémy and H. E. Stanley (2000). 'Classes of small-world networks', *Proceedings of the National Academy of Sciences USA* **97**, 11149–52.

P. W. Anderson, K. J. Arrow and D. Pines (eds) (1988). *The Economy as an Evolving Complex System*. Addison-Wesley, Redwood City, CA.

C. Andersson, A. Hellervik, K. Lindgren, A. Hagson and J. Tornberg (2003). 'The urban economy as a scale-free network', Preprint arXiv:cond-mat/0303535.

M. Anghel, Z. Toroczkai, K. E. Bassler and G. Korniss (2003). 'Competition in social networks: Emergence of a scale-free leadership structure and collective efficiency', Preprint, <http://arxiv.org/abs/cond-mat/0307740>.

W. B. Arthur (1989). 'Competing technologies, increasing returns, and lock-in by historical events', *Economic Journal* **99**, 116–31.

W. B. Arthur (1994). 'Inductive reasoning and bounded rationality (the El Farol problem)', *American Economic Review* **84**, 406–11.

W. B. Arthur (1999). 'Complexity and the economy', *Science* **284**, 107–9.

W. B. Arthur, S. N. Durlauf and D. A. Lane (1997). *The Economy as a Complex Evolving System II*. Addison-Wesley, Reading, MA.

J. Aubrey. *Brief Lives*. Ed. J. Buchanan-Brown. Penguin, London, 2000.

J. Avery (1997). *Progress, Poverty and Population*. Frank Cass, London.

R. Axelrod (1984). *The Evolution of Cooperation*. Basic Books, New York.

R. Axelrod (1997). 'The dissemination of culture', *Journal of Conflict Resolution* **41**, 203–26.

R. Axelrod (1997). 'Advancing the art of simulation in the social sciences', in R. Conte, R. Hegselmann and P. Terna (eds), *Simulating Social Phenomena*, pp. 21–40. Springer-Verlag, Berlin.

R. Axelrod (2000). 'On six advances in cooperation theory', *Analyse and Kritik* **22**, 130–51.

R. Axelrod and D. S. Bennett (1993). 'A landscape theory of aggregation', *British Journal of Political Science* **23**, 211–33.

R. Axelrod, W. Mitchell, R. E. Thomas, D. S. Bennett and E. Bruderer (1995). 'Coalition formation in standard-setting alliances', *Management Science* **41**, 1493–508.

R. Axtell (1999). 'The emergence of firms in a population of agents: local increasing returns, unstable Nash equilibria, and power law size distributions', Preprint, Working Paper No. 3, Brookings Institution, Washington, DC.

R. L. Axtell (2001). 'Zipf distribution of U.S. firm sizes', *Science* **293**, 1818–20.

R. Axtell, R. Axelrod, J. M. Epstein and M. D. Cohen (1996). 'Aligning simulation models: A case study and results', *Computational and Mathematical Organization Theory* **1**, 123–41.

P. Bak (1997). *How Nature Works*. Oxford University Press.

P. Bak, K. Chen and M. Creutz (1989). 'Self-organized criticality in the "Game of Life"', *Nature* **342**, 780–81.

P. Ball (1998). *The Self-Made Tapestry*. Oxford University Press.

P. Ball (2000). 'Jams tomorrow', *New Scientist*, 15 January, 34–8.

P. Ball (2002). 'The physical modeling of society: A historical perspective', *Physica A* **314**, 1–14.

S. C. Bankes (2002). 'Agent-based modeling: A revolution?' *Proceedings of the National Academy of Sciences USA* **99**, 7199–200.

A.-L. Barabási (2001). 'The physics of the Web' *Physics World*, July, 33.

A.-L. Barabási (2002). *Linked*. Perseus, Cambridge.

A.-L. Barabási and R. Albert (1999). 'Emergence of scaling in random networks', *Science* **286**, 509–12.

A.-L. Barabási, R. Albert and H. Jeong (1999). 'Mean-field theory for scale-free random networks', *Physica A* **272**, 173–87.

D. F. Batten (2000). *Discovering Artificial Economics*. Westview Press, Boulder, CO.

M. Batty (1997). 'Predicting where we walk', *Nature* **388**, 19–20.

M. Batty (2003). 'Agent-based pedestrian modelling', in P. Longley and M. Batty (eds), *Advanced Spatial Analysis*, Chapter 5. ESRI Press, London.

M. Batty (2003). 'The emergence of cities: Complexity and urban dynamics', Preprint.

M. Batty, J. Desyllas and E. Duxbury (2003). 'Safety in numbers? Modelling crowds and designing control for the Notting Hill Carnival', *Urban Studies* **40**, 1573–90.

M. Batty and P. Longley (1994). *Fractal Cities*. Academic Press, London.

G. S. Becker (1968). 'Crime and punishment: An economic approach', *Journal of Political Economy* **76**, 443–78.

G. S. Becker (1991). *A Treatise on the Family*. Harvard University Press, Cambridge, MA.

W. A. Bentley and W. J. Humphreys (1931). *Snow Crystals*. McGraw-Hill Book Co., New York. Reprinted in 1962 by Dover, New York.

A.T. Bernardes, D. Stauffer & J. Kertész (2001). 'Election results and the Sznajd model on Barabasi network.' Preprint, <http://arvix.org/abs/cond-mat/0111147>.

B. J. L. Berry, L. D. Kiel and E. Elliott (2002). 'Adaptive agents, intelligence, and emergent human organization: Capturing complexity through agent-based modeling', *Proceedings of the National Academy of Sciences USA* **99**, 7187–8.

C. Bicchieri, R. Jeffrey and B. Skyrms (1999). *The Logic of Strategy*. Oxford University Press.

R. Bierstedt (ed.) (1959). *The Making of Society*. Random House, New York.

D. J. Boorstin (1985). *The Discoverers*. Vintage Books, New York.

L. Borland (2002). 'Option pricing formulas based on a non-gaussian stock price model', *Physical Review Letters* **89**, 098701.

J.-P. Bouchard, P. Cizeau, L. Laloux and M. Potters (1999). 'Mutual attractions: Physics and finance', *Physics World*, January, 25.

J.-P. Bouchard and M. Mézard (2000). 'Wealth condensation in a simple model of economy', *Physica A* **282**, 536–45.

S. J. Brams (1975). *Game Theory and Politics*. Free Press, New York.

S. J. Brams (1978). *The Presidential Election Game*. Yale University Press, New Haven, CT.

L. R. Brown (2001). *Eco-Economy*. Earthscan, London.

M. Buchanan (2000). *Ubiquity*. Weidenfeld & Nicolson, London.

M. Buchanan (2000). 'That's the way the money goes', *New Scientist*, 19 August, 22–6.

M. Buchanan (2002). *Small World*. Weidenfeld & Nicolson, London.

Z. Burda, D. Johnston, J. Jurkiewicz, M. Kamiński, M. A. Nowak, G. Papp and I. Zahed (2002). 'Wealth condensation in Pareto macroeconomies', *Physical Review E* **65**, 026102.

S. Camazine, J.-L. Deneubourg, N. R. Franks, J. Sneyd, G. Theraulaz and E. Bonabeau (2001). *Self-Organization in Biological Systems*. Princeton University Press.

M. Campbell and P. Ormerod (1997). 'Social interaction and the dynamics of crime', Preprint.

A. Capocci, G. Caldarelli, R. Marchetti and L. Pietronero (2001). 'Growing dynamics of Internet providers', *Physical Review E* **64**, 035105.

R. Carvalho and A. Penn (2003). 'Scaling and universality in the microstructure of urban space', Preprint, <http://arxiv.org/abs/cond-mat/0305164>.

C. Castellano, M. Marsili and A. Vespignani (2000). 'Nonequlibrium phase transition in a model for social influence', *Physical Review Letters* **85**, 3536–9.

M. Castells (2001). *The Internet Galaxy: Reflections on the Internet, Business, and Society*. Oxford University Press.

C. Cercignani (1998). *Ludwig Boltzmann: The Man Who Trusted Atoms*. Oxford University Press.

D. Challet, M. Marsili and G. Ottino (2003). 'Shedding light on El Farol', Preprint, <http://arxiv.org/abs/cond-mat/0306445>.

D. Challet, M. Marsili and Y.-C. Zhang (1999). 'Modeling market mechanism with minority game', Preprint, <http://arxiv.org/abs/cond-mat/909265>.

D. Challet and Y.-C. Zhang (1997). 'Emergence of cooperation and organization in an evolutionary game', *Physica A* **246**, 407–18.

D. D. T. Chen (2000). 'Smart growth', *Scientific American*, December, 85–91.

D. Cohen (2002). 'All the world's a net', *New Scientist*, 13 April, 24–9.

M. C. Cohen, R. L. Riolo and R. Axelrod (2000). 'The role of social structure in the maintenance of cooperative regimes', *Rationality and Society* **13**, 5–32.

R. Cohen, K. Erez, D. ben-Avraham and S. Havlin (2000). 'Resilience of the Internet to random breakdowns', *Physical Review Letters* **85**, 4626–8.

J. J. Collins and C. C. Chow (1998). 'It's a small world', *Nature* **393**, 409–10.

R. Cont (1999). 'Modelling economic randomness: Statistical mechanics of market phenomena', in M. T. Batchelor and L. T. Wille (eds), *Statistical Physics on the Eve of the 21st Century: The James B. McGuire Festschrift*, pp. 47–64. World Scientific, Singapore.

R. Cont (1999). 'Statistical properties of financial time series'. Lecture given at the Symposium on Mathematical Finance, Fudan University, Shanghai, 10–24 August.

R. Conte (2002). 'Agent-based modeling for understanding social intelligence', *Proceedings of the National Academy of Sciences USA* **99**, 7189–90.

R. Cont and J.-P. Bouchard (2000). 'Herd behavior and aggregate fluctuations in financial markets', *Macroeconomic Dynamics* **4**, 170–96.

R. Cont, M. Potters and J.-P. Bouchard (1997). 'Scaling in stock market data: Stable laws and beyond'. In B. Dubrulle, F. Graner and D. Sornette (eds). *Scale Invariance and Beyond*. Proceedings of the CNRS Workshop on Scale Invariance, Les Houches. Springer-Verlag, Berlin.

R. N. Costa Filho, M. P. Almeida, J. S. Andrade, Jr and J. E. Moreira (1999). 'Scaling behavior in a proportional voting process' *Physical Review E* **60**, 1067–8.

F. Cramer (1993). *Chaos and Order*. VCH, Weinheim.

G. Csányi and B. Szendröi (2003). 'Structure of a large social network', Preprint, <http://arxiv.org/abs/cond-mat/0305580>.

G. Cuniberti, A. Valleriani and J. L. Vega (2001). 'Effects of regulation on a self-organized market', Preprint, <http://arxiv.org/abs/cond-mat/0108533>.

A. Czirók, A.-L. Barabási and T. Vicsek (1999). 'Collective motion of self-propelled particles: kinetic phase transition in one dimension', *Physical Review Letters* **82**, 209–12.

A. Czirók, M. Vicsek and T. Vicsek (1999). Collective motions of organisms in three dimensions. *Physica A* **264**, 299–304.

A. Czirók and T. Vicsek (1999). 'Collective motion', in D. Reguera, M. Rubi and J. Vilar (eds), *Statistical Mechanics of Biocomplexity*, Lecture Notes in Physics 527, 152–64. Springer-Verlag, Berlin.

A. Czirók and T. Vicsek (2000). 'Collective behavior of interacting self-propelled particles', *Physica A* **281**, 17–29.

P. A. David (1993). 'Path-dependence and predictability in dynamical systems with local network externalities: A paradigm for historical economics', in D. Foray and C. Freeman (eds), *Technology and Wealth of Nations: The Dynamics of Constructed Advantage*. Pinter, London.

J. Davidsen, H. Ebel and S. Bornholdt (2002). 'Emergence of a small world from local interactions: Modeling acquaintance networks', *Physical Review Letters* **88**, 128701.

C. Davidson (2001). 'E-mmune from attack', *New Scientist*, 31 March, 35–7.

G. De Fabritiis, F. Pammolli and M. Riccaboni (2003). 'On size and growth of business firms', *Physica A* **324**, 38–44.

Z. Dezso and A.-L. Barabási (2001). 'Can we stop the AIDS epidemic?', *Physical Review E* **65**, 055103(R).

P. S. Dodds, R. Muhamad and D. J. Watts (2003). 'An experimental study of search in global social networks', *Science* **301**, 827–9.

S. N. Dorogovtsev and J. F. F. Mendes (2000). 'Evolution of networks with aging of sites', *Physical Review E* **62**, 1842–5.

H. Ebel, L.-I. Mielsch and S. Bornholdt (2002). 'Scale-free topology of e-mail networks', Preprint, <http://arxiv.org/abs/cond-mat/0201476>.

V. M. Eguíluz and K. Klemm (2002). 'Epidemic threshold in structured scale-free networks', *Physical Review Letters* **89**, 108701.

V. M. Eguíluz and M. G. Zimmermann (2000). 'Transmission of information and herd behavior: An application to financial markets', *Physical Review Letters* **85**, 5659–62.

R. W. Emerson (1860). 'The conduct of life', in *The Conduct of Life and Other Essays*. J. M. Dent, London, 1908.

J. M. Epstein (2001). 'Learning to be thoughtless: social norms and individual computation', *Computational Economics* **18**(1), 9–24.

J. M. Epstein and R. Axtell (1996). *Growing Artificial Societies: Social Science From the Bottom Up*. Brookings Institution Press, Washington, DC.

K. A. Eriksen, I. Simonsen, S. Maslov and K. Sneppen (2003). 'Modularity and extreme edges of the Internet', *Physical Review Letters* **90**, 148701.

I. Farkas, D. Helbing and T. Vicsek (2002). 'Mexican waves in an excitable medium', *Nature* **419**, 131–2.

J. D. Farmer (1999). 'Physicists attempt to scale the ivory towers of finance', *Computing in Science and Engineering*, November/December, 26–39.

E. Fehr and S. Gächter (2002). 'Altruistic punishment in humans', *Nature* **415**, 137–40.

N. Ferguson (ed.) (1997). *Virtual History*. Picador, London.

R. Florian and S. Galam (2000). 'Optimizing conflicts in the formation of strategic alliances', *European Physical Journal B* **16**, 189–94.

H. Föllmer (1974). 'Random economies with many interacting agents', *Journal of Mathematical Economics* **1**, 51–62.

T. Frank (2002). 'Talking bull', *Guardian*, weekend supplement, 17 August, 21.

S. Freud (1973). *Civilization and its Discontents*, transl. J. Riviere, ed. J. Strachey. Hogarth Press, London.

D. Gauthier (1969). *The Logic of Leviathan*. Clarendon Press, Oxford.

I. Giardina and J.-P. Bouchard (2002). 'Bubbles, crashes and intermittency in agent based market models', Preprint, <http://arxiv.org/abs/cond-mat/0206222>.

R. Gimblett, C. A. Roberts, T. C. Daniel, M. Ratliff, M. J. Meitner, S. Cherry, D. Stallman, R. Bogle, R. Allred, D. Kilbourne and J. Bieri (2000). 'An intelligent agent based model for simulating and evaluating river trip scenarios along the Colorado River in Grand Canyon National Park', in H. R. Gimblett (ed.), *Integrating GIS and Agent Based Modeling Techniques for Understanding Social and Ecological Processes*, 245–75. Oxford University Press.

M. Gladwell (2001). *The Tipping Point*. Abacus, London.

N. S. Glance and B. A. Huberman (1994). 'The dynamics of social dilemmas', *Scientific American*, March, 76–81.

P. M. Gleiser and L. Danon (2003). 'Community structure in jazz', Preprint, <http://arxiv.org/abs/cond-mat/0307434>.

E. Goldsmith and J. Mander (eds) (2001). *The Case Against the Global Economy and for a Turn Towards Localization*. Earthscan Publications, London.

P. Gopikrishnan, V. Plerou, L. A. Nuñes Amaral, M. Meyer and H. E. Stanley (1999). 'Scaling of the distribution of fluctuations of financial market indices', *Physical Review E* **60**, 5305–16.

M. Granovetter (1973). 'The strength of weak ties', *American Journal of Sociology* **78**, 1360–80.

M. Granovetter (2003). 'Ignorance, knowledge, and outcomes in a small world', *Science* **301**, 73–4.

J. Guare (1990). *Six Degrees of Separation: A Play*. Vintage, New York.

C. Handy (1993). *Understanding Organizations*, 4th edn. Penguin, London.

P. M. Harman (1998). *The Natural Philosophy of James Clerk Maxwell*. Cambridge University Press.

F. A. Hayek (1962). *The Road to Serfdom*. Routledge, London.

B. Hayes (2002). Statistics of deadly quarrels. *Computing Science*, January/February.

D. Helbing (1994). 'A mathematical model for the behavior of individuals in a social field', *Journal of Mathematical Sociology* **19**, 189–219.

D. Helbing (1996). 'A stochastic behavioral model and a "microscopic" foundation of evolutionary game theory', *Theory and Decision* **40**, 149–79.

D. Helbing (2001). 'Traffic and related self-driven many-particle systems', *Reviews of Modern Physics* **73** 1067–141.

D. Helbing, I. Farkas and T. Vicsek (2000). 'Simulating dynamical features of escape panic', *Nature* **407**, 487–90.

D. Helbing, A. Hennecke and M. Treiber (1999). 'Phase diagram of traffic states in the presence of inhomogeneities', *Physical Review Letters* **82**, 4360–63.

D. Helbing and B. A. Huberman (1998). 'Coherent moving states in highway traffic', *Nature* **396**, 738–40.

D. Helbing, J. Keltsch and P. Molnar (1997). 'Modelling the evolution of human trail systems', *Nature* **388**, 47–9.

D. Helbing and P. Molnar (1995). 'Social force model for pedestrian dynamics', *Physical Review E* **51**, 4282–6.

D. Helbing, P. Molnar, I. J. Farkas and K. Bolay (2001). 'Self-organizing pedestrian movement', *Environment and Planning B: Planning and Design* **28**, 361–83.

D. Helbing and M. Schreckenberg (1999). 'Cellular automata simulating experimental properties of traffic flow', *Physical Review E* **59**, R2505–8.

D. Helbing and T. Vicsek (1999). 'Optimal self-organization', *New Journal of Physics* **1**, 13.1–13.7.

L. F. Henderson (1971). 'The statistics of crowd fluids', *Nature* **229**, 381–3.

B. Hillier and J. Hanson (1984). *The Social Logic of Space*. Cambridge University Press.

M. J. Hinich and M. C. Munger (1997). *Analytical Politics*. Cambridge University Press.

T. Hobbes (1651). *Leviathan*. Ed. C. B. Macpherson. Penguin, London, 1985.

E. Hobsbawm (1995). *Age of Extremes*. Abacus, London.

S. Hod and E. Nakar (2002). 'Self-segregation versus clustering in the evolutionary minority game', *Physical Review Letters* **88**, 238702.

Z.-F. Huang and S. Solomon (2001). 'Finite market size as a source of extreme wealth inequality and market instability', *Physica A* **294**, 503–13.

B. A. Huberman and L. A. Adamic (1999). 'Growth dynamics of the World-Wide Web', *Nature* **401**, 131.

B. A. Huberman and R. M. Lukose (1997). 'Social dilemmas and Internet congestion', *Science* **277**, 535–7.

C. Jencks (1995). *The Architecture of the Jumping Universe*. Academic Editions, London.

H. Jeong, B. Tombor, R. Albert, Z. N. Oltvai and A.-L. Barabási (2000). 'The large-scale organization of metabolic networks', *Nature* **407**, 651–4.

R. Kagan (2002). 'Power and weakness', *Policy Review* **113** June/July <http://www.policyreview.org/JUN02/kagan.html>.

S. Keen (2001). *Debunking Economics: The Naked Emperor of the Social Sciences*. Pluto Press and Zed Books, Sydney and London.

S. Keen (2003). 'Standing on the toes of pygmies: Why econophysics must be careful of the economic foundations on which it builds', *Physica A* **324**, 108–16.

S. Keen, J. Legge, G. Fishburn and R. Standish (2003). 'Aggregation problems in the non-interactive equilibrium theory of markets', Preprint.

B. S. Kerner (1998). 'Experimental features of self-organization in traffic flow', *Physical Review Letters* **81**, 3797–800.

B. S. Kerner (1999). 'The physics of traffic', *Physics World*, August, 25–30.

B. S. Kerner (2003). 'Control of spatial-temporal congested traffic patterns at highway bottlenecks', Preprint, <http://arxiv.org/abs/cond-mat/0309017>.

B. S. Kerner, S. L. Klenov and D. E. Wolf (2002). 'Cellular automata approach to three-phase traffic theory', Preprint, <http://arxiv.org/abs/cond-mat/0206370>.

B. S. Kerner and H. Rehborn (1996). 'Experimental features and characteristics of traffic jams', *Physical Review E* **53**, R1297–R1300.

B. S. Kerner and H. Rehborn (1997). 'Experimental properties of phase transitions in traffic flow', *Physical Review Letters* **79**, 4030–33.

A. Kirchner and A. Schadschneider (2002). 'Simulation of evacuation processes using a bionics-inspired cellular automaton model for pedestrian dynamics', Preprint, <http://arxiv.org/abs/cond-mat/0203461>.

A. Kirman (1996). 'Some observations on interactions in economics' Paper presented at Workshop on Structural Change, Manchester Metropolitan University, May 20–21. Available at <http://www.cpm.mmu.ac.uks/pub/workshop/kirman.html>

A. P. Kirman (1994). 'Economics with interacting agents'. Working Paper 94–05–030, Santa Fe Institute, Santa Fe, New Mexico.

K. Klemm and V. M. Eguíluz (2002). 'Growing scale-free networks with small-world behavior', *Physical Review E* **65**, 057102.

K. Klemm, V. M. Eguíluz, R. Toral and M. San Miguel (2002). 'Global culture: A noise induced transition in finite systems', Preprint, <http://arxiv.org/abs/cond-mat/0205188>.

W. Knospe, L. Santen, A. Schadschneider and M. Schreckenberg (2001). 'Human behaviour as the origin of traffic phases', *Physical Review E* **65**, 015101.

P. Krugman (1994). *Peddling Prosperity*. W. W. Norton, New York.

P. Krugman (1996). *The Self-Organizing Economy*. Blackwell, Oxford.

CRITICAL MASS

S. Kurtz (2002). 'The future of "history"', *Policy Review* **113**, June/July <http://www.policyreview.org/JUN02/kurtz.html>.

L. Laloux, M. Potters, R. Cont, J.-P. Aguilar and J.-P. Bouchard (1999). 'Are financial crashes predictable?', *Europhysics Letters* **45**, 1–5.

R. Landauer (1975). 'Inadequacy of entropy and entropy derivatives in characterizing the steady state', *Physical Review A*, **12**, 636–8.

V. Latora and M. Marchioro (2001). 'Efficient behaviour of small-world networks', *Physical Review Letters* **87**, 198701.

S. Lawrence and C. L. Giles (1998). 'Searching the World Wide Web', *Science* **280**, 98–100.

B. LeBaron (2002). 'Short-memory traders and their impact on group learning in financial markets', *Proceedings of the National Academy of Sciences USA* **99**, 7201–6.

H. Y. Lee, H.-W. Lee and D. Kim (2000). 'Phase diagram of congested traffic flow: An empirical study', *Physical Review E* **62**, 4737–41.

R. Lewin (1992). *Complexity*. Macmillan, New York.

W. Li and X. Cai (2003). 'Statistical analysis of airport network of China', Preprint, <http://arxiv.org/abs/cond-mat/0309236>.

F. Liljeros, C. R. Edling, L. A. Nuñes Amaral, H. E. Stanley and Y. Åberg (2001). 'The web of human sexual contacts', *Nature* **411**, 907–8.

F. Lillo and R. N. Mantegna (2000). 'Variety and volatility in financial markets', *Physical Review E* **62**, 6126–34.

F. Lillo and R. N. Mantegna (2000). 'Symmetry alteration of ensemble return distribution in crash and rally days of financial markets', *European Physical Journal B* **15**, 603–6.

R. E. Litan and A. N. Rivlin (2000). 'The economy and the Internet: What lies ahead?', Conference report No. 4, Brookings Institution, Washington, DC.

J. Locke (1689). *Two Treatises of Government*. Ed. M. Goldie. J. M. Dent, London, 1993.

M. P. Lombardo (1985). Mutual restraint in tree swallows: A test of the TIT FOR TAT model of reciprocity', *Science* **227**, 1363–5.

Y. Louzoun, S. Solomon, J. Goldenberg and D. Mazursky (2002). 'The risk at being unfair: World-size global markets lead to economic instability.'

Preprint.

S. Lübeck, M. Schreckenberg and K. D. Usadel (1998). 'Density fluctuations and phase transition in the Nagel–Schreckenberg traffic flow model', *Physical Review E* **57**, 1171–4.

T. Lux and M. Marchesi (1999). 'Scaling and criticality in a stochastic multi-agent model of a financial market', *Nature* **397**, 498–500.

H. A. Makse, J. S. Andrade, Jr, M. Batty, S. Havlin and H. E. Stanley (1998). 'Modeling urban growth patterns with correlated percolation', *Physical Review E* **58**, 7054–62.

H. A. Makse, S. Havlin and H. E. Stanley (1995). 'Modelling urban growth patterns', *Nature* **377**, 608–12.

B. Mandelbrot (1963). 'The variation of certain speculative prices', *Journal of Business* **35**, 394–419.

B. Mandelbrot (1977). *The Fractal Geometry of Nature*. W. H. Freeman, New York.

B. Mandelbrot (1999). 'A multifractal walk down Wall Street' *Scientific American*, February, 70–73.

R. N. Mantegna and H. E. Stanley (1995). 'Scaling behaviour in the dynamics of an economic index', *Nature* **376**, 46–9.

R. N. Mantegna and H. E. Stanley (1996). 'Turbulence and financial markets', *Nature* **383**, 587–8.

R. N. Mantegna and H. E. Stanley (1999). *An Introduction to Econophysics*. Cambridge University Press.

N. Mathias and V. Gopal (2001). 'Small worlds: How and why', *Physical Review E* **63**, 021117.

R. Matthews (1999). 'Get connected', *New Scientist*, 4 December, 24–8.

R. May (1987). 'More evolution of cooperation', *Nature* **327**, 15–17.

R. M. May and A. L. Lloyd (2001). 'Infection dynamics on scale-free networks', *Physical Review E* **64**, 066112.

A. S. Mikhailov and D. H. Zanette (1999). 'Noise-induced breakdown of coherent collective motion in swarms', *Physical Review E* **60**, 4571–5.

S. Milgram (1967). 'The small world problem', *Psychology Today* **2**, 60–67.

M. Milinski (1987). 'TIT FOR TAT in sticklebacks and the evolution of cooperation', *Nature* **325**, 433–5.

J. S. Mill (1859). *On Liberty*. Ed. G. Himmelfarb. Penguin, London, 1985.

J. S. Mill (1872). *System of Logic*. Longman, Green, Reader & Dyer, London.

J. S. Mill and J. Bentham (1987). *Utilitarianism and Other Essays*. Ed. A. Ryan. Penguin, London.

M. Mitzenmacher (2003). 'A brief history of generative models for power law and lognormal distributions', Preprint, Harvard University.

C. Moore and M. E. J. Newman (2000). 'Epidemics and percolation in small-world networks', *Physical Review E* **61**, 5678–82.

T. More (1516). *Utopia*. Everyman's Library, London, 1992.

S. Mossa, M. Barthélémy, H. E. Stanley and L. A. Nuñes Amaral (2002). 'Truncation of power law behavior in "scale-free" network models due to information filtering', *Physical Review Letters* **88**, 138701.

L. Muchnik and S. Solomon (2003). 'Statistical mechanics of conventional traders may lead to non-conventional market behaviour', Preprint.

L. Mumford (1938). *The Culture of Cities*. Secker & Warburg, London.

J. Naughton (1999). *A Brief History of the Future*. Weidenfeld & Nicolson, London.

Z. Néda, E. Ravasz, Y. Brechet, T. Vicsek and A.-L. Barabási (2000). 'Self-organizing processes: The sound of many hands clapping', *Nature* **403**, 849–50.

L. Neubert, L. Santen, A. Schadschneider and M. Schreckenberg (1999). 'Single-vehicle data of highway traffic: A statistical analysis', *Physical Review E* **60**, 6480–90.

M. E. J. Newman (2000). 'Models of the small world: A review', Preprint, <http://arxiv.org/abs/cond-mat/0001118>.

M. E. J. Newman (2001). 'Ego-centred networks and the ripple effect', Preprint, <http://arxiv.org/abs/cond-mat/0111070>.

M. E. J. Newman (2001). 'The structure of scientific collaboration networks', *Proceedings of the National Academy of Sciences USA* **98**, 404–9.

M. E. J. Newman, S. Forrest and J. Balthrop (2002). 'Email networks and the spread of computer viruses', *Physical Review E* **66**, 035101.

M. E. J. Newman and J. Park (2003). 'Why social networks are different from other types of networks', Preprint, <http://arxiv.org/abs/cond-mat/0305612>.

M. Nowak and R. M. May (1992). 'Evolutionary games and spatial chaos', *Nature* **359**, 826–9.

M. Nowak and R. May (1993). 'The spatial dilemmas of evolution', *International Journal of Bifurcation and Chaos* **3**, 35–78.

M. Nowak, R. M. May and K. Sigmund (1995). 'The arithmetics of mutual help', *Scientific American*, June, 50–55.

M. Nowak and K. Sigmund (1992). 'Tit for tat in heterogeneous populations', *Nature* **355**, 250–53.

M. Nowak and K. Sigmund (1993). 'Chaos and the evolution of cooperation', *Proceedings of the National Academy of Sciences USA* **90**, 5091–4.

M. Nowak and K. Sigmund (1993). 'A strategy of win-stay, lose-shift that outperforms tit-for-tat in the Prisoner's Dilemma game, *Nature* **364**, 56–8.

F. L. Nussbaum (1965). *The Triumph of Science and Reason 1660–1685*. Harper & Row, New York.

R. Olson (1990). *Science Deified and Science Defied*, Vol. 2. University of California Press, Berkeley.

P. Ormerod (1998). *Butterfly Economics*. Faber & Faber, London.

P. Ormerod (2000). 'The Keynesian micro-foundations of the business cycle: Some implications of globalisation', Preprint.

P. Ormerod (2002). 'The US business cycle: Power law scaling for interacting units with complex internal structure', *Physica A* **314**, 774–85.

P. Ormerod (2002). 'Sense on segregation', *Prospect*, February, 12–14.

P. Ormerod and M. Campbell (2000). 'The evolution of family structures in a social context', Preprint.

J. K. Parrish and L. Edelstein-Keshet (1999). 'Complexity, pattern, and evolutionary trade-offs in animal aggregation', *Science* **284**, 99–101.

R. Pastor-Satorras and A. Vespignani (2001). 'Epidemic spreading in scale-free networks', *Physical Review Letters* **86**, 3200–203.

R. Pastor-Satorras and A. Vespignani (2001). 'Immunization of complex networks', Preprint, <http://arxiv.org/abs/cond-mat/0107066>.

G. J. Perez, G. Tapang, M. Lim and C. Saloma (2002). 'Streaming, disruptive interference and power-law behavior in the exit dynamics of confined pedestrians', *Physica A* **312**, 609–18.

I. Peterson (1996). 'The shapes of cities', *Science News* **149**, 8–9.

I. Peterson (1996). 'The Gods of Sugarscape', *Science News*, 23 November, 332.

V. Plerou, P. Gopikrishnan, L. A. Nuñes Amaral, M. Meyer and H. E. Stanley (1999). 'Scaling of the distribution of price fluctuations of individual companies', *Physical Review E* **60**, 6519–29.

K. Popper (1997). *The Lesson of this Century*. Routledge, London.

R. Porter (2000). *Enlightenment*. Penguin, London.

T. M. Porter (1986). *The Rise of Statistical Thinking 1820–1900*. Princeton University Press.

M. Potters, R. Cont and J.-P. Bouchard (1998). 'Financial markets as adaptive ecosystems', *Europhysics Letters* **41**, 239–44.

W. K. Potts (1984). 'The chorus-line hypothesis of manoeuvre coordination in avian flocks', *Nature* **309**, 344–5.

I. Prigogine (1981). *From Being to Becoming*. W. H. Freeman, New York.

E. Ravasz and A.-L. Barabási (2002). 'Hierarchical organization in complex networks', Preprint, <http://arxiv.org/abs/cond-mat/0206130>.

C. Rawcliffe (1999). *Medicine and Society in Later Medieval England*. Sandpiper, London.

D. Read (2002). 'A multitrajectory, competition model of emergent complexity in human social organization', *Proceedings of the National Academy of Sciences USA* **99**, 7251–6.

C. Reynolds (1995). Boids, <http://www.red3d.com/cwr/boids>, last updated 6 September 2001.

C. W. Reynolds (1987). 'Flocks, herds and schools: A distributed behavioral model', *Computer Graphics* **21**(4), 25–34.

L. F. Richardson (1960). *Statistics of Deadly Quarrels*, ed. Q. Wright and C. C. Lienau. Boxwood Press, Pittsburgh.

C. A. Roberts, D. Stallman and J. A. Bieri (2002). 'Modeling complex human–environment interactions: The Grand Canyon river trip simulator', *Journal of Ecological Modelling* **153**, 181–96.

J. S. Rowlinson (ed.) (1988). *J. D. van der Waals: On the Continuity of the Gaseous and Liquid States. Studies in Statistical Mechanics* XIV.

North-Holland, Amsterdam.

B. Russell (1984). *A History of Western Philosophy*. Unwin Paperbacks, London.

C. Saloma, G. J. Perez, G. Tapang, M. Lim and C. Palmes-Saloma (2003). 'Self-organized queuing and scale-free behavior in real escape panic', *Proceedings of the National Academy of Sciences USA* **100**, 11947–52.

B. Schechter (1999). 'Birds of a feather', *New Scientist*, 23 January, 30–33.

T. C. Schelling (1978). *Micromotives and Macrobehavior*. W. W. Norton, New York.

J. B. Schneewind (ed.) (1990). *Moral Philosophy from Montaigne to Kant*, Vol. II. Cambridge University Press.

F. Schweitzer (ed.) (2002). *Modelling Complexity in Economic and Social Systems*. World Scientific, Singapore.

P. Sen, S. Dasgupta, A. Chatterjee, P. A. Sreeram, G. Mukherjee and S. S. Manna (2002). 'Small-world properties of the Indian Railway network', Preprint, <http://arxiv.org/abs/cond-mat/0208535>.

M. A. Serrano and M. Boguñá (2003). 'Topology of the world trade web', *Physical Review E* **68**, 015101(R).

V. Shvetsov and D. Helbing (1999). 'Macroscopic dynamics of multilane traffic', *Physical Review E* **59**, 6328–39.

K. Sigmund (1993). *Games of Life*. Oxford University Press.

K. Sigmund, E. Fehr and M. A. Nowak (2002). 'The economics of fair play', *Scientific American*, January, 82–7.

H. A Simon (1996). *The Sciences of the Artificial*, 3rd edn. MIT Press, Cambridge, MA.

B. Skyrms (1996). *Evolution of the Social Contract*. Cambridge University Press.

A. Smith (1776). *An Inquiry into the Nature and Causes of the Wealth of Nations*. Ed. L. Dickey. Hackett Publishing, Indianopolis, 1993.

S. Solomon (1999). 'Behaviorly realistic simulations of stock market traders with a soul', *Computer Physics Communications* **121/2**, 161.

D. Sornette (2003). *Why Stock Markets Crash*. Princeton University Press.

D. Sornette (2003). 'Critical market crashes', Preprint, <http://arxiv.org/abs/cond-mat/0301543>.

G. Soros (2002). *George Soros on Globalization*. Perseus, New York.

M. H. R. Stanley, L. A. N. Amaral, S. V. Buldyrev, S. Havlin, H. Leschhorn, P. Maass, M. A. Salinger and H. E. Stanley (1996). 'Scaling behaviour in the growth of companies', *Nature* **379**, 804–6.

G. Stix (1998). 'A calculus of risk', *Scientific American*, May, 92–7.

J. Sutton (1997). 'Gibrat's legacy', *Journal of Economic Literature* **35**, 40–59.

K. Sznajd-Weron and R. Weron (2002). 'How effective is advertising in duopoly markets?', Preprint, <http://arxiv.org/abs/cond-mat/0211058>.

D'A. W. Thompson (1942). *On Growth and Form: A New Edition*. Cambridge University Press. Reissued by Dover, New York, 1992.

B. Tjaden and G. Wasson (1997). 'The Oracle of Bacon at Virginia', <http://www.cs.virginia.edu/oracle/>.

J. Toner and Y. Hu (1998). 'Flocks, herds, and schools: A quantitative theory of flocking', *Physical Review E* **58**, 4828–58.

J. Toner and Y. Tu (1995). 'Long-range order in a two-dimensional dynamical XY model: How birds fly together', *Physical Review Letters* **75**, 4326–9.

M. Treiber and D. Helbing (2002). 'Microsimulations of freeway traffic including control measures', Preprint, <http://arxiv.org/abs/cond-mat/0210096>.

M. Treiber, A. Hennecke and D. Helbing (2000). 'Congested traffic states in empirical observations and microscopic simulations', *Physical Review E* **62**, 1805–24.

R. Tuck (2002). *Thomas Hobbes: A Very Short Introduction*. Oxford University Press.

T. Vicsek, A. Czirók, E. Ben-Jacob, I. Cohen and O. Shochet (1995). 'Novel type of phase transition in a system of self-driven particles', *Physical Review Letters* **75**, 1226–9.

S. Vines (2003). *Market Panic*. Profile, London.

M. M. Waldrop (1994). *Complexity*. Penguin, London.

K. N. Waltz (1954). *Man, the State, and War*. Columbia University Press, New York.

M. Ward (2001). *Universality*. Macmillan, London.

D. J. Watts (1999). *Small Worlds*. Princeton University Press.

D. J. Watts, P. S. Dodds and M. Newman (2002). 'Identity and search in local networks', *Science* **296**, 1302–5.

D. J. Watts and S. H. Strogatz (1998). 'Collective dynamics of "small-world" networks', *Nature* **393**, 440–42.

W. Weidlich (2000). *Sociodynamics: A Systematic Approach to Modelling in the Social Sciences*. Harwood Academic, Amsterdam.

G. S. Wilkinson (1984). 'Reciprocal food sharing in the vampire bat', *Nature* **308**, 181–4.

E. O. Wilson (1998). *Consilience*. Alfred A. Knopf, New York.

P. Winch (1958). *The Idea of a Social Science and its Relation to Philosophy*. Routledge, London.

D. Wolf (1999). 'Cellular automata for traffic simulations', *Physica A* **263**, 438–51.

J. Wu and R. Axelrod (1995). 'How to cope with noise in the iterated Prisoner's Dilemma', *Journal of Conflict Resolution* **39**, 183–9.

M. Wyart and J.-P. Bouchard (2002). 'Statistical models for company growth', Preprint, <http://arxiv.org/abs/cond-mat/0210479>.

I. Yang, H. Jeong, B. Kahng and A.-L. Barabási (2003). 'Emerging behavior in electronic bidding', *Physical Review E* **68**, 016102.

S.-H. Yook, H. Jeong and A.-L. Barabási (2002). 'Modeling the Internet's large-scale topology', *Proceedings of the National Academy of Sciences USA* **99**, 13382–6.

W.-X. Zhou and D. Sornette (2003). '2000–2003 real estate bubble in the UK but not in the USA', *Physica A* **329**, 249–63.

G. K. Zipf (1965). *Human Behavior and the Principle of Least Effort*. Hafner, New York.

Index

Achenwall, Gottfried 64–5
Ackroyd, Peter: *London: The Biography* 183–4
Adamic, Lada 482–3
African nations 436
agar gel 138–40, *139*, 148
agent-based modelling 301, 324*ff*, 420–2, 441–2
Aguilar, Jean-Pierre 282, 291, 293
Aids 451, 495, 496
Akerlof, George 263
Alberich, Ricardo 497
Albert, Prince Consort 92
Albert, Eddie 443–4, 452, 465
Albert, Réka 479–80, 481–2, 484–5, 496–7, 499, 500, 501
d'Alembert, Jean Le Rond 66
alliances 364–5, *366*, 367; and war 355–7, *358*, 359–61, 365, 368; *see also* firms and companies
altruism 514, *531*, 532, 535–7, 563, 571
Amaral, Luis Nuñes 495
Anaxagoras 40
Anderson, Philip 281, 307
Andrews, Thomas 103
ants 151, 155, 159
AOL 474, 483
applause 586–8, *588*
Aquinas, Thomas: *On the Governance of Rulers* 579
Aristotle 16, 39, 40, 159, 380, 579
ARPANET 472–3, 474, 476
Arrow, Kenneth 382
Arthur, W. Brian 254, 274–5, 340*n*; El Farol problem 416–18, 422
Ashworth, Tony: *Trench Warfare 1914–1918* 521
Asimov, Isaac: *Solaria* 459
astronomers/astronomy 41, 73–4, 77, 79; *see also* universe
AT&T 342

Athens *190*, 191, 510, 534
atomists 40–1
attendance figures 416–18, *419*
Aubrey, John: *Brief Lives* 15
audience applause 586–8, *588*
Ausloos, Marcel 293
automata 17–18, 35, 42; cellular 153–5
average man, theory of the (Quetelet) 78–9, 87, 307
Axelrod, Robert; *The Evolution of Cooperation* 533, 541, 561; landscape theory (with D. S. Bennett) 343, 345, 351–4, 356–7, 365, 367; model of cultural dissemination 427–36, 438, 440; and Prisoner's Dilemma tournament 527, 529, 530–1, 540, 541, 542, 543, 550–2, *553*
'axial maps' 189–90, *190*
Axtell, Robert 321, 323; model of firm growth 324–6, *327*, 329, *329*, *330*, 331–2, *332*, *333*, 335–6; *see also* Epstein, Joshua

Babbage, Charles 90
Bachelier, Louis 238, 240, 242, 245
Bacon, Francis 14–15, 40, 61; *New Atlantis* 22–3
Bacon, Kevin 444–6, 452–3, 463–5
Bacon Numbers/movie-actor game 443–7, 452–3, *453*, 463, 464–5, 497, 498, 501, *502*
bacterial (*Bacillus*) colonies 135–40, *136*, *139*, 142, *143*, 143–4, 146–8, *147*; automata 156–8
Bak, Per 295–301, 302, 306
Bangkok 191
Barabási, Albert-László 489, 491, 492, 496, 503, 587; (with Albert) 496–7, 499, 500, 501; (with Albert and Jeong) 479–80, 481–2, 484–5
Baran, Paul 474–7, 487
Barcelona 191
Barnes, Julian: *A History of the World in 10½ Chapters* 565

Barrett, William 108
Batty, Michael 171, 180–1, 185–6, 187
Beck, Harry: London Underground map 454, 454–5
Becker, Gary 397; A Treatise on the Family 405–9, 413
bees 48, 151, 159, 570
bell curves 49, 49–50, 77, 115–16
Bellamy, Edward: Looking Backward 567
Ben-Jacob, Eshel 146–8, 156
Bénard, Henri 128–30, 129, 140
Bennett, D. Scott see Axelrod, Robert
Bentham, Jeremy 34, 80, 226, 326, 577
Bentley, Wilson A. and Humphreys, W. J.: Snow Crystals 123, 141
Berlin 185, 188–9, 189, 391
Bernandes, Tristao 375–6 and n
Berners-Lee, Tim 473, 479
Bernoulli, Daniel 41, 46, 76, 105
Bernoulli, Jacob 76
bifurcations, non-equilibrium 131, 132, 132–3, 134, 135
Biham, Ofer 217
birds: collective behaviour 151–2
Bismarck, Prince Otto von 383
Black, Duncan 378
Black, Fischer 249, 250
Black-Scholes model 249, 250–51
Bogart, Humphrey 444
'boids' (Reynolds) 153–4, 155
Bolay, Kai 165n
Boltzmann, Ludwig 53–4, 56, 58, 59, 60, 85, 87, 101, 120, 122, 160, 161, 256; Beethoven in Heaven 58–9
Borda, Jean Charles de 381
Borges, Jorge Luis: 'The Garden of Forking Paths' 133, 362
Bornholdt, Stefan 494
Boscovich, Roger Joseph: A Theory of Natural Philosophy 41–2
Boston, Massachusetts 174, 193
Bouchard, Jean-Philippe 251, 309
Boulding, Kenneth 254
Boyle, Robert 43n, 47
Brahe, Tycho 2
Brando, Marlo 465
Brazil: election system 373–4, 375
Brecon, Wales 311–12
Brinton, Thomas, Bishop of Rochester 31
Brock, William 267–8
Brown, Robert/Brownian motion 51–2, 137, 238
Buccleuch, Duke of 69
Buchanan, Mark: Ubiquity 303–5
Buckle, Henry Thomas 80–81, 92, 93; History of Civilization in England 81–2, 83, 84, 85

Burda, Zdzislaw 278
Burke, Edmund 70, 277, 554; Reflections on the Revolution in France 70–71
Bush, George W. 370–71
business cycles 231, 233, 235–6, 238; see also real business cycle theory
businesses see firms and companies

Cagniard de la Tour, Charles, Baron 102–3
Campbell, Michael 398–400, 400
Candolle, Alphonse De 90
Cantillon, Richard: Essai sur la nature du commerce en général 574
capillarity, theory of 104
capitalism 222–3, 228–9, 231, 234, 236, 258, 278–9, 280, 313, 315, 316–17, 580
Cardiff 187
Cardoso, Gustavo 467
Carey, John 566
Carlyle, Thomas 254
Carnot, Nicolas Sadi 42, 43
'Carnot's cycle' 43–4
cartels 319
Carvalho, Rui 189–91
Castells, Manuel: The Internet Galaxy 467, 470–1
catastrophe theory (Thom) 4, 101
Catholic Church 12, 63
censuses 52, 63–4
CERN 473
Challet, Damien, and Zhang, Yi-Cheng: 'minority-game' 418–20, 419, 498
chaos theory 4, 238, 301
Charles I 8, 9, 10, 16, 20, 62, 509, 577
Charles II 9, 21, 31, 32, 62
Charles, Jacques 47
chemotaxis 150
Chen, Donald 184–5
Chen, Kan 301
chess 347–8, 349, 354, 523
choice 338, 371ff, 522, 577; see also free will
choice theory 381–2
Churchill, Winston 360
Cicero: De legibus 579, 580
cities see urban growth; urban planning
Civil War, English 8–9, 10, 20–21, 221, 425–6, 565
clapping 586–8, 588
Clark, Wesley 476–7
Clausius, Rudolph 45, 47, 54, 85, 86, 122
Clinton, Bill 487, 535
Coase, Ronald 327
Cobb, G. Belton: Stand to Arms 520, 531
Coconut Grove nightclub fire, Boston 174
Cohen, Michael 440
coin tossing 75, 76
Cold War 475–6, 508, 525, 540, 557–9, 562

collective behaviour 5, 36, 106, 135, 148–58, 159, 324, 372–3, 440–1; *see also* crowd behaviour; herding; audience applause 586–8, *588*; *see also* Schelling, Thomas
companies *see* firms and companies
complexity theory 4–5, 155
computer hackers and crackers 488
computers 154, 338, 341–3, 344–5, 523; and cyber-terrorism 487–92; viruses 468, 469, 492–6
Comte, Auguste 71, 73, 362; *Cours de philosophie positive* 71
Condorcet, Marie-Jean-Antoine-Nicolas Caritat de 65, 66, 68, 88; *Esquisse. . .* 65–8, 69; 'Essay on . . . Majority Decisions' 66, 380, 381
conformity *see* conventions, social
conscientious objection 561
convection patterns *129*, 129–31, 134, 140
conventions, social 383–5, *386*, 387–9, *388*
Conway, John Horton: Game of Life 154–5
cooperation 514–15, 521, 527–9, 530–4, 536, 537, 543, 544–9, 550–4, *555*, 556–7, 560, 562, 563; *see also* alliances
Cootner, Paul 243
Copernicus, Nicolaus 11
Cornhill Magazine 92
correlation functions 248–9
Costa Filho, Raimundo 373, 374*n*, 376
Cotgrove, Stephen: *The Science of Society* 7
Counter-Reformation, the 11
Cournot, Antoine Augustin 88
'crackers' 488 *and n*
crime and criminals 89, 91–2, 369, 396–401, *400*, 403–4, 413, 415, 425, 511, 514, 562–3; Internet 471, 487–9
critical exponents 284–5, 286, 289–91
critical opalescence 283, 289
critical points 103, 106, 110, 114, 115, 116–17, 135, 282–4, 285; and fluctuations 286–9, *288*, 295; and market crashes 291, *292*, 293–4
Cromwell, Oliver 8–9, 10, 32, 62, 355, 565
crowd behaviour 160–6, 174–82, 202, 571, 586–8
cultural drift 433*n*
cultures 424–7; dissemination of 427–36 *see also* racial segregation
Curie, Pierre 108, 268
Curie point 108, 109–10, 113, 114, 118, 133, 134, 285, 289
cyber-terrorism 487–9
Cyples, William 92
Czirók, András 156–7, 158

Dallas, Texas 197, 217
Dalzell, Judge Stewart 472
Darwin, Charles 63, 86–7, 192; *Descent of Man* 86, 513–14; *On the Origin of Species* 86, 125

Darwinism 34, 87, 513, 514, 544, *548*
Davenant, Charles 277
Davies, Cecil 561
DBM *see* dielectric breakdown model
Dean, James 444
decision-making 369, 372, 378, 383, 384, 385, 387; *see also* elections
Defoe, Daniel 223
democracies 32, 34, 371, 377, 379, 380, 381, 382, 509–10, 513, 534–5, 580
Democritus 40
demographic statistics 52, 60, 62, 63–4, 68, 77, 81–2, 90–1
Denver, Colorado 193
derivatives trading 250, 251
Desaguliers, Jean Théophile: *The Newtonian System of the World. . .* 69
Descartes, René 11, 15, 19, 21, 41
determinism 18–19, 42, 76, 91, 92–3, 577
Devonshire, Earl of 14
Dezso, Zoltan 496
Dhaka 191
Dickens, Charles 82
Dictyostelium discoideum (slime mould) 149–50, *150*
dielectric breakdown model (DBM) 186
diffusion 50–1, 52
diffusion-limited aggregation (DLA) 136–8, 140, 142, 186, 187, 188
Diggers, the 10, 565
Disraeli, Benjamin 83
divergences 284–6
divine right of kings 10, 509, 513, 573
DLA *see* diffusion-limited aggregation
doorways: and crowd behaviour 166, *166*, *167*, 174–80, *177*,
Dostoevsky, Fyodor ; *Crime and Punishment* 397; *Notes from the Underground* 92–3
Dow, Charles 234
Downs, Anthony 378
Dresher, Melvin 523, 524
driving conventions 369, 383–4
Dugas, Gaetan 451
Duisburg, Germany 196–7, 212, 217
Durkheim, Emile: *Revue philosophique* 443
Durlauf, Steven 267–8

earthquakes 299, 303
economics 220, 253, 573–4; free market economy 222–4, 275–80, 580–3; fundamentals 259–60, 261–2; income distribution 307–10, *308*, 438–9; and interaction 266–70, 277–8; and the Internet 468, 478; market fluctuations and crashes 231, 232, 233–7, 238–52, 259, 261, 270–5, 280, 291–5; and marriage rates 405–9,

411, *412*, 413, *414*; and the minority game 415–17, 420, 321; and physics 223, 225–6, 238, 254, 256–7, 266–8, 282; and rationality and irrationality 262–5, 274–5; recessions 229–30, 231, 275–6 ;and self-organized criticality 300–2; theories and models 69–70, 224–7, 234, 238–44, 249–52, 254–69, 270–1, 277–8, 281–2, 574–7; *see also* business cycles; capitalism; Keynes, John Maynard; Smith, Adam; traders, stock market econophysics/econophysicists 225–6, 251, 278, 504

Edelstein-Keshet, Leah 151, 182

Eden, M. 138

Edgeworth, Francis 254, 255–6, 258, 264 *Mathematical Psychics* 227, 254–5, 256–7

efficient market hypothesis 248, 260

Ehrenfest, Paul 59

Eindhoven 191

Einstein, Albert 38, 51, 52, 238, 260, 445

elections/voting systems 370–2, 373–83

electrodeposition 136, 137

Elephant and Castle, London 173

Elizabeth I 184

Elliot, Ralph 233–4

emergence 155

Emerson, Ralph Waldo 26; *The Conduct of Life* 99, 223–4, 253, 564; 'Fate' 82–3; 'On Politics' 577–8

energy 46 *and n*, 49, 127 *and n*

energy landscapes 347–8, *349, 350*, 350–2, *358*, 359

Engels, Friedrich (with Marx): *Communist Manifesto* 230, 566

Enlightenment, the 11, 34, 39, 65, 66, 70, 71, 222, 223, 305, 572–3, 574, 576

entropy 45, 54–6, 95–6, 123, 127, 128

Epicurus 40

epidemiology 493–5

Epstein, Joshua 369, 383, 384–5, *386*, 387–9, *388*; (with Robert Axtell) *Growing Artificial Societies*/Sugarscape model 277–8, 279, 393*n*, 437–41

equilibrium and non-equilibrium states 123, 124, 125–35, 142, 144, 156, 159, 187; *see also* Nash equilibria

Erdös, Paul 445 *and n*, 452; Erdös Numbers 445; (with Rényi) 452, 455, 456, *486*, 499

error curves 74–7, 75, 80, 85, 239

Euclid: *Elements of Geometry* 15

eugenics 87, 566, 583

Euler, Leonhard 445*n*

European Union 426

Fama, Eugene 244*n*

Farkas, Illés 164*n*, 175

Farr, William 90

fatalism 91, 92, 96

Fehr, Ernst 536–7

Ferguson, Adam 576–7, 578, 579

Ferguson, Niall: *Virtual History* 362, 363

Fibonacci sequence 234

Filmer, Robert: *Patriarcha. . .* 509

firms and companies 315–17; and competition 317–18, 319, 339–41; court actions 534; growth rates/sizes 313–14, 320, 321–3, 324–9, *325, 329, 330*, 331–2, *334*; mergers and alliances 338, 342–3, 344–6, 352–4; monopolies 318, 319; oligopolies 318–20; and production 316 *and n*; and profits 317, 320, 334–5; reasons for failure 333–4 ; worker interactions 335–6

fish: collective behaviour 150, *150*

Fisher, Irving 225, 235, 238

Flood, Merrill 523, 524

Florian, Razvan 353*n*

fluids 286*n*; *see also* critical points; phase transitions; Waals, Johannes van der; density 105–6, 114

Flynn, Errol 444

Föllmer, Hans 267

Fontenelle, Bernard: *Conversations on the Plurality of Worlds* 7

forest fires 299

Fourier, Joseph 77

Fox, J. J. 88

fractals *136*, 136–8, 142, 143–4, 186, *187*, 242, 246

free will 5, 18, 42, 60, 88, 91, 92, 93–4, 95, 96, 97, 135; and economics 257–9

freezing 100, 108, 126, 201–3, 286, 403

Freud, Sigmund 59

Civilization and its Discontents 506, 507

friendships *see* networks, social

futures 250

Gächter, Simon 536–7

Galam, Serge 353*n*

Galileo Galilei 11, 16–17, 39, 40

galleries, designing 171–2

Galton, Francis 87; *Hereditary Genius* 87

Game of Life (Conway) 154–5

game theory 268, 398, 508–9, 517, 531–2, 533, 539, 544, 549–50, 559, 562; *see also* Prisoner's Dilemma

games 347–8, *349*, 354–5, 523–4; *see* game theory

gases, kinetic theories of 41, 46–51, 52–4, 84–6, 100–5

Gassendi, Pierre 21, 40

Gaulthier, David 532–3

Gauss, Carl Friedrich 74–5

gaussian statistics 75, 239–42, 243, 246, 251, 271

Gay-Lussac, Joseph Louis 47

Geller, Uri 60
General Motors, Warren, Michigan 198
geometry 15–16, 125
George, Susan 423
Gibbs, Josiah Willard 123, 126, 127, 133, 193, 256,
    413; *Elementary Principles in Statistical
    Mechanics* 122–3
Gibrat, Robert 321–3, 324, 331
Giddings, Franklin: *The Scientific Study of
    Human Society* 584
Gilday, Cindy 423
Gladwell, Malcolm: *The Tipping Point* 101,
    369–70, 403–4
God 12, 20, 60, 79, 91, 92, 96, 523, 532
Goethe, Johann Wolfgang von 220
Goodstein, David: *States of Matter* 59
Gopal, Venkatesh 503
government(s) 20–1, 68–9, 80, 82, 505, 506–8,
    509–10, 511–12, 513, 515–17, 532, 534–5, 539,
    570, 579–80; and the economy 275–6, 279–80;
    see also democracies; monarchial rule
Granovetter, Mark 447
graphs 452; ordered 456, 457; random 452, *455*,
    455–6, 481, 484, *484*; and random rewiring
    460–63; relational 452–5, *454*
Graunt, John 62
Observations upon the Bills of Mortality 62, 64
Graves, Robert: *Goodbye To All That* 520
Grim Trigger (strategy) 548
Grotius, Hugo: *The Laws of War and Peace* 23–4,
    27
Guare, John: *Six Degrees of Separation* 450
Gutenberg, Beno 299

Hague, The 191
Haldane, J. B. S. 311
Hales, Stephen: *Vegetable Staticks* 121
Halley, Edmond 73
Hamilton, William 305
Handy, Charles: *Understanding Organizations*
    311
Hanson, Julienne 172, 173–4, 189
Hare, Maurice Evan 94
Harrington, James: *Oceana* 221
Havlin, Shlomo 187–8
Hay, Ian: *The First Hundred Thousand* 505–6
Hayek, Friedrich von: *The Road to Serfdom* 580–2
heat, theories of 42–4, 46; *see* thermodynamics
heat capacity 284
Hegel, Friedrich 543
Heilbroner, Robert: *The Worldly Philosophers*
    222, 253, 255, 257, 258
Hein, Piet: *Grooks* 38
Heisenberg, Werner 445; Uncertainty Principle
    98

Helbing, Dirk 161–2, 163–71, 172, 175–80, 211,
    212–15, *216*, 217–19, 257
helium 117, 118, 284
Helmholtz, Hermann von 46, 95
Henderson, L. F. 160–1, 162
Hennecke, Ansgar 213, 215
Heraclitus 124
Herapath, John 46
herding 164n, 179, 269–70, 271
Herschel, John 58, 79, 85, 103
high-rise estates 173–4
Hillier, Bill 172, 173–4, 189
Hillsborough disaster (1989) 174–5
Hirn, Gustave-Adolphe 105
history 80–2, 93–4, 361–3, 367, 424–5
Hitler, Adolf 360, 580
Hobbes, Thomas 5, 8, 9, 13, *14*, 14–16, 17, 20–1,
    25, 32, 70, 146, 507, 510, 511, 532–3, 570,
    573–4, 584; *Behemoth. . . 33*; *De cive* 20, 21;
    Leviathan 11–13, 17, 18–19, 21, 23, 26–33, *30*,
    34, 35, 61, 149, 159, 220–21, 222, 439, 468,
    469, 515–16, 539, 580; *Philosophical
    Rudiments Concerning Government and
    Society* 18, 20, 509
Hobsbawm, Eric: *Age of Extremes* 337, 360, 518
Hod, Shahar 420–1
Hölderlin, Friedrich 567
Hollar, Wenceslas 31
Hong Kong 191; Hang Seng index 146, 292
Hopkinson, John 108 *and n*
Hopper, Dennis 464
Horney, Karen 162
Household Words (periodical) 82
Houston, Texas 194
html (Hypertext Mark-up Language) 479
Huang, Zhi-Feng 309
Huberman, Bernardo 217–18, 483
Humanism 11, 24
Humboldt, Wilhelm von: *Ideas towards a
    Definition of the Limits of State Action* 512
Hume, David 33, 69; *Treatise of Human Nature*
    69
Hunt, Frederick 82
Huxley, Aldous: *Brave New World* 566
Huxley, Julian 566
hyperlinks 473, 478, 479–81, *480*, 499
hysteresis 204

IBM 342, 353
impossibility theorem (Arrow) 382
Industrial Revolution 42, 43, 64, 173, 227, 234,
    315
Inquisition, the 11, 18
interaction 266–70, 277–8, 335–6, 372–3; *see also*
    alliances; collective behaviour

Internet, the 467, 468, 469–71, 472–3, 474–8, 475, 484–7, 485, 504; chat rooms 460; and cyber-terrorism 486–92; viruses 468, 469, 492–6
Irving, Larry 468
Ising, Ernst 111
models of magnetism 111, 112, 113–14, 119, 134, 267, 286, 288, 290, 351, 376, 346, 346–7, 348, 351, 364
Israel 541, 550, 556

James II 1, 62
Jefferson, Thomas 70
Jencks, Charles 101
Jeong, Hawoong (with Albert and Barabási) 479–80, 481–2, 484–5
Jevons, Stanley 235–6
Johnson, Samuel 220
Jones, David 561

Kagan, Robert 516–17
Kahn, Hermann: *Thinking About the Unthinkable in the 1980s* 557–8
Kamerlingh Onnes, Heike 117–18
Kant, Immanuel 39, 81, 94, 305, 362, 507, 512; *On History* 64, 91; *Perpetual Peace* 516, 556
Kapitsa, Pyotr 118
Kapteyn, Jacobus 321
Karinthy, Frigyes 450n
Kay, John 225, 262, 266–7, 335
Keltsch, Joachim 168
Kepler, Johannes 2, 142
Kerner, Boris 207, 210–11
Kett, Henry 184
Keynes, John Maynard 237, 255, 258, 263, 269–70, 275–6, 281–2
Khrushchev, Nikita 558
Kinnock, Neil 372, 377
Kipling, Rudyard 193
Kirman, Alan 265–6, 267, 268–9, 270, 271
Kleist, Heinrich von 145
Klemm, Konstantin 433n
Kochen, Manfred 447–8
Kondratieff, Nikolai: cycles 234, 235
Kraines, David and Vivien 542
Kropotkin, Peter: *Mutual Aid* 514, 515, 537
Krugman, Paul: *Peddling Prosperity* 226, 227, 233n, 256, 276, 279n
Kubrick, Stanley: *Dr Strangelove* 540, 541, 559
Kuhn, Thomas 101

laissez-faire 80, 82, 222, 277
Lambert, John 9
landscape theory 343–5, 351–4, 356–7, 359–67
languages 427 *and n*, 428

Laplace, Pierre-Simon 42, 73–4, 76–7, 84, 86n, 88, 104
Las Vegas 190
Lavoisier, Antoine: *Traité élémentaire de chimie* 125
law of large numbers (Poisson) 76, 78, 96, 98, 310
Leeuwenhoek, Antony van 146
Leibniz, Gottfried 20
Lenz, Wilhelm 109, 110–11
Leucippus 40
Levellers, the 10
Levi, Primo: 'For a Good Purpose' 469
Lévy, Paul: Lévy flights 243 *and n*, 244–6
Lewin, Kurt 162–3
Lichtenberg, Georg Christoph 121, 145
Lidl supermarkets 312–13
Lighthill, James 197–8, 213
Lillo, Fabrizio 247
Lind, Michael 395–6
liquids 99–107, 112–13, 116–17, also *see* fluids; freezing
Locke, John 33, 34, 517, 562; *Two Treatises of Government* 505, 506–7, 509, 510–12, 514, 539
London 173, 180, 180–81, 183–4, 185, 186, 187, 188–9, 191, 192, 194
London Underground map 454, 454–5
Longfellow, Henry Wadsworth 79n
Longley, Paul 186, 187
Los Angeles 194
Love Bug computer virus 468, 494
Lucretius: *De rerum natura* 40
Lundberg, George: *Foundations of Sociology* 564
Luther, Martin 10, 11
Lux, Thomas 271, 272, 273

Mach, Ernst 51
Machiavelli, Niccolò: *The Prince* 538
MacIver, Robert M.: *The Web of Government* 7, 25
McNamara, Robert 559
McNaughton, John 559
magnetism/magnetization 99, 108–11, 110, 133–4, 158, 268, 286–7, 350; *see also* Ising models
Makse, Hernán 187–8
Malik, Kenan 583
Malthus, Thomas 68, 90, 282, 513; *Essay on the Principle of Population* 63, 68–9
Mandelbrot, Benoit 242–3, 244, 245, 246, 304n; *The Fractal Geometry of Nature* 307
Mandeville, Bernard: *The Fable of the Bees* 570–1
Mannheim, Karl: *Ideology and Utopia* 369
Mantegna, Rosario 243, 247
Marchesi, Michele 271, 272, 273
Markowitz, Harry 263
marriage 404–11, 412, 413, 414, 415, 424
Marshall, Alfred 255, 256–7, 327

Marvel Comics 497–8

Marx, Karl/Marxism 34, 63, 79–80, 227, 228–30, 315, 317, 362, 549, 578; (with Engels) *Communist Manifesto* 230, 231, 235, 566

Mathias, Nisha 503

Matsushita, Mitsugu 135–6, 137–8, 140, 146, 148, 186

Maurois, André 362

Maxwell, James Clerk 38, 47, 48, 48–51, 52, 53–4, 56, 83–5, 86, 96–8, 99, 101, 106*n*, 108, 120, 122, 160, 161, 239, 254, 256, 413

May, Robert 552, 554

mean-field theories 268, 289–90

mechanistic theories 5, 17–21, 23, 33, 35, 39–40, 41, 42, 53, 56, 71, 86, 226

Merchant, Carolyn: *The Death of Nature* 35

mercury 118

Mersenne, Marin 21

Merton, Robert 249

metabolism 489–91, 490

metastable states/metastability 201–2, 203–4, 204, 365, 402, 402, 403, 404

microeconomics 255, 260, 261, 266–7, 276

Microsoft 344, 474

Milgram, Stanley 448–50, 457

Mill, John Stuart 34; *On Liberty* 512; *A System of Logic* 80, 337, 568, 576, 583

minimization 122

minority game 418–20, 419, 498

Moe, Richard 195

Moelwyn-Hughes, Emyr Alun 49

Moivre, Abraham de: error curve 75, 76, 77, 239

Molnár, Péter 163–5, 168, 170

monarchial rule 9–10, 29, 31–2

Monbiot, George: *Captive State* 312, 313, 319

Monk, George 9

Montesquieu, Baron de (Charles Secondat de la Brède) 570, 575; *The Spirit of the Laws* 69

moral laws 576–7, 578–9

More, Thomas: *Utopia* 22, 566–7

Morgenstern, Oskar: *Theory of Games and Economic Behavior* 524

morphology diagrams 139, 140, 141, 147

Morris, William: *News from Nowhere* 567

movie-actor game *see* Bacon Numbers

Moynihan, Daniel Patrick 394

Mumford, Lewis: *The Culture of Cities* 26, 185, 186, 572–3

Musil, Robert 59

The Man Without Qualities 59–60

Nagel, Kai 197, 200, 217

Nagel-Schreckenberg (NaSch) model 197, 199–200, 203, 204, 207, 210, 213

Nakar, Ehud 420–1

Nakaya, Ukichiro 141–2

Napier, John 17

Napoleon Bonaparte 253, 282, 339, 564

Nash equilibria 328–9, 345

NATO 355, 367, 540, 549

natural laws/natural rights 23–4, 25, 27–8, 89, 305, 224, 305, 505, 511, 576, 577–9

natural selection 86–7, 125, 513, 531, 545, 548

negative feedback 230

nematode worms 489, 491

Netscape 474

networks 443, 446–7, 450–1; and clustering process 368, 448, 458–9, 459, 463, 498; studies; and theories 447–50, 455–64, 465–6; *see also* Bacon Numbers; *see also* scale-free networks

Neumann, John von 154, 523

New Lanark, Scotland 565–6

New Orleans 191

New York City: crime rates 403–4

Newman, Mark 443, 445*n*, 493

Newmarch, William 82, 414–15

Newton, Isaac 2, 11, 16, 20, 39, 40, 41, 53, 66, 73, 74, 94, 96, 197, 305 ; *Opticks* 99; *Principia mathematica* 125, 574*n*

Nietzsche, Friedrich 83

Nightingale, Florence 79 *and n*, 90

Nixon, Richard 394

Nobel Prize winners 52, 108, 128, 249, 263, 290, 327, 406

noise 131, 157, 157–8, 238, 239, 289*n*

non-equilibrium *see* equilibrium and non-equilibrium states

normal distribution 74

Notting Hill Carnival, London 180, 180–81

Nowak, Martin 544, 545–7, 552, 554

nucleation 403, 411

Nussbaum, Frederick: *The Triumph of Science and Reason 1660–1685* 31

Oakeshott, Michael: *Experience and its Modes* 362

Olson, Richard: *Science Deified and Science Defied* 32–3, 578–9

Onion, The (magazine) 370

Ono, S. 113

Onsager, Lars 113, 126, 127, 128, 290

Open Software Foundation (OSF) 342, 343, 353

options pricing 249–50, 251

Ormerod, Paul 244, 273–4, 391, 398–401, 410–11, 413 ; *Butterfly Economics* 578

Osborne, M. F. M. 236*n*

OSF *see* Open Software Foundation

Osiander, Andreas 11

Ostwald ripening 344

Owen, Robert 565–6

Owen, Wilfred 561

Paczuski, Maya 200
Paine, Thomas 67; *The Rights of Man* 67
paradigm shifts (Kuhn) 101
Pareto, Vilfredo 236, 307–9, *308*, 439
Park, Robert E. 569
Parrish, Julia 151, 182
Pascal, Blaise 17
Pasteur, Louis 136
Pastor-Satorras, Romualdo 494, 495–6
path dependence 340*n*
Paul, St 362
Pavlov (strategy) 542, 546–7, 548, 556
Peierls, Rudolf 113*n*
Peirce, Charles 86
Penn, Alan 189–91
Pericles 534–5
Perrin, Jean 52
Petty, William 1, 61, 62–3; *Political Arithmetick* 1–3, 5, 61–2, 571–2
phase boundaries 140
phase diagrams 140
phase transitions 100–3, 106–7, 108–10, *110*, 115, 118–20, 133, *157*, 158, 364; critical (second-order) 110, *110*, 113–14, 134–5, 203, 286, 404; first-order 110, *110*, 203, 401–2, *402*
physics: and economics 223, 225–6, 238, 254, 256–7, 266–8, 282; and society 3–4, 5–6, 13, 35–7, 56–7, 60, 61, 567–9, 571, 572, 576, 577–9, 583–5
Plato: *The Republic* 12, 21, 22, 509, 510, 566
Pleasence, Donald 464, 465
Poincaré, Henri 238
Poisson, Siméon-Denis 74, 76, 78
poker (game) 523–4
political arithmetic (Petty) 1–3, 5, 61–2, 571–2
politics/political science 3, 12–13, 69, 70, 578; parties 338–9, 378, 535, *see also* democracies; elections; government(s)
polygyny 407
Pool, Ithiel de Sola 447–8
Popper, Karl: *The Lessons of This Century* 507, 534, 535, 538, 556, 557, 558, 563
positivism 71, 80, 83, 414
Potts, Wayne 151–2
power 13, 25–9, 33, 221, 222, 285
power laws 285, 295, 296, 324; and growth of firms 329, *339*, 331, *332*; and income distribution 307–9, *308*; and networks 481, 497 *and n*, 498, 501–2, *502*
probability distributions 243*n*, 296–300, *297*, 302–3, 304, 306–7; and voting patterns 374–5, *375*, 376; and the World Wide Web 479–81
Prigogine, Ilya 128, 131, *132*, 132–3

Prisoner's Dilemma, the 524–7, 529, 530–1, 532–3, 534, *536*, 543, *545*, *548*, 549, 550, *553*, 556, 559–61, 562, 563
probabilities/probability theory 53, 56, 65–6, 74, 88, 90, 91, 98
probability distribution 53, 75–6, 241, *241*, 242–3, 246*n*, 247; *see also* power laws

quantum chromodynamics 119
quantum mechanics 98, 118–19
Quesnay, François 69, 70; *Tableau économique* 69–70
Quetelet, Adolphe 71–3, 72, 77–8, 79, 84, 87, 92, 161, 239, 307; *Social Physics* 79–80
Quincke, Georg Hermann 129*n*

racial segregation 391–5, 427
RAND Corporation 475, 523, 525, 558
random graphs 452, *455*, 455–6, 460–3, 481, 484, *484*
random walks 50, *50*, 51, 137, 148, 238–9, 242, 243 *and n*
randomness 74, 75–7, 83, 94–5, 134, 157–8
Rapoport, Anatol 455, 528, 542–3
Rayleigh-Bénard convection patterns *129*, 130–31, 140
Reagan, Ronald 444
real business cycle (RBC) theory 261, 270, 273
recessions 229–30, 231, 275–6
Redner, Sidney 302, 303
Reformation, the 10, 24
Rehborn, Hubert 207, 210–11
religion 10–11, 12, 32, 40, 63, 83, 92, 191; *see also* God
renormalization 290
Rényi, Alfred 452, 455
Reynolds, Craig 152–3, 155
Ricardo, David 228, 282
Richardson, Lewis Fry 304, 305
Richter, Charles 299
Ross, Edward A.: *Social Control* 145
Rousseau, Jean-Jacques 33, 67, 423, 512–13, 516, 575; *Discourse on the Origin of Inequality* 525
Royal Institution 43*n*
Royal Society, London 62, 146
Rudolf, Crown Prince of Austria 60

Safeway supermarkets 312
Saint-Simon, Claude-Henri de 69
Salinger, Michael 324
Sander, Len 137, 140
Santa Fe: El Farol (bar) 416–18, 422
Savoy Court, London 383*n*
scale-free distributions 283, *288*, 289, 295
scale-free networks 480, 484–6, 489, 494–504

Scheinkman, Jose 300–1
Schelling, Thomas: *Micromotives and Macrobehavior* 193, 389, 390–5, 398, 410, 441, 582, 583
Schoenberg, Arnold 59
Scholes, Myron 249, 250–1
Schreckenberg, Michael 197, 199, 212, 217; *see also* Nagel-Schreckenberg model
Schrödinger, Erwin 44*n*, 445
Schumacher, E. F.: *Small Is Beautiful* 311
Schumpeter, Joseph 233, 258; *Business Cycles* 233*n*, 234, 235
Seattle, Washington 191, 195
segregation, racial 391–5, 427
self-organized criticality (SOC) 295–300, 306, 307; and economics 300–302; and history 303–5
Senior, Nassau 82
Shaftesbury, Ashley Cooper, 3rd Earl of 511
Shannon, Claude 97
Shaw, George Bernard 220, 404–5, 427
Shiller, Robert 261, 270
Shirras, George Findlay 309
Short, Thomas: *A Complete History of the Increase and Decrease of Mankind* 63
Sigmund, Karl 544, 545–7; *Games of Life* 560
SimCity (computer game) 437
Simmel, Georg 569
Simon, Herbert: *The Sciences of the Artificial* 191, 281, 441, 564
Sinatra, Frank 465
Sinclair, John: *Statistical Account of Scotland* 64–5
Skyrms, Brian: *Evolution of the Social Contract* 558
slime mould *see Dictyostelium discoideum*
Smart Growth America 184–5
Smith, Adam 220–1, 574*n*, 576; *Wealth of Nations* 34, 70, 80, 221–4, 226, 227, 228 *and n*, 236, 315, 317, 389, 405, 569–70, 574, 580
Smith, John Maynard 542
Smoluchowski, Marian 98
Snow, C. P.: *The Two Cultures* 44–5
snowflakes 122, 123, 124, 127 *and n*, *141*, 141–3, 146
Snyder, Carl: *Capitalism the Creator* 309
Snyder, Glenn 356
social contract 33, 506, 513
social networks *see* networks
Solomon, Sorin 278, 309
Sornette, Didier 292, 293–4 *and n*
Soros, George 265
Spence, Michael 263
Spencer, Herbert 58, 87; *The Principles of Sociology* 121, 186
spin glass 347 *and n*, 348, *349*, *350*, 350–1

spinodal points 365, *366*, 367, 402, 404
Spinoza, Baruch 525
Stalin, Joseph 359, 360, 580
Stanley, Gene 187–8, 189, 245, 324, 495, 501 *and n*, 502
statistical mechanics 56, 59, 60, 86, 120, 122
statistical physics *48*, 52, 120, 225
Statistical Society of London 82, 88, 89–90
statistics 49, 52–3, 58, 60, 64–5, 81–4, 86, 87–92, 98, 99
steam engines 42–3, 44 *and n*, 46
Steiger, Rod 464, 497
Stiglitz, Joseph 263
stochastic processes 75
Strogatz, Steven, and Watts, Duncan 458–9, 460–3, 464, 481, 484, 489, 493–4, 4 96, 499, 503
Stuttgart University 168, *169*
Sugarscape *see* Epstein, Joshua
Sun Microsystems 342, 344
superconductors 118
superfluids 118, 284
supermarkets 312–13, 318, 319
susceptible-infected-susceptible (SIS) models 493, 494
Süssmilch, Johann Peter 63, 64, 91
Sutherland, Donald 464–5
Sutton, John 323
Swift, Jonathan; *Gulliver's Travels* 69; 'A Modest Proposal' 572
symmetry breaking 134
Szilard, Leo 516

Taillandier, A. 89
Tait, Peter Guthrie 85–6
Tang, Chao 295–8, 300, 302
Tate Gallery, London 171–2
Temple, Sir William 505
temptation 523; *see* Prisoner's Dilemma
Tennyson, Alfred, Lord; *In Memoriam* 513; 'Locksley Hall' 303; 'Lucretius' 96
thermodynamics 44, 46, 120, 123, 126; First Law 44, 97; Second Law 44–5, 54, 56, 95–7, 98, 122–3, 127; Third Law 44
Thirty Years' War 10, 63
Thom, René 4, 101
Thompson, Benjamin (*later* Count Rumford) 43 *and n*
Thompson, D'Arcy Wentworth: *On Growth and Form* 125, 130
Thomson, William (*later* Lord Kelvin) 46, 95, 97
Thoreau, Henry David 121–2
Thucydides 346, 510
tipping point (Gladwell) 101, 404
Tit For Tat (TFT) 528–30, 531, 535, 537, 539,

540-3, 544-8, 550-2, 554, 556-7, 558-9, 561, 562-3; Generous Tit For Tat (GTFT) 541-2, 545, 546, 547, 548, 556
Tokyo *190*, 424
Tolstoy, Leo: *War and Peace* 93-4, 337
trade 220, 221-2, 223, 224, 226, 227, 314-15, 438-9, 573-4; *see also* firms and companies; traders, stock market
traders, stock market 247-8, 257-9, 261-2, 264-5, 268-70; chartists and fundamentalists 265, 271, 273, 276
traffic jams 115, 193-5, 204-5, *206*, 207-8, *208*, 210-12, 217, 218, 219, 577
traffic planning 52, 195-6, 203-4; flow models 196-201, 204-5, *206*, 207, 212-15, *214*, *216*, 217-19; *see also* traffic jams
trails/trail patterns 168-71, *169*, *170*, 257
Treiber, Martin 213, 215, 218, 219
Trevor-Roper, Hugh 363
Tucker, Josiah, Dean of Gloucester 223
Twain, Mark 83
typewriters 340

Ulam, Stanislaw 154
Uncertainty Principle 98
Uniform Resource Locator (URL) 478
United Nations 426, 516
universality 114-15, 159
universe, the: theories 19-20, 46, 53, 119, 235
Unix International Incorporated (UII) 342-3, 352, 353
urban growth 183-92, *187*, *189*, *190*
urban planning 172-4; *see also* walkways
URL *see* Uniform Resource Locator
Usenet 473
Utilitarianism 34, 80, 226, 326, 512, 577
utility 259, 326
utopianism 12, 21-3, 33, 34, 35, 37, 65-8, 565-7, 576

Veblen, Thorstein 264, 265
Verschaffelt, Jules 286
Vespignani, Alessandro 432-3, 494, 495-6
Vicsek, Tamás 156-7, *158*, 164 *and n*, 175, 179, 587
Voltaire 66, 565

Vonnegut, Kurt: *Cat's Cradle* 100, 126, 202
vortex growth patterns *147*, 147-8
vortex motions 150, *150*
voting *see* elections

Waals, Johannes Diderik van der 103-8, 115, 116-17, 122, 163, 282, 283-4, 286, 289, 364, 411, *412*
walkways, pedestrian 165-6, *166*, *167*, 168, 172, 173
Wall Street Crash (1929) 225, 233*n*, 275
Walras, Léon 236
Waltz, Kenneth; *Man, the State, and War* 3, 538; *Theory of International Politics* 355
wars 63, 68, 91, 303-5, 337, 367, 380-1, 426-7, 439, 505-6, 516, 517, 538, 556; *see also* Civil War, English; World Wars
Washington DC 193
Watts, Duncan 465, 466, 493; *see also* Strogatz, Steven
wealth distribution 307-10, *308*, 438-9
Weiss, Pierre 108-9, 268, 289
Wells, H. G. 516
Whewell, William 90
Whitham, Gerald 197, 213
Wiedemann, Rainer 198
Wiesenfeld, Kurt 295-8, 300, 302
Wilde, Oscar 427*n*
William III 1
Wilson, Edward O. 155, 395, 531
Wilson, Kenneth 290
Winstanley, Gerrard 565
Witten, Tom 137, 140
Wittgenstein, Ludwig 59, 60
Woodford, Michael 300-1
World Wars; First 304, 355, 517-22, 531, 533-4; Second 304, 355-7, *358*, 359-61, 523, 561
World Wide Web (WWW) 468, 470, 471, 473-4, 478-84, *480*, 499; search engines 482-3
Wundt, Wilhelm 58

Zhang, Yi-Cheng *see* Challet, Damien
Zhou, Wei-Xing 294*n*
Zipf, George Kingsley: *Human Behavior and the Principle of Least Effort* 281, 305-7, 310, 389, 501, 577*n*